Handbook of Forestry

Handbook of Forestry

Edited by **Malcolm Fisher**

R CALLISTO
REFERENCE

New York

Published by Callisto Reference,
106 Park Avenue, Suite 200,
New York, NY 10016, USA
www.callistoreference.com

Handbook of Forestry
Edited by Malcolm Fisher

International Standard Book Number: 978-1-63239-397-5 (Hardback)

Printed in the United States of America.

Contents

Preface

Trees are an integral part of human existence. They provide numerous environmental, social and economic benefits to man. A forest is an area which is densely covered with trees. Forests hence can be considered a major source of production of these benefits to man. However, the increasing exploitation of forests for such commercial practices has led to global warming. As a result there have been drastic climatic changes leading to many natural disasters. It is these changes that has caught the attention of man and made him aware of the need for forest management.

Forests are an important component of a biosphere which is known to support many life forms and hence their preservation and management is now emerging as a vital applied science. Forest management, also known as Forestry, is the science which deals with the study and practice of creation, utilization and maintenance of forests. Its concerns are mainly the fulfillment of the demands of human existence while simultaneously engaging in responsible and judicial use of natural resources and their conservation. In areas that are topographically affected, proper forestry is extremely important so as to avoid severe soil erosion or even landslides. In areas that are most prone to disasters like landslides, creation and maintenance of forests, which is a part of forestry, can help stabilize soils and prevent damage or loss of property and human life.

This book is concerned with the emergence of forestry as a science and the practice of forestry in different ways in various parts of the world. I would like to thank all our researchers and enthusiasts of the field for their efforts and contributions to this book.

Editor

Foliar Litter Decomposition: A Conceptual Model with Focus on Pine (*Pinus*) Litter—A Genus with Global Distribution

Björn Berg

Department of Forest Sciences, University of Helsinki, FIN-00014 Helsinki, Finland

Correspondence should be addressed to Björn Berg; bb0708212424@gmail.com

Academic Editors: J. Kaitera and S. F. Shamoun

The genus *Pinus* encompasses c 120 species and has a global distribution. Today we know more about the decomposition of pine needle litter than litter from any other genus. This paper presents a developed conceptual three-phase model for decomposition, based on pine needle litter, starting with newly shed litter and following the process until a humus-near stable residue. The paper focuses on the mass-loss dynamics and factors regulating the process in the early phase, the late one, and the humus-near phase. For the late phase, the hampering influence of N and the rate-enhancing effect of Mn on the decomposition are given extra attention. Empirical factors related to the limit value/stable residue are discussed as well as the decomposition patterns and functions for calculating limit values. The climate-related litter concentrations of N and Mn are discussed as well as their possible influence on the size of the stable residue, which may accumulate and sequester carbon, for example, in humus layers. The sequestration of carbon in humus layers is discussed as well as the effect of tree species on the process. Although the paper focuses on litter of pine species, there are comparisons to studies on other litter genera and similarities and differences are discussed.

1. Introduction

The process "plant litter decomposition" is quantitatively as large as the photosynthesis. The process is necessary for the recirculation of nutrients and a continued buildup of plant biomass as well as for the maintenance of food webs through the energy released by the degradation of organic compounds.

The main process which we, a bit simplified, call "plant litter decomposition" is extremely complex and can be subdivided into a multitude of subprocesses. These include not only release of nutrients but also a stepwise degradation of the main chemical compounds present in the shed litter. Also, synthesis processes are included in the concept, resulting in new compounds, which in turn may recombine, often creating a recalcitrance of the remaining litter mass. In spite of its importance, this system of subprocesses is mainly unknown.

The concept "plant litter" is wide both regarding components and their chemical composition and foliar litter appears to be the litter fraction that has been most studied. Still, with foliar litter of different species having very different chemical composition (e.g., DELILA III data base; http://www.eko.uj.edu.pl/deco/) we may expect different decomposition patterns among species or at least among genera. Also, bark, twig, branch, root, and foliar litter from the same plant may have different chemical composition and thus follow different decomposition patterns.

To create a first comprehensive image of the decomposition process as such, we may need to create at least a conceptual model, ideally based on one litter species or one genus. Such a model should include influences of litter nutrients and organic components as well as rate-regulating factors and the litter's contribution to accumulation of soil organic carbon. To do this we need to formulate basic questions about the process, simple questions that are reflecting what we need to know.

An example; for a long time there was a generally accepted assumption that litter decomposed completely and thus decomposition rates were considered to be not only of high interest but also a main parameter as discussed by Prescott [1]. However, questions based on such assumptions were of less value as it has shown that rates may range from a high one for newly shed litter to close to zero at a later stage, well before the litter is completely decomposed. At present, we may describe such changes in rate in just general terms and for just a few species.

By tradition, there appears to be a general opinion and image that climate rules decomposition on a regional scale, whereas litter chemical composition dominates the process on a local one. The picture may be more complicated, though, and is not entirely correct. Meentemeyer [2] and later Berg et al. [3] and Kang et al. [4] demonstrated a large-scale effect of climate on decomposition rate of newly shed plant litter. However, such an effect is not general to all litter species [5, 6] and not to all decomposition stages [7], but the changed substrate composition may dominate decomposition rates—at least for some foliar litter species.

The influence of the substrate on the decomposition process may thus be such that the proposed climate change and a suggested raise in temperature may be without influence on the mineralization rate of soil organic matter (SOM). Such a substrate influence is due to the changed chemical composition during decomposition. What may cause the increasing recalcitrance of partly decomposed litter and the decrease in decomposition rate is today not clear, although the well-established decrease in rate has been related to concentration of gravimetric lignin in litter (Acid Unhydrolyzable Residue—AUR). This relationship was observed by Fogel and Cromack [8] and later developed [2, 7] to different climate situations. However, new analytical techniques have made it clear that the concept AUR may be complex and appears to encompass native lignin, waxes, tannins, cutins, as well as newly synthesized compounds [9, 10].

A subdivision of AUR into different components may be of help to distinguish what compounds that accumulate in litter and what factors that influence their degradation. We cannot exclude that the developing ^{13}C-NMR technique may change the concept of native lignin as a recalcitrant compound.

In most Nordic/boreal coniferous forests, the number of soil animals is low [11], which means that the decomposition is mainly microbial and more than 90% is carried out by microorganisms. The dominant microbial degradation may facilitate the interpretation of decomposition data as well as the application of microbial mechanisms to the decomposition process. Further, the lack of burrowing soil animals creates intact humus layers, which facilitates the determination of carbon (C) sequestration.

A help parameter for decomposition studies was found by Howard and Howard [12] as they estimated a level of the accumulated mass loss with the decomposition rate zero, which later has been called a "limit value" for decomposition. They used a function giving an asymptote. Their study was confirmed by Wieder and Lang [13] who introduced a more handable function (Equation 3; Section 3.3.4). Later the concept was developed [14, 15], including the suggestion of a negative influence of nitrogen (N) concentration on the limit value. The pattern described by (3) means that the accumulated mass loss approaches a limit value, which can be described by an asymptotic function.

So far such limit values have been described mainly for litter in forest systems in boreal and temperate areas and recently also for subtropical and tropical ones [16, 17]. Litter at this stage may be assumed to be close to SOM as discussed by Berg et al. [18], who used litter N concentration as an internal marker.

Limit values for 106 sets of 21 species of decomposing foliar litter from natural forest systems were estimated [19], using litters representing a wide range in chemical composition. A highly significant negative relationship between limit values and initial litter N concentrations was found [19]. Limit values for pine (*Pinus*) species have also been positively related to initial concentrations of manganese (Mn) [20, 21].

With the massive information in the data base from the Long-term Intersite Decomposition Experiment Team (LIDET) confirming that decomposing foliar litter leaves a stable residue [16, 17] we may expect that the focus of interest for decomposition studies may change to be more directed towards (i) explaining the retardation of the decomposition process, (ii) developing the concept limit value that defines the stable fraction, and (iii) explaining the stability of the residue. Hopefully, future studies will be more directed towards single species or genera as was suggested in a recent synthesis paper [22].

That long-term humus buildup and storage are possible was shown by Wardle et al. [23] when they determined such a buildup to have taken place for close to 3000 years. The buildup was reconstructed quantitatively [24] using foliar litter fall and the remaining stable fraction. Such a buildup is not uncommon and the property of humus-layer depth has been presented as a dominant property of a humus form "Tangelhumus" [25]. This humus form, with up to c. 1 m deep humus layers, has been described for the Alps and for mountains in Central Europe.

The reconstruction of humus layers has been extended to encompass the sequestration of N [26, 27]. The approach of using limit values seemed to make it possible to quantify the remaining, recalcitrant mass in the very late stages and may allow a further evaluation. Such a relative stability may be important for the sequestration of carbon dioxide (CO_2) in humus. As regards the long-term stability of humus, the information is scarce and to some extent contradictory. Thus, Berg and Matzner [28] report that increasing N concentrations in humus decrease the decomposition rate. That report was based on a study of Bringmark and Bringmark [29] covering a gradient along Sweden. In contrast, there are reports on extremely active humus disintegration that may be related to humus N concentrations [30].

The aim of this paper is to review and organize existing knowledge of the decomposition of foliar litter with focus on pine species (*Pinus*) needle litter. The information will be organized into a structure to create a system of influencing factors on the decomposition process starting with newly shed litter and following it to SOM and C sequestration. To this purpose, I have reviewed existing information on Scots pine (*Pinus sylvestris*) needle litter and other pine species. A reason to this choice is that we seem to have more information about pine needle litter and its decomposition than of any other litter species or genus. Further, with c. 105 to 125

TABLE 1: Some abbreviations, acronyms, and specific terms used in this paper.

Acronym	Meaning
MAT	Mean annual temperature (°C)
MAP	Mean annual precipitation (mm)
AET	Annual actual evapotranspiration (mm) acc. to [2]
PET	Potential evapotranspiration (mm) acc. to [2]
AUR	Acid Unhydrolyzable Residue or gravimetric lignin (refers to different methods, e.g., sulfuric acid lignin, Effland lignin, acid detergent lignin). See; for example, [9, 10].

identified pine species, now with global distribution and occurring in boreal, temperate, and subtropical forests, we appear to have a model substrate, decomposing in several environments, and climate situations. Pine litter already has been suggested as a reasonable model substrate [22]. Throughout the paper, I will focus on the components N, Mn, AUR, and native lignin. As far as we know today, these are key components in the decomposition process, not excluding that further factors may be identified.

Further, I discuss (i) what parameters that may be of importance for a characterization of a given decomposition pattern, (ii) if the presented conceptual model may be of use for further genera of foliar litter, and (iii) if developing analytical tools may change basic findings about the decomposition process.

Data and information is taken from the literature with focus on foliar litter of Scots pine and other pine species. I repeatedly quote the DELILA database (DELILA II and DELILA III; http://www.eko.uj.edu.pl/deco/).

2. Terminology

Abbreviations, acronyms, and specific terms are collected in Table 1 and in part explained below.

Accumulated Mass Loss. The total amount of litter mass lost from a decomposing litter substrate, normally expressed as a percentage of initial mass.

AUR (Acid Unhydrolyzable Residue) refers to gravimetric lignin such as sulfuric-acid lignin and Effland lignin.

Decomposition Pattern refers to the general development of accumulated litter mass loss with time. Often the decomposition rate slows down and ceases, with the consequence that the accumulated mass loss does not increase. At the rate zero, we may estimate a "limit value" with a so far known range from c. 42 to 100%. See also [21].

Holocellulose. A term covering cellulose plus hemicelluloses, and thus the polymer carbohydrates in litter.

Limit Value is the calculated value for the extent of decomposition of a given litter type, at which the decomposition rate approaches zero. Limit value may be given as accumulated

mass loss (%) or as a fraction. It may also be calculated, using remaining amount and then given as percentage remaining amount.

3. Discussion

3.1. The Chemical Composition of Pinus Foliar Litter with Focus on Nutrients and AUR

3.1.1. Organic-Chemical Composition

Some Comments. There is limited information on organic compounds in foliar litter. Cellulose, hemicellulose, and AUR are the quantitatively dominant ones (Tables 2 and 3) with a variation in proportions between annual litter falls and among species. The developing ^{13}C-NMR technique has given us a tool that may open new possibilities to work on plant litter, although so far this technique gives the quantity of specific bonds, rather than compounds and we will discuss both the new method and the more traditional ones.

Several studies on chemical composition have been made using extractions of hemicellulose and cellulose based on solubility in alkaline solutions (e.g., acid detergent lignin) and although useful, they are not included in this review as they are less specific.

Traditional Techniques. Available detailed data on organic components in foliar litter is old today and few studies have been carried out, which separates the specific components [31]. Separating solid substance and different soluble fractions, the authors identified 44 organic compounds in Scots pine needle litter.

We may see that there is not much variation in the main compounds among the investigated five litter species (Table 2), still there is a certain variation in soluble substances and AUR. Some variations in proportions of hemicelluloses and cellulose have been observed with glucans and mannans having higher concentrations in the three coniferous species. In contrast, silver birch (*Betula pendula*) has higher levels of xylans and rhamnans.

As regards AUR, Johansson et al. [7] found a concentration that ranges from 223 to 288 $mg\,g^{-1}$ when analyzing newly shed Scots pine needle litter collected in the same stand over 17 consecutive years. Although the concept AUR is not very specific and includes native lignin, waxes, cutins, and tannins [9, 10], it still deserves to be discussed as it has been widely used as an index of recalcitrance for litter. Available data [34] shows a range in AUR between 234 $mg\,g^{-1}$ in Chir pine (*Pinus roxburghii*) and Scots pine to 432 in Khesi pine.

An empirical, general, and positive relationship has been found between concentrations of AUR and N in pine needle litter (Figure 1) as well as for specific groups like coniferous and broadleaf litter species [39]. A similar observation was made in N fertilization experiments for Scots pine and Norway spruce (*Picea abies*) [32, 40, 41].

^{13}C-NMR Technique. New analytical techniques, such as ^{13}C-NMR, may in part replace the traditional, time consuming ones and provide new information. So far ^{13}C-NMR rather

TABLE 2: Initial composition of organic-chemical components and nutrients in five litter species. Data from [14, 31–33].

| | \multicolumn{9}{c}{Concentrations of components (mg g^{-1})} | | | | | | | | |
| | \multicolumn{2}{c}{Substances soluble in} | AUR | \multicolumn{6}{c}{Relative proportions of carbohydrates} | | | | | |
	Water	Ethanol		Rhamnan	Xylan	Galactan	Mannan	Araban	Glucans
Scots pine	164	113	231	3	23	32	75	36	245
Scots pine[1]	92	120	240	1	23	30	64	43	214
Lodgepole pine	103	42	381	6	34	46	90	48	254
Norway spruce	32	48	318	7	33	28	105	40	288
Silver birch	241	57	330	16	77	44	14	49	166
Grey alder	254	39	264	9	30	32	10	44	116

[1] Recalculated from [31].

TABLE 3: Concentrations of seven main nutrients and AUR in foliar litter of a few pine species. Available data. From [34].

| Litter sp. | \multicolumn{9}{c}{Concentration (mg g^{-1})} | | | | | | | | |
	N	P	S	K	Ca	Mg	Mn	AUR	Lit. ref.
Scots pine[a]	4.2	0.23	0.44	0.79	5.3	0.45	1.03	249	[7]
Lodgepole pine	3.9	0.34	0.62	0.56	6.4	0.95	1.79	360	[15]
Maritime pine	6.8	0.54	1.01	1.95	3.1	1.90	0.59	326	*
Red pine	6.0	0.36	0.73	1.40	8.9	2.00	0.73	265	**
White pine	5.9	0.21	0.68	0.70	7.2	1.10	0.80	256	**
Jack pine	7.8	0.64	0.77	2.30	4.0	2.10	0.25	329	**
Limber pine	4.3	0.43	0.52	1.10	5.3	1.10	0.21	258	**
Ponderosa pine	5.5	0.45	0.56	1.5	3.8	1.22	0.20	294	**
Stone pine	3.0	0.57	1.36	5.9	7.1	2.4	0.19	312	[35]
Corsican pine	4.7	0.54	0.71	3.5	7.8	1.3	0.50	276	**
Monterey pine	5.6	0.22	0.70	1.3	1.9	0.93	0.47	406	**
Aleppo pine	4.3	0.38	1.3	1.73	25.2	2.33	0.03	341	[58]
Virginia pine	6.5	0.76	1.37	1.8	5.1	1.05	1.48	347	**
Black pine	6.1	0.35	1.12	0.8	10.6	1.6	0.86	280	[43]
Chir pine	6.7	nd	nd	nd	nd	nd	nd	234	[49]
Khasi pine	9.8	nd	nd	nd	nd	nd	nd	432	[49]

[a] Average values over 17 years at one Scots pine stand.
* DELILA III data base (http://www.eko.uj.edu.pl/deco).
** B. Berg and C. McClaugherty (unpubl.).

gives information about specific bonds than compounds, which may open up for new interpretations of data. At the present development stage, however, the concentration of a chemical compound is given as that of a certain C-bond. A bond is identified as the response in a certain frequency interval and we may expect such responses to be complex [10] (Table 4).

Some recent papers [10, 42, 43] have reported initial chemical composition of whole litter (Table 4). We may see rather wide ranges in response signals among species. For example, a range factor of more than 4 for Methoxy-C, a factor of c. 2 for Aromatic-C, and Phenolic-C and c. 3 for Carbonyl-C. Among the three pine species, there was a variation too, but smaller. These studies encompassed whole litter, thus including extractives.

Alkyl-C or *Aliphatic-C*. This frequency interval (0–50 ppm) indicates long chains with -CH$_2$-units. A side chain in hemicellulose, namely, an acetate group belongs here as well as C in side chains of lignin. Available data (Table 4) gives a clear variation among litter species (from 15.7 to 25.8%).

Methoxy-C (50–60 ppm) shows the methoxyl carbon in lignin. However, this frequency interval also includes the alkyl carbon bound to N in proteins. A clear variation is seen, ranging from 1.5 to 6.61%.

O-Alkyl-C (60–93 ppm) encompasses mainly carbohydrate carbon, namely, that in cellulose and hemicelluloses, but also the side chains of lignin going from carbon 3. Further, some signals from tannins come in this interval.

Di-O-Alkyl-C (93–112 ppm) mainly encompasses cellulose plus hemicelluloses, and thus carbohydrate carbon but shows no difference between the different carbohydrates. We may see (Table 4) that the highest frequencies are found for O-Alkyl-C, followed by Di-O-Alkyl-C, which mainly reflect cellulose plus hemicellulose. These appear to be highest for Scots pine and beech. The intensity for O-Alkyl-C ranges from 39.4 for black spruce (*Picea mariana*) to 61.4 for black pine (*Pinus nigra*) litter.

Aromatic-C or *Aryl-C*. The intensity in this interval (112–140 ppm) comes from the aromatic carbon in both lignin and condensed tannins. It may also show the guaiacyl group of lignin.

TABLE 4: Relative contribution (% of total area) of different groups of C-bonds as derived from ^{13}C CP-NMR spectra of newly shed foliar litter.

Litter species	Alkyl-C 0–50	Methoxy-C 50–60	O-alkyl-C 60–93	Di-O-alkyl-C 93–112	Aromatic-C 112–140	Phenolic-C 140–165	Carbonyl-C 165–190	Lit. ref.
Scots pine	15.8	5.28	50.9	12.5	7.32	5.82	2.39	[42]
Scots pine	16.4	5.46	51.4	12.4	6.80	5.35	2.20	[42]
Jack pine	23.4	2.2	44.8	10.5	8.6	5.9	4.6	[10]
Black pine[a]	16.1	5.0	61.4	—	—	5.0	3.6	[43]
Norway spruce	17.2	6.21	44.9	11.6	9.65	7.30	3.13	[42]
Black spruce	20.5	4.4	39.4	9.9	12.4	6.7	6.7	[10]
Douglas fir	23.2	1.5	45.2	8.7	9.4	6.4	5.5	[10]
Tamarack	15.9	1.7	43.3	16.5	8.0	11.1	3.5	[10]
White birch	25.8	2.4	43.7	11.7	7.1	6.3	2.9	[10]
Silver birch	23.0	5.58	43.6	12.1	6.13	5.56	3.98	[42]
Aspen	22.8	1.5	39.4	12.2	7.3	7.5	6.6	[10]
Beech	15.7	3.2	48.7	11.9	9.6	6.0	5.0	[10]

[a]The investigated intervals were somewhat different.

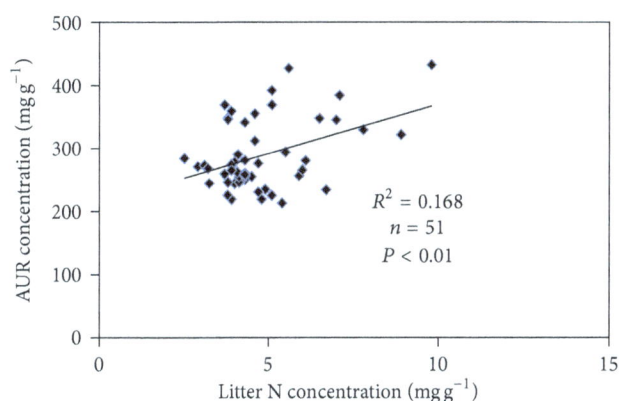

FIGURE 1: A positive relationship between N and AUR concentrations in needle litter of 51 samples of 10 pine (Pinus) species. Data from [39] and the DELILA database; http://www.eko.uj.edu.pl/deco/.

Phenolic-C. The intensity in this interval comes from phenolic C (140–165 ppm) in both lignin and condensed tannins. It may also show the syringyl group of lignin. Aromatic-C and Phenolic-C, reflecting aromatic carbon show a clear variation among litter species, with a range from 12.4 for black spruce to 6.13 for silver birch litter. Phenolic-C ranged from 11.1 for tamarack (*Larix laricina*) to 5.0 for black pine.

Carboxyl-C or Carbonyl-C (165–190 ppm). This region includes carboxylic acids, amides and esters. The intensity for Carbonyl-C ranged from 6.7 for black spruce to 2.2 for Scots pine.

So far relatively few studies have been carried out, which limits the possibilities to draw conclusions.

3.1.2. Two Main Nutrients and Heavy Metals

What Factors Influence Litter Concentrations of N and Mn? It appears that concentrations of nutrients in pine litter and several other genera vary along at least three axes, one being species/genus, another climate, and a third one soil

properties at the site of growth. We cannot exclude that these are linked and when a given species is dependent on both climate and soil properties, both influence the litter chemistry simultaneously.

Pine Species—Available Data. Newly shed needle litter has been analyzed for just a few pine species and the analyses are rather incomplete, encompassing mainly the seven main nutrients. In general, the nutrient levels are low, for example, for N, with a range from 3.0 to a highest value of 9.8 for Khesi pine (Table 3). Pines normally grow on nutrient-poor soil, which in part may explain this generally low level.

The natural variation over 17 years at one stand of Scots pine ranged from 2.9 to 4.8 mg g^{-1} [7] with an occasional 10.4 mg g^{-1} possibly due to an interruption in the retranslocation process. Still, occasionally high N levels (23.0 mg g^{-1}) have been observed, for example, for black pine [43], when the trees grew in a system supporting the dinitrogen-fixing Mount Etna broom (*Genista aetnensis*).

The variation among species/genera increases when we leave the genus *Pinus*. So far there has not been any systematic grouping of genera/species versus nutrient concentrations, so what nutrients such a variation encompasses, in addition to N, is not possible to say. The large global approach of Kang et al. [33] indicates a difficulty already for N and P. Although N and P are mainly correlated, there were differences both in concentrations and the N-to-P ratio among continents and main functional groups. An alternative approach is that of Tyler [47] using no less than 58 macro- and micronutrients as well as rare earth metals, possibly over a limited area.

Variation with Climate. The chemical composition of the newly shed pine litter varies considerably with climate. In 1995, Berg et al. [48] reported positive relationships between N, P, S, and K and the climate indices MAT and AET. They used mainly Scots pine needle litter but also those of 7 further pine species, collected over Western Europe. Later, a positive relationship was reported between N and MAT for eight species of pine litter ($n = 56$) in a gradient from the Equator

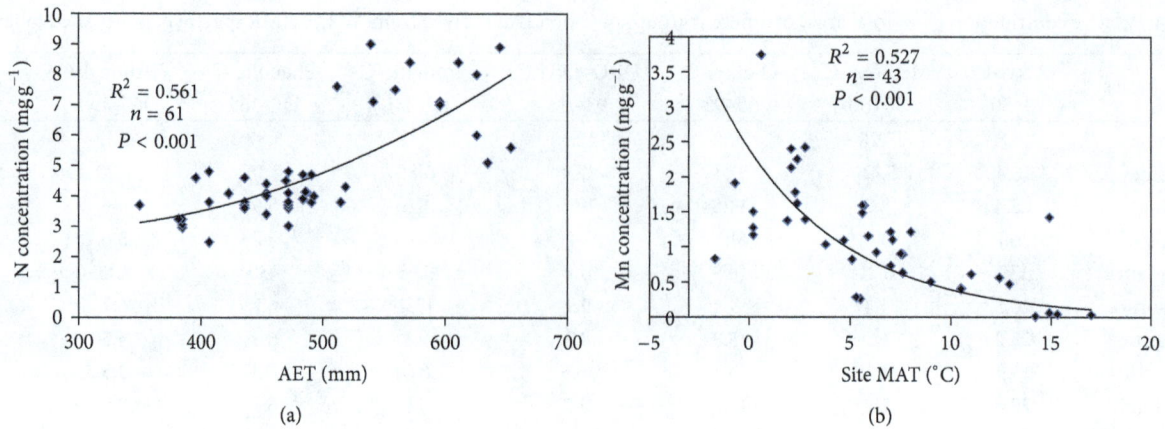

(a)   (b)

FIGURE 2: All available data for newly shed pine (*Pinus*) needle litter in a climatic gradient ranging from northernmost Fennoscandia to south Spain and south Italy. Nitrogen and Mn were analyzed on the same litter samples. (a); increasing N concentration with increasing site AET (b); decreasing Mn concentration with increasing site MAT. Data from [21, 48], and the DELILA database; http://www.eko.uj.edu.pl/deco/.

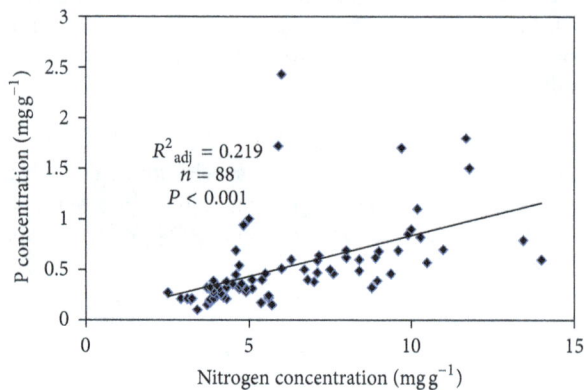

FIGURE 3: Relationship between concentrations of N and P in pine (*Pinus*) species needle litter, collected in stands over Europe, Asia, and North America. The data encompasses 22 pine species. B. Berg, unpublished. Data from [7, 48] and the DELILA database; http://www.eko.uj.edu.pl/deco/.

in Southeast Asia to the Arctic Circle in Scandinavia with a MAT range from 0 to 28°C [49]. Using standardized data, they could confirm that MAT ($P < 0.001$) and not MAP gave a highly significant relationship. We may see (Figure 2) that the available data for pine needle litter gives a highly significant relationship ($R^2 = 0.561$; $n = 61$; $P < 0.001$) to AET.

That the main nutrients are related not only to climate factors but to each other may be seen over the climate gradient as discussed by Berg et al. [48]. An example is that of N and P. For 22 pine species ($n = 88$) in the northern hemisphere, the concentrations of N and P are well correlated (Figure 3). The variation may be due to the different species or environmental factors, possibly MAT. For one species, namely, Scots pine with 38 samples, the correlation was considerably better ($R^2_{adj} = 0.830$; $P < 0.001$) as compared

to $R^2_{adj} = 0.219$ ($P < 0.001$) for the 22 different species (Figure 3). This emphasizes that also for litter, the concept species is important. This kind of relationship (N versus P) has been confirmed on a global level for several litter species [33].

A further nutrient that may be related to climate indices is Mn. Negative relationships between Mn concentrations and AET/MAT have been reported for pine litter [21, 48] (Figure 2(b)). We may see that there is a highly significant negative relationship to MAT for eight pine species ($n = 43$) covering the area from northernmost Scandinavia to North Africa.

The relationship between Mn concentration and MAT was clearly stronger than that to AET. Both relationships were negative, indicating that the higher the average temperature, the lower was the Mn concentration in the litter. The highest Mn levels were found in Scots pine in north Scandinavia ($3.67 \, \text{mg} \, \text{g}^{-1}$) and the lowest ($0.03 \, \text{mg} \, \text{g}^{-1}$) in Aleppo pine (*Pinus halepensis*) in North Africa (Figure 2(b)). This relationship is still empirical.

With the relationship for N being positive to MAT and that for Mn being negative, there was a negative relationship between N and Mn in the pine needle litter with $R^2 = 0.145$; $n = 42$; $P < 0.05$ (Figure 4). Thus, with increasing MAT (AET), there is an increasing N-to-Mn quotient in the litter. The concentrations of these two nutrients, important for lignin degradation thus vary strongly over a climatic gradient with a range in MAT from −1.7 to 17°C and a geographical range of 4200 km.

A certain variation among species, possibly due to place of growth or deviating site conditions may be seen for Ca and Mg (Table 3). In their analyses, Berg et al. [48] found no relationship to climate factors for Ca and Mg and concluded that soil factors were ruling their concentrations in litter.

3.1.3. A Covariation in AUR and Nitrogen Concentrations in Newly Shed Litter. There appears to be a positive relationship between concentrations of N and AUR in foliar litter for

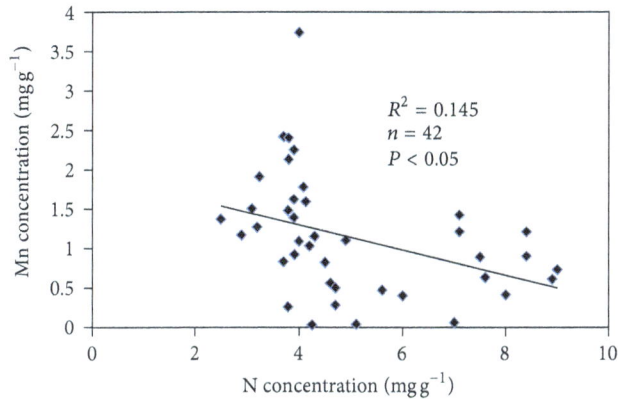

FIGURE 4: Within the genus *Pinus* there appears to be a negative relationship between concentrations of N and Mn in newly shed needle litter along a climate gradient ranging from 1.7 to 17°C. B. Berg unpublished. Data from [48] and the DELILA database; http://www.eko.uj.edu.pl/deco/.

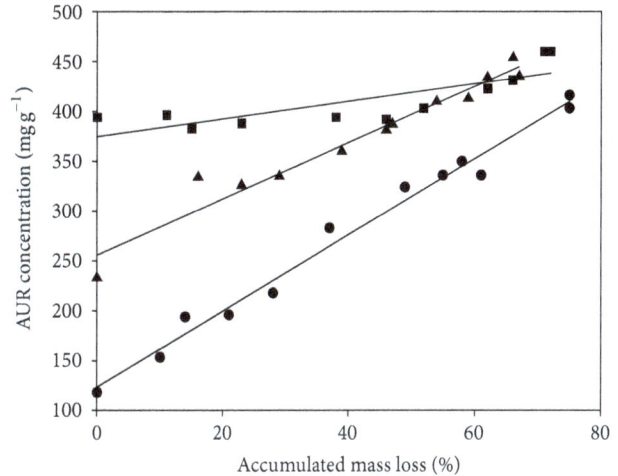

FIGURE 5: During decomposition, the concentration of AUR increases. The AUR concentration may be plotted versus accumulated litter mass loss. By doing that, we obtain a linear relationship, which is readily comparable for litter incubated in different climate zones and among different litter species. Scots pine (●), lodgepole pine (■), Norway spruce (▲). Data from [31, 52, 53], figure from [32].

both coniferous and broadleaf litter (both at $P < 0.0001$) as well as for pine needle litter [39] (Figure 1). The latter relationship was based on 10 pine species ($n = 51$) with $P < 0.01$. This relationship is empirical and we cannot exclude that it will be subdivided as AUR will be, using ^{13}C-NMR. It may be mentioned that after N fertilization, both Scots pine and Norway spruce shed needle litter with increased AUR concentrations [40, 41]. The concentration of AUR also has been shown to vary with climate. Thus, Berg et al. [48] reported positive relationships between AUR in pine needle litter and both MAT and potential evapotranspiration (PET). Also this relationship is empirical.

3.1.4. Relative Amounts of N and Mn in Litter Fall.
The relative composition of foliar litter fall for pine species changes with MAT and AET as does the amount of litter fall. Thus, litter fall is positively related to AET and in northern Europe also to MAT, where temperature rather than precipitation is limiting [50]. This was seen also by Liu et al. [51] for pine litter over Eurasia.

As the concentration of N in needle litter fall increases with, for example, MAT and AET, the amounts of N that are returned to the soil in litter fall are predictable and there is a heavy increase with increasing AET.

For Mn, the situation is less clear but largely the concentration decreases with MAT and AET. Still, there is a high variation in concentration among years and sites which makes the amounts less predictable. We may, however, estimate proportions between N and Mn for a few cases. Using real data for pine needle litter gives for a stand with a MAT of 0°C an N-to-Mn ratio of 2.5, at a MAT of 10°C the ratio becomes 25.6, and at a MAT of 17°C it becomes 133. Thus, the relative amount of N increases considerably with MAT as compared to that of Mn.

3.2. The Long-Term Decomposition Process

3.2.1. How Do Concentrations and Amounts of Solubles, Holocellulose, AUR, and Native Lignin Develop with Accumulated Mass Loss?

Long-Term Development of Some Main Processes. Some litter species allow us to follow the decomposition process until rather high values of accumulated mass loss or until the decomposition process appears to go very slowly. Several studies using pine needle litter have recorded values for accumulated mass loss of 80% and above. Such mass losses may be reached in shorter or longer periods, to some extent depending on climate. We discuss the concept long term with reference to accumulated mass loss.

The concept long term is not directly related to the three identified stages [28] (below). The three stages were introduced to describe some main steps in litter degradation, namely, what components and factors that dominate and in part regulate the decomposition rate. The stages may also help to distinguish what nutrients that are limiting the decomposition in the different steps.

In the course of decomposition, when the more easily degradable compounds are decomposed, AUR remains relatively intact for a long time and the amount may even increase. This means that the litter becomes enriched in AUR, its concentration increases and may reach even above 50% (Figure 5).

The AUR analysis gives normally a clear increase in concentration and is a useful index for litter recalcitrance. However, in decomposing litter, AUR integrates the most hydrolysis-resistant organic structures and is an analytical fraction derived from lignin, condensed tannins, cutins, and waxes [10]. As an index of litter recalcitrance and age it may be useful. However, we must call it an empirical index.

Soluble Substances. Foliar litter may contain considerable levels of soluble substances. For example, concentrations of water-solubles ranging between 7% in Scots pine needles [7] and 30% in grey alder (*Alnus incana*) leaves have been reported [14]. So far, four principal groups of soluble organic material in pine needle litter have been identified, namely, sugars, phenolics, hydrocarbons, and glycerides. The soluble sugars are predominantly mono- and oligosaccharides that were involved in metabolic processes of the plant. The soluble phenolics are low-molecular weight compounds that serve either as defensive agents against herbivory, are lignin precursors, or waste products; hydrolysable tannins are a common example of soluble phenolics. Phenolics are highly variable in their solubility and many have a tendency to condense into less soluble forms or to react with larger molecules.

Part of these substances may be leached out of the litter [54, 55] and part may be degraded in the litter structure. Few attempts have been made to follow the degradation of simple soluble components in litter and it should be pointed out that most studies describe net disappearance only. The soluble fraction is challenging to study, due to the complexities of tracing the formation of new solubles during decomposition and the disappearance of the same solubles due to leaching or metabolism.

Compounds such as triglycerides and hydrocarbons that are soluble in light petroleum ether may disappear quickly, whereas fatty acids and diterpene acids remain for longer periods. Simple sugars, for example, glucose and fructose, or compounds related to simple sugars such as glycosides and pinitol, are also degraded very early and at a high rate [31]. Still, as an example, glucose, which is present initially in newly shed litter is also produced from decomposing cellulose and is thus found even in the later stages of decomposition. The same applies to the simple sugars of hemicelluloses. Also several phenolic substances that are found in newly shed litter are produced, but later during the degradation of, for example, native lignin.

Solubles are mainly studied as compounds soluble in water and ethanol/acetone and generally the water soluble fraction is quickly decomposed, and consequently it decreases in concentration. This has been generally observed for the few litter species that have been studied in some detail. The ethanol/acetone soluble fraction, which contains phenolics and higher fatty acids may remain more constant or even increase in concentration although there appears to be a clear variation among litter species.

Cellulose and Hemicelluloses. Based on traditional analytical techniques [31, 32], the components cellulose, hemicelluloses, and AUR have been shown to be degraded at different rates—in the early stages of decomposition. Still, throughout the decomposition process, there are no drastic changes in the concentration of cellulose. Berg et al. [31] found for Scots pine needle litter a slight increase followed by a decrease to about the initial concentration. For pine litter, part of the cellulose is

lignified when the needle litter is newly shed and part is not. This may be common among litters although it has been little investigated. Some litters such as Norway spruce needle litter and that of common oak (*Quercus robur*) appear to have very little, if any cellulose that is not lignified.

The most common hemicelluloses decompose in a similar fashion in litter. For the most part, they behave like cellulose, although they may have different positions in the decomposing fiber tissue. The concentrations of, for example, xylans, mannans, arabinans, and galactans with smaller variations remain about constant as far as the decomposition process has been followed. Considering the structure and complexity of the hemicelluloses, we may simplify our discussion and regard them as a group. When considered together, we may see that in Scots pine needle litter their concentration at 70% accumulated mass loss is about the same as at the start of the incubation [31]. Like cellulose, hemicelluloses appear to be in part lignified and in part not.

Using a long-term incubation of needle litter of Japanese cedar (*Cryptomeria japonica*) and Hinoki cypress (*Chamaecyparis obtusa*), Ono et al. [56] showed an initially fast loss of Alkyl-C and O-Alkyl-C bonds corresponding to carbohydrates. It appears that the decomposition rates for O-Alkyl-C were higher in the first incubation year. They compared the rates to those of Aromatic-C representing lignin and Carbonyl-C (below).

"Lignin" Is Often Determined as Acid Unhydrolyzable Residue (AUR). Native lignin is not a very clear concept either in fresh or decomposing litter and so far lignin has been defined on the basis of proximate (gravimetric) analytical methods rather than purely chemical criteria. When applied on newly shed litter, some proximate methods yield results which may be close to chemically defined lignin. However, in decomposing plant litter, lignin is modified by the humification processes, including condensation reactions, and by partial degradation by microorganisms. The formation of such decay products, which are included in the AUR fraction, may raise arguments about the extent to which true lignin is measured in decomposing litter when analyzed as AUR. In addition, gravimetric lignin (AUR) will contain, among other compounds, cutin, waxes, and tannins [10]. Further, an inorganic fraction (ash) can be of considerable magnitude. Although the latter fraction for, for example, Scots pine litter is about 1% of the total litter mass, it may amount to some percent in the gravimetric lignin analysis. The ash content of newly shed deciduous litter can be much higher, going above 10% in some cases [57]. Furthermore, ash concentrations may increase during decay and should be considered when reporting AUR contents of decomposing litter. In an extreme case [36], fine particles of mineral soil were found to penetrate the litter and "ash" increased from 5.3 to 15.6%.

There are some clear differences between AUR and NMR-determined lignin, which need to be clarified. Although what is determined as AUR is not true lignin but a mixed chemical fraction with compounds, some of which have similarities to true lignin. It is important to note that even native lignin is highly variable among and even within species. Thus, native

TABLE 5: Mean intensity distributions (% of total area) as well as mean ratio Alkyl-to-O-Alkyl. Data, which originates from [10] give ratios for 75 MHz ^{13}C CP NMR of spectra of newly-shed litter and after 2 and 6 years of decomposition. Mean values refer to 10 foliage litter.

Litter species	Alkyl-C 0–47	Methoxy-C 47–58	O-Alkyl-C 58–92	Di-O-Alkyl-C 92–112	Aromatic-C 112–140	Phenolic-C 140–160	Carbonyl-C 160–185	Alkyl/O-Alkyl
Mean newly shed	18.9	4.5	42.3	12.3	9.8	6.6	5.6	0.32
Mean 2 years	23.0	5.5	36.6	10.0	10.9	6.5	7.6	0.45
Mean 6 years	22.2	6.6	33.7	10.1	12.4	6.9	8.1	0.44

lignin cannot be described with the same chemical precision as cellulose or other plant polymers.

Lignin is not a well-defined compound when it is produced and it remains a poorly defined compound as it is decomposing. The nomenclature for lignin that has been modified during decomposition is still in question and even misleading terms like "Acid-Insoluble Substance" are seen in the literature. One suggestion is "Non-Hydrolyzed Remains" (NHR) [58]. The more recent suggestion "Acid Unhydrolyzable Residue" (AUR) [10] appears to have won a more general acceptance.

It should be pointed out that although the terminology sometimes is misleading, the gravimetric lignin or AUR that contains chemically recalcitrant matter, is still today an important concept to litter decomposition. It has turned out that AUR also represents a biologically recalcitrant unit. Although we also need methods to identify native lignin as well as to identify the compounds included in AUR, we may still use AUR as an index for degradability [34].

Two approaches have been made to describe the change in concentrations of cellulose and hemicellulose versus that of AUR. In the long term, the concentration of holocellulose decreases, whereas that of AUR increases, and as observed, there is a level at which the relative amounts remain constant. Berg et al. [59] suggested the term holocellulose-to-lignin quotient (HLQ). Two such quotients are as follows:

$$HLQ = \frac{holocellulose}{(AUR + holocellulose)} \quad (1)$$

(see [59]). Another was suggested by Melillo et al. [60] as follows:

$$LCI = \frac{AUR}{(AUR + holocellulose)}. \quad (2)$$

The upper relationship [59] approaches a minimum value asymptotically, which may be different for different litter types. For example, Berg et al. [59] found a clear difference between the HLQ values for Scots pine and silver birch indicating a difference in the quality of the carbon source.

The purpose of the two approaches was to obtain a quality measure for the litter carbon source. With a new analytical tool, namely, the ^{13}C-NMR, we cannot exclude that the indices will be more useful.

True Lignin. The old concept that lignin concentration increases during decomposition has been efficiently questioned [10, 61]. We cannot exclude that an increase may be different among litters and depend on the main degrading

organism (e.g., white rot versus brown rot). According to some papers, it thus appears that concentrations of native lignin itself do not increase during the decomposition process, but rather the sum of components that may be recalcitrant as seen when analyzing for AUR (e.g., sulfuric-acid lignin). On the other hand, Ono et al. [56] found an increase in concentrations of Aromatic-C when following the decomposition dynamics over four years of Japanese cedar needle litter and that of Hinoki cypress. Considering the faster loss of O-Alkyl-C and Alkyl-C such an increase is reasonable. We cannot exclude that contradictory reports reflect a difference in composition of decomposing organisms (e.g., white rot versus brown rot).

When using ^{13}C-NMR analysis, bonds such as Methoxy-C, Aromatic-C, and Phenolic-C may indicate the concentration of lignin. Aromatic-C and Phenolic-C are also found in condensed tannins [10]. In their CIDET study, Preston et al. [10] reported intensity spectra for decomposition of 10 foliar litter species with measurements at 0, 2, and 6 years (Table 5). For Aromatic-C and Phenolic-C, there were clear increases in concentration.

Considering the discussion in the above papers [10, 61], we cannot exclude that in a near future it will be possible to distinguish litter decomposition by white rot versus that by brown rot and their relative participation in lignin degradation.

3.2.2. Dynamics of Two Main Nutrients: N and Mn

Some Comments. In decomposition studies just total concentrations of nutrients are normally used. The information value of total concentrations is limited for evident reasons and does not tell us how these nutrients are bound in litter and the decomposing material, nor what fraction that is available. Some nutrients are in part bound with covalent bonds to organic molecules and thus belong to organic complexes, and some of those in ion form are readily leachable, for example, potassium (K). Thus, N, P, and S are bound in proteins and the nucleic acids that remain in the litter when shed. During decomposition also the developing microbial biomass needs the N, P, and S for building proteins and nucleic acids and when following their total concentrations with accumulated mass loss, we may see that their concentrations increase in linear proportion. Their concentrations need to be in proportion in the microbial biomass and this has been noted especially for the relationship between N and P which has been well studied [62].

Nitrogen dynamics appears to have been more studied than that of other nutrients and there are several synthesis

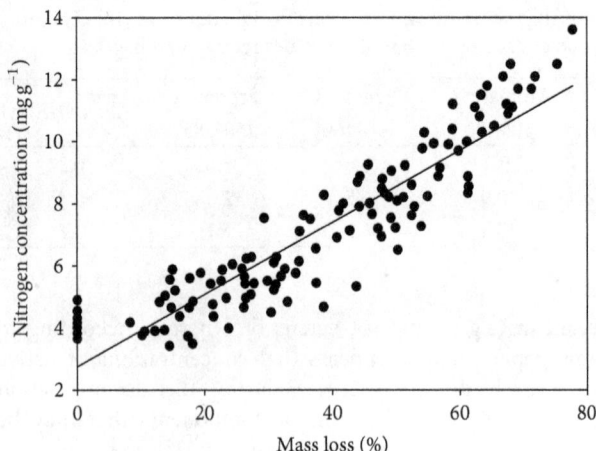

FIGURE 6: Linear relationship between increasing concentrations of N and accumulated mass loss for decomposing Scots pine needle litter. Incubations were made at one site, a nutrient-poor Scots pine forest. Data are pooled from 14 studies, each representing an incubation of local litter sampled in a different year ($R^2 = 0.843$; $n = 131$; $P < 0.001$). Figure from [32], data from [44, 67].

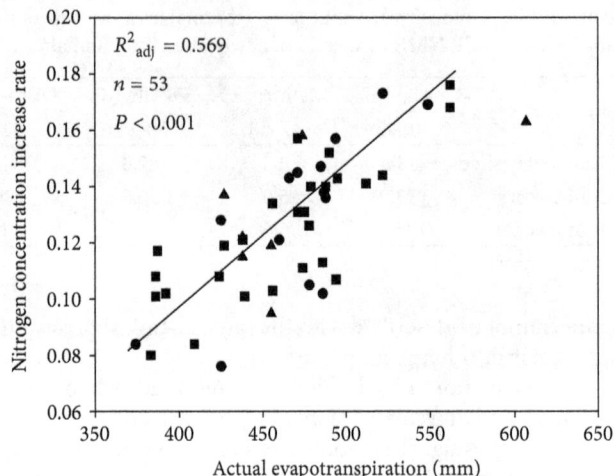

FIGURE 7: Linear relationship between the climatic index actual evapotranspiration (AET) and Nitrogen Concentration Increase Rate (NCIR) for decomposing needle litter of (●) Scots pine, (■) Norway spruce, and (▲) other pines (lodgepole pine and white pine). Figure from [32], data from [67].

works [63, 64]. Basic syntheses may also be found in textbooks [65, 66].

The discussion below is limited to the main nutrients N and Mn, which have been found to be important for the long-term decomposition of litter. Heavy metals in natural and clean environments have been suggested to be important for regulating decomposition at the very late stages [9, 34]. Still, the literature on, for example, Fe, Pb, Zn, and Cu dynamics is very limited.

Nitrogen Concentrations Increase as Decomposition Proceeds. That concentrations of total N increase in decomposing litter is well known. Berg and McClaugherty [32] related the increasing N concentrations to litter accumulated mass loss for several litter types, resulting in a linear increase [32, 44, 67] (Figure 6), possibly until the limit value for decomposition is reached [18]. This type of relationship is useful and may be used for analytical purposes. Such a linear increase has been found for many species including foliar litter of pine species, Norway spruce as well as for broadleaf litters [67, 69].

For one Scots pine stand, there was a limited variability among decomposition studies (Figure 6). The litter was naturally produced from a Scots pine monocultural system, and the variation in initial N concentration was the observed annual variation. Similar comparisons were made for needle litter of lodgepole pine and Norway spruce litter with rather little variation. The increase in N concentration during decomposition may be considerable. Thus, for Scots pine, a linear increase was found in concentration from an initial $4\,mg\,g^{-1}$ up to 12 to $13\,mg\,g^{-1}$ N at about 75% accumulated mass loss ($R^2 = 0.99$; $n = 16$; $P < 0.001$) [70].

Some deciduous litter species, such as silver birch, also give linear relationships, although much of the mass is lost initially, resulting in a fast increase in N relative to mass loss. This linearity is empirical and the reasons for the linear relationship are far from clear, given the simultaneous in- and outflows of N during the decomposition process [66, 71]. This relationship has also been elaborated [18, 67].

Some Influences on N Dynamics May Be Systematized. Using the linear relationship between N concentration and accumulated mass loss, Berg and McClaugherty [32, 34] compared the slopes of the linear relationships among several litter species and among several studies of decomposing Scots pine needle litter in one forest system (Figure 7). They called the slope of the relationship Nitrogen Concentration Increase Rate (NCIR). An advantage with linear relationships is that they may be readily and simply compared and Berg and McClaugherty [32, 34] found a set of factors influencing the NCIR. One factor appeared to be litter species, another was the influence of initial litter N concentrations, and a third the influence of climate. Thus, the NCIR increased with increasing initial N concentration, a property which appeared to be in common for different species. They found this to be valid for at least a few pine species and for Norway spruce. Berg and Cortina [69] also noticed this when comparing NCIR for seven very different litter types incubated in one system.

One mechanism for conserving N in decomposing litter could be via covalent bonds to macromolecules during the humification process. A first step is the ammonium fixation described by Nömmik and Vahtras [72]. When the initial amount of N in the litter is higher, there will be more N available for fixation, giving a higher NCIR. Such a conclusion is reasonable since Axelsson and Berg [73] found that the N availability is limiting the rate of the process.

Earlier studies suggested that quinones were formed with N in the heterocyclic rings [74]. Two recent papers give further compounds [45, 75]. The study of Knicker [75]

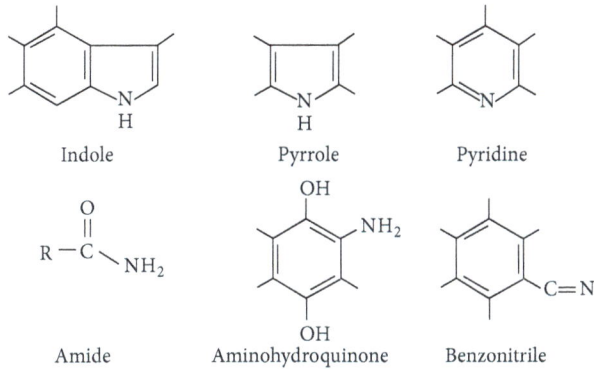

FIGURE 8: Products detected by ^{15}N-NMR after reaction of ^{15}N-labeled ammonium hydroxide with humic material after oxidative ammonolysis of lignin model compounds. Figure from [75].

suggested several heterocyclic components (Figure 8), which may be part of the recalcitrant complexes formed.

Influence of climate on NCIR. For local Scots pine needle litter and a unified Scots pine needle litter preparation, the relationship between NCIR and AET was investigated across a climatic gradient, with AET ranging from 380 to 520 mm. There was a highly significant positive relationship ($R^2_{adj} = 0.640$; $n = 31$; $P < 0.001$) indicating that the N concentration will increase faster relative to accumulated mass loss under a warmer and wetter climate. This correlation was significant when local and unified needle litters were used in combination as well as when using only local needle litter ($R^2_{adj} = 0.517$; $n = 18$; $P < 0.001$). Also, for litter of Norway spruce, the NCIR values increased with increasing AET values and the relationship was well significant ($R^2_{adj} = 0.534$; $n = 14$; $P < 0.01$). Combining data for brown coniferous litter resulted in a highly significant linear relationship with $R^2_{adj} = 0.569$; $n = 53$; $P < 0.001$ (Figure 7).

Thus, climate as indexed by AET is a significant factor in affecting the rate of N concentration increase in decomposing leaf litter. As the increases were calculated on the basis of accumulated mass loss rather than time, the results mean that, at a given accumulated mass loss, a particular litter decaying in an area with higher AET will contain more N than the same litter decaying in an area with lower AET.

Manganese Concentrations Change with Accumulated Mass Loss. There are few reviews and/or syntheses on Mn dynamics in decomposing litter. As seen from a comparison of Scots pine and Norway spruce needle litter, the former has not only lower Mn concentration in newly shed litter but there is also a clear difference in Mn dynamics during decomposition. Berg et al. [76] compared the two genera in two approaches, (i) by using information from 8 paired stands with Scots pine and Norway spruce and (ii) by using available data for Mn dynamics for 3 pine species and Norway spruce. They used 63 decomposition studies in which Mn was analysed in each litter sampling (546 data points) and related concentrations and remaining amounts of Mn in litter to accumulated mass loss.

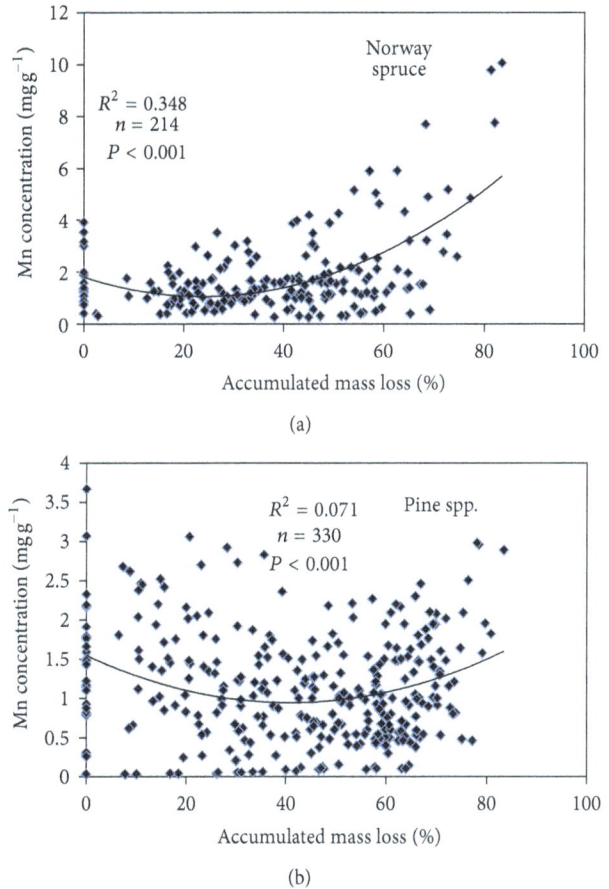

FIGURE 9: Manganese concentration in decomposing needle litter of Norway spruce (a), Scots pine, lodgepole pine, and Aleppo pine (b). Data originates from a climatic gradient. A quadratic function of the type $X^2 - X$ fitted both data sets reasonably well. Data from [58, 69, 76]. Please note the different scales on the Y axis.

For both the pine (37 studies) and the spruce litter (26 studies), the pattern for Mn concentration with accumulated mass loss varied among the single studies. The variation may depend on initial concentration, ranging from an increase at low initial concentrations to a clear decrease at high initial levels.

For both litters, a high initial Mn concentration resulted in a decrease in concentration as decomposition proceeded. Mainly it appeared that the concentration decreased and at c. 15–20% accumulated mass loss Mn concentrations reached a minimum after which the concentrations increased following accumulated litter mass loss (Figure 9).

Scots Pine versus Norway Spruce in Paired Stands as well as Available Data for Pine spp. and Norway Spruce. In their synthesis, Berg et al. [76] investigated how litter Mn concentration changed with accumulated mass loss in 8 paired stands with Norway spruce and Scots pine (pairwise the same soil and the same climate). The resulting patterns showed a clear difference between them with a not very clear pattern for Scots pine (cf. Figure 9). As decomposition proceeded, the Mn concentration reached a minimum, followed by a slight increase, which was significant, but not very pronounced. In contrast, the Norway spruce litter showed a very clear and strong increase in concentration, which was significantly stronger than for litter of Scots pine ($P < 0.0001$). For each group of litter, Mn concentration followed a significant positive quadratic function ($X^2 - X$). The average Mn concentration at 80% accumulated mass loss was 6.26 mg g^{-1} for Norway spruce litter and 1.47 mg g^{-1} for that of Scots pine. The quotient in Mn concentration at 80% mass loss between Norway spruce and Scots pine was 4.3.

In the same study, they combined all available data for pine species and Norway spruce and the pattern (Figure 9) was similar to that for the paired stands. Pine litter did not give a very clear pattern for changes in concentration and at 60–70% accumulated mass loss, the range in concentrations was about as wide as for the newly shed litter. A quadratic function was highly significant ($R^2 = 0.070$; $n = 330$; $P < 0.001$) but indicated a very low increase in concentration. For Norway spruce litter, on the other hand, there was a clear increase in concentration (Figure 9) and significantly higher than for pine species.

Manganese Release Patterns during Decomposition. Manganese release from litter was generally linear to accumulated mass loss. This was investigated for 63 decomposition studies encompassing Scots pine, lodgepole pine, Aleppo pine, and Norway spruce, and found to be linear for each individual study [76]. This linearity was used for comparing the Mn release rates for some litters with different initial concentrations.

The slope of such a linear relationship gives the release rate, and the slopes for the 63 relationships were related to litter initial Mn concentration. It appeared that the release rate was in proportion to the litters' initial concentration of Mn ($P < 0.001$) and this relationship was highly significant in spite of the different genera. Although the species groups fitted a common function, there were significant differences among them. Thus, the functions for the groups Norway spruce and pine species were significantly different ($P < 0.001$); [76] with a clearly lower release rate for Mn in Norway spruce litter (Figure 10).

3.3. A Conceptual Model Based on Three Identified Stages

3.3.1. Some Introductory Comments

Two Phases Have Developed to Three. Based on traditional analytical techniques, Berg and Staaf [40] set up a conceptual two-phase model for decomposition of Scots pine needle litter and included N as a rate-retarding factor for AUR decomposition in the late stage. Later, the model was developed to encompass three identifiable stages [28]. In addition to N, this modified model included Mn as an influencing factor for the late stage.

The *early phase* was based on the decomposition of non-lignified carbohydrates and mass-loss rate was enhanced by higher levels of the main nutrients N, P, and S. Further, in this phase, there was a direct effect of climate on decomposition. The *late phase* started when a net loss of AUR was observed

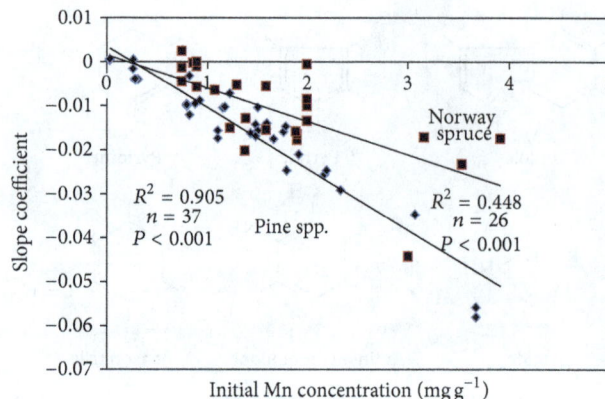

FIGURE 10: Net release/uptake in foliar litter of pine species and Norway spruce using Mn release coefficients for linear functions. Available data for pine species was used ($n = 37$) as well as for Norway spruce ($n = 26$). The resulting linear function for pine and spruce litter combined was highly significant ($R^2 = 0.635$; $n = 63$; $P < 0.001$). The figure shows that the data set could be subdivided after genus with the two functions significantly different at $P < 0.001$. Pine species (◆), Norway spruce (■). Data recalculated from [76].

and in contrast to the early phase, a raised N concentration would have a rate-suppressing effect, whereas a higher Mn concentration would increase the litter mass-loss rate. The effect of climate would decrease and possibly disappear. The *third stage* or humus-near stage, [28] was defined by the limit value, which identified a stable litter fraction.

Both new data and additional analytical work on lignin in litter [61] have made it necessary to clarify and develop the definition of the phases, considering the new information [34].

In a new approach, Hobbie et al. [77] followed the development of enzymatic activities during decomposition of leaf litter of white pine (*Pinus strobus*) and pin oak (*Quercus ellipsoidalis*) providing further support to the model. A further study [78] encompassed flowering dogwood (*Cornus florida*), red maple (*Acer rubrum*), and red oak (*Quercus rubra*). Also, in this case, studies on enzyme activities supported the model.

3.3.2. Early Stage: What Factors May Regulate the Decomposition Rate?

Substrate Chemical Composition. In the early phase, amounts and concentrations of water-soluble substances decrease quickly before reaching relatively similar and stable levels [31]. Also, free unshielded holocellulose is degraded in this phase. In the first work, based on AUR, no net loss of AUR was seen in the early phase. Recently, Klotzbücher et al. have found [61] that also in the early phase, there is some lignin degradation (cf. Figure 11). Their data set included Scots pine needle litter. Still, although there is some degradation of lignin, this appears not to influence the effects of the main nutrients and the degradation of carbohydrates appears to dominate the early phase.

The recent finding of Klotzbücher et al. [61] has shown that part of the native lignin thus may be degraded in

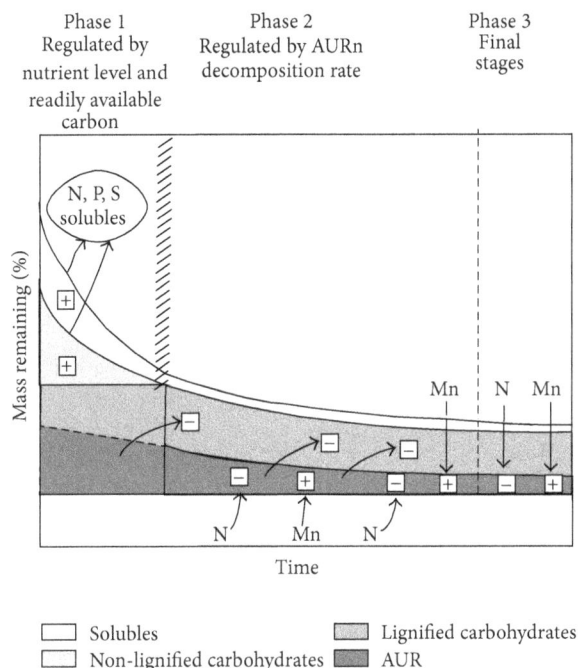

FIGURE 11: Conceptual model for rate-regulating factors and chemical changes during decomposition, modified from [28, 34]. The decomposition of water-soluble substances and unshielded cellulose/hemicellulose is stimulated by high levels of the major nutrients such as N, P, and S (early stage—phase 1). When the main part of all unshielded holocellulose is decomposed, lignin-encrusted holocellulose and lignin remain. The early phase has been suggested to last for even up to c. 40% accumulated mass loss for some pine litters [34]. For other foliar litters, for example, spruce and oak, the early phase has been found to be very short or possibly nonexisting. In the late stage—phase 2, the degradation of lignin dominates the litter decomposition rate. Nitrogen hampers the degradation of lignin and higher N concentrations suppress the decomposition, whereas Mn appears to have a stimulating effect on the degradation of lignin and thus on litter. Finally, in the humus-near stage (phase 3), the litter decomposition rate is (close to) zero and the accumulated mass loss reaches its limit value. The model is modified according to a suggestion of Klotzbücher et al. [81]. They found that there is a loss of lignin from the start of the incubation. Still it appears that for, for example, Scots pine and other pine litters the early phase is not dominated by lignin degradation.

the early stage. We may speculate that the lignin in the foliar litter tissue is not evenly distributed. As the lignification of the living tissue goes more slowly than the growth of the cellulose and hemicellulose, part of the carbohydrates is not lignified or only to a low extent when the litter is shed. It would be reasonable to assume that Klotzbücher et al. found that lignin in less lignified tissue was degraded [61].

Although this lignin mainly appears not to have any dominant role for the decomposition or for the phases, we may accept new results [42] suggesting that there was a negative correlation between the frequency of bonds related to, for example, lignin and tannins (Aromatic-C and Phenolic-C) and respiration rate from newly incubated whole litter.

This study, based on ^{13}C-NMR, needs further confirmation but appears to be a good example that all rate-suppressing effects in the early phase simply cannot be measured using gravimetric determinations (e.g., using litterbags). Further, it indicates an effect encompassing all bonds including both those in solubles and in solid components.

In the early phase, the mass-loss rate still may be positively related to total concentrations of the major nutrients, such as N, P, and S, which often are limiting for decomposition rates over several species [14], among them Scots pine [40]. We may note a recent discovery by Kaspari et al. [79] that even the highly soluble sodium (Na) has been found to be limiting for litter decomposition in areas at inland sites, namely, at a distance from sea-spray. Such an effect may apply to at least the early stage in addition to that of the main nutrients.

The stimulating effects of N in the early stage and its suppression of decomposition in the late one has been confirmed experimentally in three recent studies increasing the number of litter species for which the model is applicable. Using green leaves and leaf litter of white pine and pin oak with different initial concentrations of N, Hobbie et al. added inorganic N and organic N as well as a mix of the nutrients P, K, calcium (Ca), magnesium (Mg), and iron (Fe) to incubated litter [77]. They also incubated litter in a stand with long-term N additions. Calculating k_{init} (k_A) using an asymptotic function adapted for remaining amount, they obtained a significantly higher rate after N addition. We may see that decomposition of litter that had received N additions was stimulated and significantly faster. Effects of addition of inorganic N were not quite significant.

Perakis et al. [80], using Douglas fir (*Pseudotsuga menziesii*) needle litter with different concentrations of N confirmed also for this species that N is a limiting nutrient in the early stage. By adding N fertilizer (ammonium nitrate and urea), they found that the mass loss in the first 8 months increased as compared to the unfertilized litter. They obtained a significant relationship between mass loss and initial litter Mn concentration in N-fertilized plots but not in unfertilized. The initial P concentrations were similar among their eight litter preparations and appear not to have been limiting.

In their study, Carreiro et al. [78] used flowering dogwood, red maple, and red oak and confirmed the stimulating effect of added N in the early stage and a repressing effect in the late one.

Appearance Pattern of Enzyme Activities. Hobbie et al. [77] determined both cellulolytic and lignolytic enzyme activities after 6 months and 1, 2, and 3 years. We may see (Figures 12 and 13) that after six months' incubation, cellulolytic activity was well measurable and that positive effects of added N were observed indicating a higher production of these enzymes by the decomposing microorganisms. The activity of β-glucosidase and cellobiohydrolase increased and reached a maximum after one year of incubation after which a clear decrease took place. As regards phenol oxidase and peroxidase, the activity was just measurable at the first sampling (after 6 months) but increased with incubation time (Figure 13).

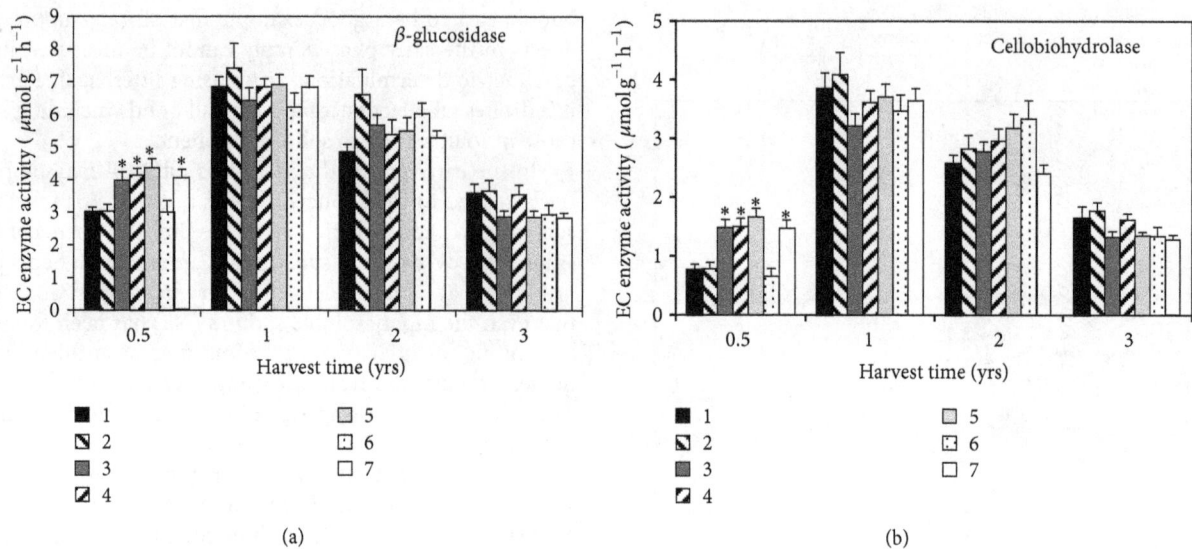

(a)   (b)

FIGURE 12: Hydrolytic enzyme activity (β-glucosidase and cellobiohydrolase) in white pine and pin oak foliar litter harvested during the first three years of a decomposition experiment. Values are given by treatment as averaged over sites and substrates. An asterisk indicates that a particular treatment differed significantly from the control treatment at a particular harvest time. Values are means with standard error bars. Overall model R^2 values from 3-way analysis of variance (ANOVA) including treatment, site, and substrate as main effects were done separately for each harvest date and ranged from 0.25 to 0.58 for β-glucosidase, 0.24 to 0.69 for cellobiohydrolase. (1) Control, (2) carbon addition (25.5 g C m^{-2} y^{-1} as glucose), (3) addition of inorganic N (10 g N m^{-2} y^{-1} as NH$_4$NO$_3$), (4) addition of carbon (25.5 g C m^{-2} y^{-1} as glucose) and inorganic N (10 N m^{-2} y^{-1} as NH$_4$NO$_3$), (5) long-term N additions (10 g N m^{-2} y^{-1} as NH$_4$NO$_3$ since 1999), (6) addition of non-N nutrients (P, K, Ca, Mg, S, Fe), (7) addition of organic N (10 g N m^{-2} y^{-1} as amino acids). From [77].

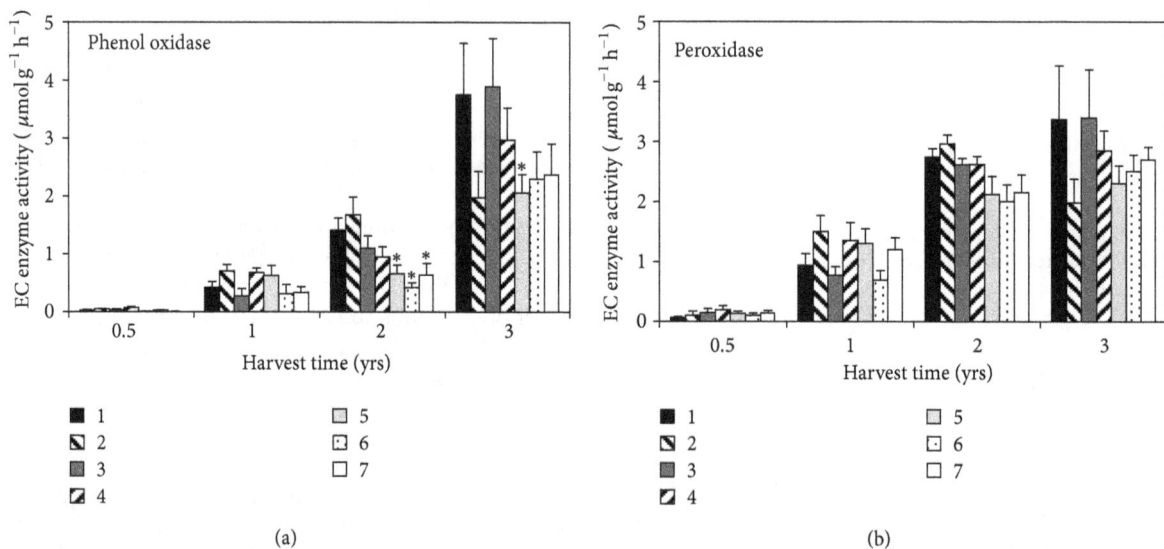

(a)   (b)

FIGURE 13: Oxidative enzyme activities (phenoloxidase and peroxidase) in white pine and pin oak foliar litter harvested during the first three years of a decomposition experiment. Values are given by treatment and averaged over sites and substrates. Statistical comparisons were only done for the 2- and 3-year harvests because of high numbers of zero values in prior harvests. An asterisk indicates that a particular treatment differed significantly from the control treatment at a particular harvest time. Values are means with standard error bars. Overall model R^2 values from 3-way analysis of variance (ANOVA) including treatment, site, and substrate as main effects were 0.34 and 0.24 for phenoloxidase and 0.22 and 0.23 after 2 and 3 years of decomposition, respectively. (1) Control, (2) carbon addition (25.5 g C m^{-2} y^{-1} as glucose), (3) addition of inorganic N (10 g N m^{-2} y^{-1} as NH$_4$NO$_3$), (4) addition of carbon (25.5 g C m^{-2} y^{-1} as glucose) and inorganic N (10 N m^{-2} y^{-1} as NH$_4$NO$_3$), (5) long-term N additions (10 g N m^{-2} y^{-1} as NH$_4$NO$_3$ since 1999), (6) addition of non-N nutrients (P, K, Ca, Mg, S, Fe), (7) addition of organic N (10 g N m^{-2} y^{-1} as amino acids). From [77].

Climate Influence. For newly shed litter, it appears that climate may influence litter mass-loss rate. For local needle litter of Scots pine, it has been possible to demonstrate a clear influence of climate on decomposition rate, using the range in climate within a 2000 km-long gradient [7]. The mass loss in the first year ranged from about 10.9% in northern Finland (close to Barents Sea) to about 43.7% in northern Germany. The dominant rate-regulating factor was the climate, as indexed by annual actual evapotranspiration (AET) or by mean annual temperature (MAT), and none of the substrate-quality factors alone was significant. Using unified Scots pine litter, Berg et al. [3] showed the climate relationship in a gradient from northern Finland to southern United States (southern Georgia). Such effects of climate could thus be recorded for local and unified pine needle litter in pine forests with their relatively open canopy covers.

Still, a more general relationship was demonstrated [4]. By combining available data for broadleaf and coniferous litter, the authors found a relationship between first-year mass loss and MAT. The geographic range was considerable and extended from the Equator to north Scandinavia. However, when investigating separate functional groups, the authors found clear differences among pine species, spruce, and oak species. Thus, mass-loss rate for pine litter was in highly significant and positive linear relationship to MAT, whereas that of spruce showed no relationship.

The Extent of the Early Phase and a Possible Transition Stage between the Early and the Late Stages. The extent of the early stage has been suggested to be c. 25–27% accumulated mass loss [40, 82] and later Berg and McClaugherty [34] suggested that it may extend to c. 40% accumulated mass loss. These results were based on the response of the decomposition to nutrient concentration versus that to AUR concentration. We cannot exclude, however, that both the extent of a well-defined early phase and a less clear transition phase to the late stage are more unclear and possibly there is, in addition, a temporal variation over a wide range even within one species. First, the lignification in the green leaf, even within one species may be variable among years and place of growth, which may be reflected in the litter. We do not know the size of this variation. Further, we cannot exclude that the transition between the early and the late stage will be less distinguishable with a less clear response to both climate and to concentrations of N, P, and S. Also, a negative response to N and a positive one to Mn may be less distinguishable (B. Berg and J. Kjønaas, unpubl.). Such a transition stage is reasonable to expect and we can expect that new analytical techniques may improve the possibility to distinguish it.

3.3.3. The Late Stage. What Factors May Regulate Decomposition Rate? Berg and Staaf [40] defined a late stage as the one in which the decomposition of gravimetric lignin/AUR dominated the mass loss of litter. This definition was later improved by Berg and Matzner [28]. In both studies, AUR was used instead of lignin. Although AUR is not really an acceptable replacement to native lignin, we may use it as an index of increasing recalcitrance. In their model, Berg and

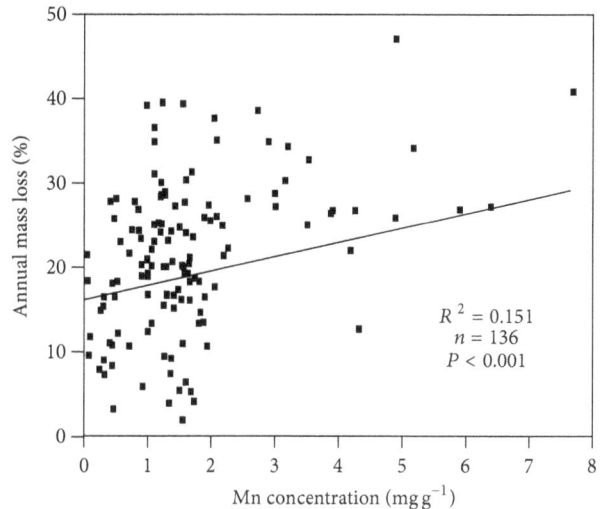

FIGURE 14: All available data (Norway spruce, lodgepole pine, Scots pine, Aleppo pine, silver birch, grey alder) for annual mass loss of foliar litter in late stages were related to litter Mn concentration at the start of each year. Mass-loss data originate from sites distributed over Sweden plus from two sites in northern Libya. From [85].

Matzner [28] included the effect of N concentration as a rate-retarding agent of lignin decomposition (above). The positive effect of Mn on production of lignolytic enzymes [83, 84] was followed by a report on positive relationships between Mn concentration and litter mass-loss rate [85] (Figure 14). Such effects were found for litter of lodgepole pine and Norway spruce as well as for a mix of litter species [85].

Incubating leaves and leaf litter of white pine and pin oak in N-fertilized plots, Hobbie et al. followed the incubated litter over time. After 1 year, they found clearly reduced rates for the litter that had received N additions [77]. In their study on the development of cellulolytic and lignolytic enzymes in the sampled litter, they noted (i) a heavy increase in lignolytic enzymes after one year of incubation and (ii) a decrease of lignolytic enzymes after N additions. That study [77] supports the proposed phases.

In a study using needle litter of Douglas fir [80], a significant decrease (late stage) was found in mass-loss rates after additions of ammonium nitrate.

3.3.4. The Very Late Stages and the Concept "Limit Value." First in the study of Howard and Howard [12] and later in two independent ones [13, 14], the concept limit value was described (Figure 15). Further, it was found that the limit values were different among litter species. A "limit value" is estimated using a function that gives an asymptotic value for the accumulated mass loss. The following one has been suggested [13, 14]:

$$L_t = m\left(1 - e^{-kt/m}\right), \qquad (3)$$

where L_t is the accumulated mass loss (in percent), t time in days, k the decomposition rate at the beginning of decay, and m the asymptotic level that the accumulated mass loss will ultimately reach, normally not 100% and often considerably

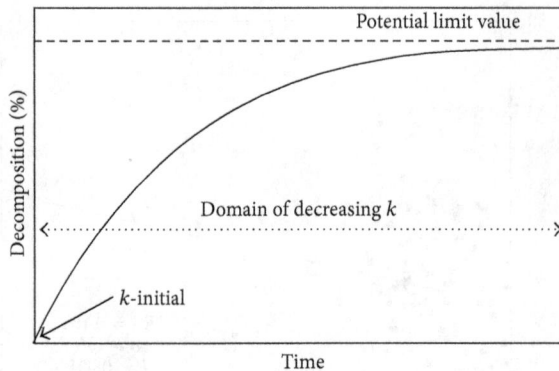

FIGURE 15: Decomposition rates for decomposing litter gradually decrease and the accumulated mass loss may approach an asymptote, a limit value. The process may be described using (3), which may be used to estimate the decomposition rate at any point in time. The initial decomposition rate (k-initial) gives the rate at time 0 from [34].

less. From a high initial value at time zero its value decreases as the process proceeds and the rate may reach the value zero at the limit value (Figure 15). The k of this function should not be directly compared to rate constants estimated with other models.

Available literature data [19] allowed the estimation of all 106 limit values for foliar litter decomposing in natural systems. When regressing these against the initial N concentrations, a highly significant and negative relationship was obtained (R^2 = 0.323; n = 106; P < 0.001). Later, [76] a highly significant relationship was found between 149 limit values and initial N concentration. The possible causal background to this relationship between limit values and N has been discussed (above). The fact that in this large data set the relationship to N concentration was significant may indicate a general effect of N over a good number of species, in that case 16 ones. Thus it may encompass both deciduous and coniferous litter and ecosystems in boreal and temperate forests.

However, in a study on limit values for just one litter genus, namely, pine (*Pinus* spp.) needle litter, mainly that of Scots pine, Berg et al. [41], using backward elimination and local litter in a climatic gradient, found that initial Mn concentration was the only significant factor for limit values. The backward elimination procedure simply removed the non-significant factors stepwise and out of ten ones, namely, MAT, MAP, water solubles, AUR, and six main nutrients only Mn was selected as a significant factor. Using the same data set, Berg and McClaugherty [34] applied a quadratic function of the type $-X^2 + X$ and obtained a highly significant and improved relationship. That relationship was based on litter from four pine species and a range in litter Mn concentrations from 0.03 mg g^{-1} to 3.1 mg g^{-1}. That data set originated from a climate gradient with MAT ranging from −0.7 to 17°C. In contrast to the above finding for 16 deciduous and coniferous species [76], there was no significance for N in that data set using exclusively pine litter. Of the eight substrate-quality

factors and two environmental ones, Mn concentration was left as the single factor.

The limit value as such is still mainly an empirical finding although we have both causal relationships and highly significant relationships to Mn concentrations. Further, using a single genus (*Pinus*) supported the relationship to Mn.

In order to determine specific factors that influence the limit values it appears reasonable to study separate genera or possibly litter species. Thus, for common oak leaf litter, there was no relationship to litter Mn concentration but to that of Ca [6]. An attempt to relate limit values for Norway spruce to litter Mn concentration was not successful. However, Berg [86] found a significant and positive relationship for limit values and Ca concentration for Norway spruce litter. Thus, with two studies giving similar results, we cannot exclude that an effect of Ca on limit values may be related to at least these litter species.

Still, using available data, we may see some general trends. Berg et al. [76] using a data set with 149 limit values to initial litter concentrations of both N and Mn obtained highly significant relationships in both cases (Figure 17).

We have mentioned (above) that the litter becomes increasingly enriched in the AUR complex, including recombination products, the stability of which appears to be related to both N and Mn. However, of substrate-quality factors, we cannot neglect the possible role of the increasing concentrations of heavy metals such as Cu, Pb, Fe, and Zn. There are few data revealing the dynamics of these four heavy metals, but some data has been published for decomposing Scots pine needle litter [32, 34]. For Scots pine litter, all four heavy metals increased in concentration when related to accumulated mass loss. Such concentration increases may be correlated to that of N and could explain part of the significant relationships.

Pine litter appears in general to have linear net release of Mn [76]. With Mn release being linear to accumulated mass loss, it may well be correlated to the increasing concentrations of N and to the increasing concentrations of the above heavy metals. For all cases (Mn, N, and heavy metals), we may see potential causal relationships to a retarded decomposition. Still, the roles of the main heavy metals need to be clarified, although we may keep in mind that such a role may be related to site properties, too.

3.4. Decomposition Patterns

Patterns, Functions and Influences. The decomposition patterns or the shape of accumulated mass loss plotted versus time may develop following different functions. One has been discussed, namely (3), a further one is (4) [88, 89], which describes total decomposition as follows;

$$\ln\left(\frac{M_t}{M_0}\right) = -kt, \qquad (4)$$

in which M_0 is initial mass and M_t mass at time t, k is the rate constant and t is time.

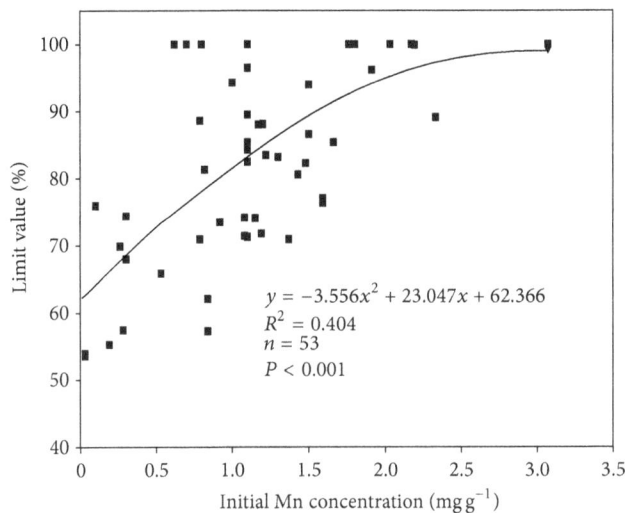

FIGURE 16: The level of the limit value, given as accumulated mass loss. The limit value for needle litter of pine (Pinus) species appears to be related to litter Mn concentration. Decomposition studies from four pine litter species, mainly Scots pine, were combined into a climatic gradient and limit values estimated using litter accumulated mass loss. A backward elimination procedure removed factors that were not significant. Of MAT and MAP and eight substrate-quality factors, namely, water solubles, AUR, N, P, K, Ca, Mg, and Mn, manganese was selected as the single significant factor. Figure from [34] based on data from [21].

Equation (3) (above) allows the calculation of a stable fraction, identified by the mass-loss rate of zero, thus separating the readily decomposed organic matter from the stabilized fraction, which may be of different sizes. This means that a graph describing the accumulated mass loss may approach very different limit values (Figure 15). Thus, the litters with different limit values (Figure 16) may be considered to have different patterns.

To discuss decomposition pattern, we may use an index and I have used the limit value as provisional index. The decomposition pattern thus differs depending on the level of the limit value. A complete decomposition with a limit value of 100% and one with a lower at, say, 45% would give different patterns or differing curvatures of the graph.

Pattern versus Litter Nutrients. It appears that the decomposition patterns for pine litter may be related to initial Mn concentrations. Manganese appears to be the dominant factor determining limit values, at least in pine ecosystems (Figure 16). In a study [21] on local, transplanted, and experimental pine needle litter, in all 56 decomposition studies ("sequences"), the role of Mn versus limit value was determined. Using backward elimination, Berg et al. investigated the eight substrate-quality variables N, P, K, Ca, Mg, Mn, AUR, and water solubles plus MAT and MAP. The litter was mainly that of Scots pine but the set included lodgepole pine, stone pine, and Aleppo pine.

The positive relationship between initial concentrations of Mn and limit values for pine needles (Figure 16) is empirical, as is the more general relationship between N and limit values found in earlier studies [19] although both are

based on causal relationships. In addition, the effect of Mn on the degradation of lignin and thus on lignified tissue as well as on the degradation of secondary products such as humic acid [84, 90] has been demonstrated with highly decomposed litter (Figure 14) [85].

We may speculate that the positive relationship between the limit value and the total concentration of Mn in highly decomposed litter arises because higher Mn concentrations enable the decomposition to proceed further before a recalcitrant fraction is developed. The formation of a very slowly decomposing fraction may occur when litter approaches its limit value because the remaining Mn at this stage is bound in a form that is relatively unavailable to the degrading microorganisms. The dynamics of total Mn in decomposing foliar litter are variable over litter types but have been little studied [69, 76].

In earlier investigations [19, 20], N was significantly ($P <$ 0.001) and negatively related to limit values and the causal relationships are well established [72, 84]. These relationships were based on available data, including 16 species (Figure 17). Although Mn was selected for pine litter [21], Berg et al. found that N was the last variable to be eliminated. The reason for N not being included in the final model for pine litter could be that N has no essential relationship with the limit value, and alternatively that the concentration range in N was too limited in the studied material.

Thus, we cannot exclude that N simply may have a lower influence on limit values than Mn in decomposition of pine needle litter, possibly because the range of initial litter N concentrations was relatively narrow among substrates. By contrast, the negative relationship between N and limit values observed by Berg [19] using 106 limit values was based on 21 different litter species/types with a wide range in initial N concentrations (2.9 to 30.7 mg g^{-1}). However, in both cases, there are strong general causal relationships between Mn or N and the microbial degradation of lignin (and lignified tissue) that could explain these empirical relationships [84, 91].

Can climate influence the pattern? The parameter MAT may deserve to be commented upon. There is a relationship between litter Mn concentration in foliar litter and MAT as well as one between MAT/AET and N concentration. Already in 1995, a highly significant and negative relationship was found between Mn concentration in the newly shed pine needle litter and MAT, suggesting that litter Mn concentration may also reflect the local climate [48]. The corresponding relationship to N concentration was positive.

These results suggest that higher MAT is associated with lower Mn and higher N concentrations in the needle litter that is formed at a given site. Although the mechanism underlying these relationships needs further studies, an effect of MAT on Mn and N concentrations in litter may influence the fraction of organic matter that becomes recalcitrant, leading to a regional and climate-related variation in the formation of recalcitrant organic matter from *Pinus*.

We discussed the strong variation in the quotient N-to-Mn in newly shed litter over a climatic gradient and gave the ratio of 2.5 at a MAT of 0°C, at 10°C it was 10 times as high and at 17° more than 50 times higher. Using the

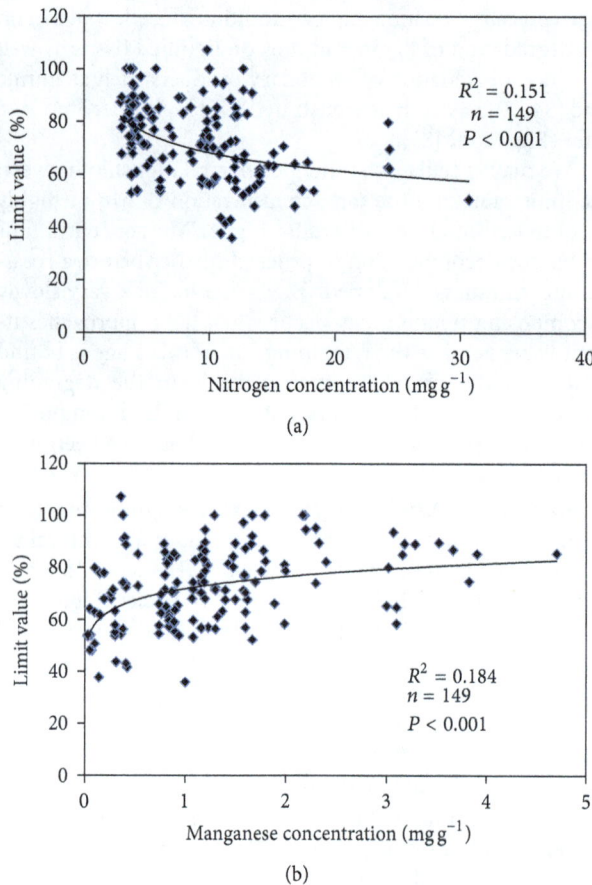

FIGURE 17: Relationships (power function) between initial litter concentration of N and Mn and limit values. (a) Nitrogen, (b) Manganese. In both cases, the same set of data was used with 149 decomposition studies and 16 species included (DELILA II database, http://www.eko.uj.edu.pl/deco/). Figure from [87].

available information of the roles of N and Mn, the litter chemical composition thus would influence the pattern and give a less degradable litter at a warmer climate—unless so far unknown factors influence the litter's degradability. So far the decomposition studies appear to have confirmed this for pine litter.

3.5. Carbon Sequestration: Amounts and Patterns

3.5.1. Can We Use Limit Values to Estimate C Sequestration Rates in Humus Layers? We defined the limit value based on accumulated litter mass loss (3). Although we have defined the limit value by estimating the accumulated mass loss at which the rate becomes 0 for a given fraction, it does not mean that the "stable" fraction is completely stable or biologically undegradable. We may only conclude that the biological degradation is slow [82]. Although we appear to have very little direct information, we may speculate about how slow it may be and if this rate may change. Coûteaux et al. [82] comparing decomposition rates of Scots pine humus and far-decomposed litter (close to the limit value) obtained a rate of "less than 0.0001% d^{-1}" or less than 0.037% per year.

For our discussion, we may call the remaining stable fraction "humus" and ask about what factors that influence its stability. Berg and Matzner [28], quoting Bringmark and Bringmark [29], reported that a high N concentration was rate-suppressing for respiration from humus-layer samples collected over a region. Still, total N concentration was just one factor and we have no answer to how many further factors there are.

If a sample of a humus layer is taken out from its environment, we may measure its respiration rate and several studies give rates that sometimes correspond to 100% decomposition in a short time, even less than a year. Still, with the humus undisturbed, the humus layer will last and grow, and undisturbed humus layers have been found to develop for close to 3000 years reaching a depth of almost 1.5 meters [23].

Although foliar litter is not the only litter component, forming humus layers, it is a major one and in some ecosystems even the dominant one. Using data from a Scots pine monoculture c. 120 years old, with a humus layer developed on ashes from the latest forest fire, it has been possible to reconstruct the amount of stored carbon in the humus layer [24, 26] using extensive litter-fall data and an average limit value of 89.0% ($n = 8$). A similar calculation was made for the close to 3000 year-old humus layers described by Wardle et al. [23] and a quantitative reconstruction was made of three groups of humus layers with the average age of 1106, 2081, and 2984 years [24, 26].

3.5.2. Tree Species Influence the C Sequestration Rate. Several attempts have been made to study the accumulation rate of humus in humus layers, the "primary sequestration" [46]. In most cases, this appears to have been made in monocultures or tree species trials. The experiment of Ovington [92] showed clearly that there was a linear relationship between stand age and the increase in mass (and C) in the LFH layer on top of mineral soil.

The rate of C sequestration in a well-developed humus layer is easily determined using gravimetric measurements of the humus-layer C and a comparison to the time for its buildup. Still, there are relatively few such values reported in the scientific literature, possibly due to the long accumulation time and the absolute request for good information about stand history and the humus layer when the buildup started.

To evaluate the effect of tree species on C sequestration rates, Berg and McClaugherty [34] made an evaluation of existing data and focused on two larger studies, one in Denmark [37] and one in the UK [38], together encompassing 10 trials with c. 16 tree species in paired stands. Each of these studies was made with the purpose of evaluating several tree species, which also were occurring in both studies. With a similarity in climate, this allowed an evaluation of the species ability to sequester C in the humus layer. Site history was given [38, 93].

An analysis of variance gave a rather clear response in spite of a high variation in sequestered carbon. Berg and McClaugherty [34] distinguished that coniferous species/ genera had a significantly higher sequestration rate—twice as high—as deciduous species (Figure 18). They found three main groups with pine and spruce species having the highest

Pine Spruce
_____
Fir Larch Beech
_____ Grand fir Other decid. Oak spp
_____
481 355 266 283 227 196 112 103

Sequestration rate in LFH layers (kg C ha^{-1} yr^{-1})

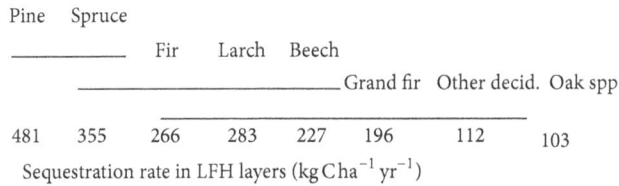

FIGURE 18: Data from 7 tree species trials in Denmark [37] and 3 in England [38]. Each of the studies reported a carefully determined amount of soil organic matter on top of the mineral soil in monocultural stands, "primary sequestration" [46]. Data were combined to analyse and determine the variation among genera [34]. The full line underlining genus names gives C accumulation rates that are not significantly different. Ash was subtracted and 50% C was assumed for the organic matter of Ovington's study [38]. Compiled data and figure are from [34].

sequestration rate, followed by fir, larch, and beech and in the third place "other deciduous." A similar observation was the evaluation of Berg [94], based on existing data [95, 96] comparing sequestration rates over 30 years in paired stands and observing about twice as high a rate in a Norway spruce stand as compared to one with common beech (*Fagus sylvatica*).

3.5.3. Regional Approach to Determine C Sequestration Rates in Well-Developed Humus Layers. We may argue that sequestration rates should be related to both foliar litter fall, and the size of the stable residue. However, in a comparison of litter fall from Scots pine and Norway spruce over Sweden [50, 97], a higher foliar litter fall was found for Norway spruce than for Scots pine. Still, the C sequestration rate in humus layers was higher in Scots pine stands when rates were compared for Sweden as a region [98].

In two attempts to calculate the C sequestration in the humus layer over a region using countrywide (Sweden), directly measured data, Berg et al. obtained sequestration rates [98] that were close to those estimated for the same region using limit values [97]. In a comparison [99], three different approaches were compared for the same region (the forested land of Sweden), providing similar results.

Although the numbers obtained by a calculation using limit values can be verified by direct measurements, we cannot exclude the possibility that calculations using litter fall data and limit values rather give an index for C sequestration. We may consider that there are several litter components that may be degraded with different decomposition patterns, which additionally may differ among species and ecosystems.

Conflict of Interests

The author declares that there is no conflict of interests regarding the publication of this paper.

References

[1] C. E. Prescott, "Do rates of litter decomposition tell us anything we really need to know?" *Forest Ecology and Management*, vol. 220, pp. 66–74, 2005.

[2] V. Meentemeyer, "Macroclimate and lignin control of litter decomposition rates," *Ecology*, vol. 59, pp. 465–472, 1978.

[3] B. Berg, M. P. Berg, P. Bottner et al., "Litter mass loss rates in pine forests of Europe and Eastern United States: some relationships with climate and litter quality," *Biogeochemistry*, vol. 20, no. 3, pp. 127–153, 1993.

[4] H. Kang, B. Berg, C. Liu, and C. J. Westman, "Variation in mass-loss rate of foliar litter in relation to climate and litter quality in eurasian forests: differences among functional groups of litter," *Silva Fennica*, vol. 43, no. 4, pp. 549–575, 2009.

[5] B. Berg, M.-B. Johansson, and V. Meentemeyer, "Litter decomposition in a transect of Norway spruce forests: substrate quality and climate control," *Canadian Journal of Forest Research*, vol. 30, no. 7, pp. 1136–1147, 2000.

[6] M. P. Davey, B. Berg, B. A. Emmett, and P. Rowland, "Decomposition of oak leaf litter is related to initial litter Mn concentrations," *Canadian Journal of Botany*, vol. 85, no. 1, pp. 16–24, 2007.

[7] M. B. Johansson, B. Berg, and V. Meentemeyer, "Litter mass-loss rates in late stages of decomposition in a climatic transect of pine forests. Long-term decomposition in a Scots pine forest. IX," *Canadian Journal of Botany*, vol. 73, no. 10, pp. 1509–1521, 1995.

[8] R. Fogel and K. Cromack, "Effect of habitat and substrate quality on Douglas fir litter decomposition in western Oregon," *Canadian Journal of Botany*, vol. 55, pp. 1632–1640, 1977.

[9] C. M. Preston, J. R. Nault, J. A. Trofymow, C. Smyth, and The CIDET Working Group, "Chemical changes during 6 years of decomposition of 11 litters in some Canadian forest sites. Part 1: elemental composition, tannins, phenolics, and proximate fractions," *Ecosystems*, vol. 12, no. 7, pp. 1053–1077, 2009.

[10] C. M. Preston, J. R. Nault, J. A. Trofymow, and The CIDET Working Group, "Chemical changes during 6 years of decomposition of 11 litters in some Canadian forest sites. Part 2: ^{13}C abundance, solid-state ^{13}C NMR spectroscopy and the meaning of 'lignin'," *Ecosystems*, vol. 12, no. 7, pp. 1078–1102, 2009.

[11] T. Persson, E. Bååth, M. Clarholm et al., "Trophic structure, biomass dynamics and carbon metabolism of soil organisms in a Scots pine forest," *Ecological Bulletins*, vol. 32, pp. 419–462, 1980.

[12] P. J. A. Howard and D. M. Howard, "Microbial decomposition of tree and shrub leaf litter. I. Weight loss and chemical composition of decomposing litter," *Oikos*, vol. 25, no. 3, pp. 341–352, 1974.

[13] R. K. Wieder and G. E. Lang, "A critique of the analytical methods used in examining decomposition data obtained from litter bags," *Ecology*, vol. 63, no. 6, pp. 1636–1642, 1982.

[14] B. Berg and G. Ekbohm, "Litter mass-loss rates and decomposition patterns in some needle and leaf litter types. Long-term decomposition in a Scots pine forest. VII," *Canadian Journal of Botany*, vol. 69, no. 7, pp. 1449–1456, 1991.

[15] B. Berg and G. Ekbohm, "Decomposing needle litter in *Pinus contorta* (lodgepole pine) and *Pinus sylvestris* (Scots pine) monocultural systems—is there a maximum mass loss?" *Scandinavian Journal of Forest Research*, vol. 8, no. 4, pp. 457–465, 1993.

[16] W. S. Currie, M. E. Harmon, I. C. Burke, S. C. Hart, W. J. Parton, and W. Silver, "Cross-biome transplants of plant litter show decomposition models extend to a broader climatic range but lose predictability at the decadal time scale," *Global Change Biology*, vol. 16, no. 6, pp. 1744–1761, 2010.

[17] M. E. Harmon, W. L. Silver, B. Fasth et al., "Long-term patterns of mass loss during the decomposition of leaf and fine root litter:

an intersite comparison," *Global Change Biology*, vol. 15, no. 5, pp. 1320–1338, 2009.

[18] B. Berg, R. Laskowski, and A. Virzo De Santo, "Estimated nitrogen concentrations in humus based on initial nitrogen concentrations in foliar litter: a synthesis. XII. Long-term decomposition in a Scots pine forest," *Canadian Journal of Botany*, vol. 77, no. 12, pp. 1712–1722, 1999.

[19] B. Berg, "Litter decomposition and organic matter turnover in northern forest soils," *Forest Ecology and Management*, vol. 133, no. 1-2, pp. 13–22, 2000.

[20] B. Berg, G. Ekbohm, M. Johansson, C. McClaugherty, F. Rutigliano, and A. V. De Santo, "Maximum decomposition limits of forest litter types: a synthesis," *Canadian Journal of Botany*, vol. 74, no. 5, pp. 659–672, 1996.

[21] B. Berg, A. De Marco, M. P. Davey et al., " Limit values for foliar litter decomposition—pine forests," *Biogeochemistry*, vol. 100, no. 1, pp. 57–73, 2010.

[22] B. Berg, "Scots pine needle litter—can it give a mechanism for carbon sequestration?" *Geografia Polonica*, vol. 85, no. 2, pp. 13–23, 2012.

[23] D. A. Wardle, O. Zackrisson, G. Hörnberg, and C. Gallet, "The influence of island area on ecosystem properties," *Science*, vol. 277, no. 5330, pp. 1296–1299, 1997.

[24] B. Berg, C. McClaugherty, A. Virzo De Santo, and D. Johnson, "Humus buildup in boreal forests: effects of litter fall and its N concentration," *Canadian Journal of Forest Research*, vol. 31, no. 6, pp. 988–998, 2001.

[25] K. E. Rehfüss, *Waldböden, Entwicklung, Eigenschaften und Nutzung*, vol. 29 of *Pareys Studientexte*, Parey, Hamburg, Germany, 2nd edition, 1990, (German).

[26] B. Berg and N. Dise, "Calculating the long-term stable nitrogen sink in northern European forests," *Acta Oecologica*, vol. 26, no. 1, pp. 15–21, 2004.

[27] B. Berg and N. Dise, "Validating a new model for N sequestration in forest soil organic matter," *Water, Air, and Soil Pollution*, vol. 4, no. 2-3, pp. 343–358, 2004.

[28] B. Berg and E. Matzner, "Effect of N deposition on decomposition of plant litter and soil organic matter in forest systems," *Environmental Reviews*, vol. 5, no. 1, pp. 1–25, 1997.

[29] E. Bringmark and L. Bringmark, "Large-scale pattern of mor layer degradation in Sweden measured as standardized respiration," in *Humic Substances in the Aquatic and Terrestrial Environments. Proceedings of an International Symposium, Linköping, Sweden, August 1989*, vol. 33 of *Lecture notes in earth sciences*, pp. 255–259, Springer, Berlin, Germany, 1991.

[30] G. Guggenberger, "Acidification effects on dissolved organic matter mobility in spruce forest ecosystems," *Environment International*, vol. 20, no. 1, pp. 31–41, 1994.

[31] B. Berg, K. Hannus, T. Popoff, and O. Theander, "Changes in organic chemical components of needle litter during decomposition. Long-term decomposition in a Scots pine forest: I," *Canadian Journal of Botany*, vol. 60, no. 8, pp. 1310–1319, 1982.

[32] B. Berg and C. McClaugherty, *Plant Litter. Decomposition. Humus Formation. Carbon Sequestration*, Springer, Berlin, Germany, 2nd edition, 2008.

[33] H. Kang, Z. Xin, B. Berg et al., "Global patterns of leaf litter nitrogen and phosphorus stoichiometry in woody plants with latitude and climatic factors," *Annals of Forest Science*, vol. 67, pp. 811–818, 2010.

[34] B. Berg and C. A. McClaugherty, *Plant Litter. Decomposition. Humus Formation. Carbon Sequestration*, Springer, Berlin, Germany, 3rd edition, 2014.

[35] B. Berg, A. Virzo De Santo, F. A. Rutigliano, A. Fierro, and G. Ekbohm, "Limit values for plant litter decomposing in two contrasting soils—influence of litter elemental composition," *Acta Oecologica*, vol. 24, no. 5-6, pp. 295–302, 2003.

[36] B. Wessén and B. Berg, "Long-term decomposition of barley straw: chemical changes and ingrowth of fungal mycelium," *Soil Biology and Biochemistry*, vol. 18, no. 1, pp. 53–59, 1986.

[37] L. Vesterdal and K. Raulund-Rasmussen, "Forest floor chemistry under seven tree species along a soil fertility gradient," *Canadian Journal of Forest Research*, vol. 28, no. 11, pp. 1636–1647, 1998.

[38] J. D. Ovington, "Studies on the development of woodland conditions under different trees. II. The forest floor," *The Journal of Ecology*, vol. 42, pp. 71–80, 1954.

[39] B. Berg, C. Liu, R. Laskowski et al., "Relationships between nitrogen, AUR, and climate among tree foliar litters," *Canadian Journal of Forest Research*, vol. 43, pp. 103–107, 2013.

[40] B. Berg and H. Staaf, "Decomposition rate and chemical changes of Scots pine needle litter. II. Influence of chemical composition," in *Structure and Function of Northern Coniferous Forests. An Ecosystem Study*, T. Persson, Ed., vol. 32, pp. 373–390, Ecological Bulletins, Stockholm, Sweden, 1980.

[41] B. Berg and C. O. Tamm, "Decomposition and nutrient dynamics of litter in long-term optimum nutrition experiments. II.Nutrient concentrations in decomposing *Picea abies* needle litter," *Scandinavian Journal of Forest Research*, vol. 9, no. 2, pp. 99–105, 1994.

[42] B. Erhagen, M. Öquist, T. Sparrman et al., "Temperature response of litter and soil organic matter decomposition is determined by chemical composition of organic material," *Global Change Biology*, 2013.

[43] A. De Marco, R. Spaccini, P. Vittozzi et al., "Decomposition of black locust and black pine leaf litter in two coeval forest stands on Mount Vesuvius and dynamics of organic components assessed through proximate analysis and NMR spectroscopy," *Soil Biology and Biochemistry*, vol. 51, pp. 1–15, 2012.

[44] B. Berg, H. G. W. Booltink, A. Breymeyer et al., "Data on needle litter decomposition and soil climate as well as site characteristics for some coniferous forest sites. 2nd ed. Section 2. Data on needle litter decomposition," Report 42, Departments of Ecology and Environmental Research. Swedish University of Agricultural Sciences, 1991.

[45] K. A. Thorn and M. A. Mikita, "Ammonia fixation by humic substances: a nitrogen-15 and carbon-13 NMR study," *Science of the Total Environment*, vol. 113, no. 1-2, pp. 67–87, 1992.

[46] B. Berg, C. McClaugherty, and A. Virzo De Santo, "Practicalities of estimating carbon sequestration," *CAB Reviews*, vol. 3, no. 84, pp. 1–15, 2008.

[47] G. Tyler, "Changes in the concentrations of major, minor and rare-earth elements during leaf senescence and decomposition in a *Fagus sylvatica* forest," *Forest Ecology and Management*, vol. 206, pp. 167–177, 2005.

[48] B. Berg, R. Calvo de Anta, A. Escudero et al., "The chemical composition of newly shed needle litter of Scots pine and some other pine species in a climatic transect. X. Long-term decomposition in a Scots pine forest," *Canadian Journal of Botany*, vol. 73, no. 9, pp. 1423–1435, 1995.

[49] C. Liu, B. Berg, W. Kutsch et al., "Leaf litter nitrogen concentration as related to climatic factors in Eurasian forests," *Global Ecology and Biogeography*, vol. 15, no. 5, pp. 438–444, 2006.

[50] B. Berg and V. Meentemeyer, "Litter fall in some European coniferous forests as dependent on climate: a synthesis," *Canadian Journal of Forest Research*, vol. 31, no. 2, pp. 292–301, 2001.

[51] C. Liu, C. J. Westman, B. Berg et al., "Variation in litterfall-climate relationships between coniferous and broadleaf forests in Eurasia," *Global Ecology and Biogeography*, vol. 13, no. 2, pp. 105–114, 2004.

[52] B. Berg and J. E. Lundmark, "Decomposition of needle litter in *Pinus contorta* and *Pinus sylvestris* monocultures—a comparison," *Scandinavian Journal of Forest Research*, vol. 2, pp. 3–12, 1987.

[53] B. Berg and C. O. Tamm, "Decomposition and nutrient dynamics of Norway spruce needle litter in a long-term optimum nutrition experiment. I. Organic matter decomposition," Tech. Rep. 39, Department of Ecology and Environmental Sciences. Swedish University of Agricultural Sciences, 1991.

[54] C. A. McClaugherty, "Soluble polyphenols and carbohydrates in throughfall and leaf litter decomposition," *Acta Oecologia*, vol. 4, pp. 375–385, 1983.

[55] L. Bogatyrev, B. Berg, and H. Staaf, "Leaching of plant nutrients and total phenolic substances from some foliage litters. A laboratory study," Tech. Rep. 33, Swedish Coniferous Forest Project, 1983.

[56] K. Ono, S. Hiradate, S. Morita, K. Ohse, and K. Hirai, "Humification processes of needle litters on forest floors in Japanese cedar (*Cryptomeria japonica*) and Hinoki cypress (*Chamaecyparis obtusa*) plantations in Japan," *Plant and Soil*, vol. 338, no. 1, pp. 171–181, 2011.

[57] J. D. Aber, C. A. McClaugherty, and J. M. Melillo, *Litter Decomposition in Wisconsin Forests-Mass Loss, Organic-Chemical Constituents and Nitrogen*, vol. R3284 of *University of Wisconsin Research Bulletins*, University of Wisconsin, Madison, Wis, USA, 1984.

[58] M. Y. Faituri, "Soil organic matter in Mediterranean and Scandinavian forest ecosystems and dynamics of nutrients and monomeric phenolic compounds," *Silvestra*, vol. 236, 136 pages, 2002.

[59] B. Berg, G. Ekbohm, and C. McClaugherty, "Lignin and holocellulose relations during long-term decomposition of some forest litters. Long-term decomposition in a Scots pine forest. IV," *Canadian Journal of Botany*, vol. 62, no. 12, pp. 2540–2550, 1984.

[60] J. M. Melillo, J. D. Aber, A. E. Linkins et al., "Carbon and nitrogen dynamics along the decay continuum: plant litter to soil organic matter," in *Ecology of Arable Lands*, M. Clarholm and L. Bergström, Eds., pp. 53–62, Kluwer, Dordrecht, The Netherlands, 1989.

[61] T. Klotzbücher, T. R. Filley, K. Kaiser, and K. Kalbitz, "A study of lignin degradation in leaf and needle litter using ^{13}C-labelled tetramethylammonium hydroxide (TMAH) thermochemolysis: comparison with CuO oxidation and van Soest methods," *Organic Geochemistry*, vol. 42, no. 10, pp. 1271–1278, 2011.

[62] S. Güsewell and J. T. A. Verhoeven, "Litter N : P ratios indicate whether N or P limits the decomposability of graminoid leaf litter," *Plant and Soil*, vol. 287, no. 1-2, pp. 131–143, 2006.

[63] B. Berg and H. Staaf, "Leaching, accumulation and release of nitrogen in decomposing forest litter," in *Terrestrial Nitrogen Cycles. Processes, Ecosystem Strategies and Management Impacts*, F. E. Clark and T. Rosswall, Eds., vol. 33, pp. 163–178, Ecological Bulletins, Stockholm, Sweden, 1981.

[64] S. Manzoni, R. B. Jackson, J. A. Trofymow, and A. Porporato, "The global stoichiometry of litter nitrogen mineralization," *Science*, vol. 321, no. 5889, pp. 684–686, 2008.

[65] J. D. Aber and J. M. Melillo, *Terrestrial Ecosystems*, Cengage Learning. Emea, 1991.

[66] B. Berg and R. Laskowski, *Litter Decomposition: A Guide to Carbon and Nutrient Turnover*, vol. 38 of *Advances in Ecological Research*, Elsevier, San Diego, Calif, USA, 2006.

[67] B. Berg, C. McClaugherty, and M. B. Johansson, "Chemical changes in decomposing plant litter can be systemized with respect to the litter's initial chemical composition," Reports from the Departments in Forest Ecology and Forest Soils, Swedish University of Agricultural Sciences. Report 74, 1997.

[68] J. D. Aber and J. M. Melillo, "Nitrogen immobilization in decaying hardwood leaf litter as a function of initial nitrogen and lignin content," *Canadian Journal of Botany*, vol. 60, no. 11, pp. 2263–2269, 1982.

[69] B. Berg and J. Cortina, "Nutrient dynamics in some decomposing leaf and needle litter types in a *Pinus sylvestris* forest," *Scandinavian Journal of Forest Research*, vol. 10, no. 1, pp. 1–11, 1995.

[70] H. Staaf and B. Berg, "Accumulation and release of plant nutrients in decomposing Scots pine needle litter. Long-term decomposition in a Scots pine forest: II," *Canadian Journal of Botany*, vol. 60, no. 8, pp. 1561–1568, 1982.

[71] B. Berg, "Dynamics of nitrogen (^{15}N) in decomposing Scots pine (*Pinus sylvestris*) needle litter. Long-term decomposition in a Scots pine forest. VI," *Canadian Journal of Botany*, vol. 66, no. 8, pp. 1539–1546, 1988.

[72] H. Nömmik and K. Vahtras, "Retention and fixation of ammonium and ammonia in soils," in *Nitrogen in Agricultural Soils*, F. J. Stevenson, Ed., vol. 22 of *Agronomy Monographs*, pp. 123–171, Agronomy Society of America, Madison, Wis, USA, 1982.

[73] G. Axelsson and B. Berg, "Fixation of ammonia (^{15}N) to Scots pine needle litter in different stages of decomposition," *Scandinavian Journal of Forest Research*, vol. 3, pp. 273–280, 1988.

[74] M. R. Lindbeck and J. I. Young, "Polarography of intermediates in the fixation of nitrogen by p-quinone-aqueous ammonia systems," *Analytica Chimica Acta*, vol. 32, pp. 73–80, 1965.

[75] H. Knicker, "Stabilization of N-compounds in soil and organic-matter-rich sediments—what is the difference?" *Marine Chemistry*, vol. 92, pp. 167–195, 2004.

[76] B. Berg, B. Erhagen, M. B. Johansson et al., "Manganese dynamics in decomposing foliar litter," *Canadian Journal of Forest Research*, vol. 43, pp. 1–10, 2013.

[77] S. Hobbie, W. C. Eddy, C. R. Buyarski et al., "Response of decomposing litter and its microbial community to multiple forms of nitrogen enrichment," *Ecological Monographs*, vol. 82, pp. 389–405, 2012.

[78] M. M. Carreiro, R. L. Sinsabaugh, D. A. Repert, and D. F. Parkhurst, "Microbial enzyme shifts explain litter decay responses to simulated nitrogen deposition," *Ecology*, vol. 81, no. 9, pp. 2359–2365, 2000.

[79] M. Kaspari, S. P. Yanoviak, R. Dudley, M. Yuan, and N. A. Clay, "Sodium shortage as a constraint on the carbon cycle in an inland tropical rainforest," *Proceedings of the National Academy of Sciences of the United States of America*, vol. 106, no. 46, pp. 19405–19409, 2009.

[80] S. Perakis, J. J. Matekis, and D. E. Hibbs, "Interactions of tissue and fertilizer nitrogen on decomposition dynamics of lignin-rich litter," *Ecosphere*, vol. 3, no. 6, article 54, 2012.

[81] T. Klotzbücher, K. Kaiser, G. Guggenberger, C. Gatzek, and K. Kalbitz, "A new conceptual model for the fate of lignin in

decomposing plant litter," *Ecology*, vol. 95, no. 5, pp. 1052–1062, 2011.

[82] M. M. Coûteaux, K. B. McTiernan, B. Berg, D. Szuberla, P. Dardenne, and P. Bottner, "Chemical composition and carbon mineralisation potential of Scots pine needles at different stages of decomposition," *Soil Biology and Biochemistry*, vol. 30, no. 5, pp. 583–595, 1998.

[83] J. Perez and T. W. Jeffries, "Roles of manganese and organic acid chelators in regulating lignin degradation and biosynthesis of peroxidases by *Phanerochaete chrysosporium*," *Applied and Environmental Microbiology*, vol. 58, no. 8, pp. 2402–2409, 1992.

[84] A. Hatakka, "Biodegradation of lignin," in *Biopolymers, Volume 1: Lignin, Humic Substances and Coal*, M. Hofman and A. Stein, Eds., pp. 129–180, Wiley, Weinheim, Germany, 2001.

[85] B. Berg, K. T. Steffen, and C. McClaugherty, "Litter decomposition rate is dependent on litter Mn concentrations," *Biogeochemistry*, vol. 82, no. 1, pp. 29–39, 2007.

[86] B. Berg, "Initial rates and limit values for decomposition of Scots pine and Norway spruce needle litter: a synthesis for N-fertilized forest stands," *Canadian Journal of Forest Research*, vol. 30, no. 1, pp. 122–135, 2000.

[87] B. Berg, "Decomposition patterns for foliar litter—a theory for influencing factors," manuscript submitted.

[88] J. S. Olson, "Energy storage and the balance of producers and decomposers in ecological systems," *Ecology*, vol. 44, pp. 322–331, 1963.

[89] H. Jenny, S. P. Gessel, and F. T. Bingham, "Comparative study of decomposition rates of organic matter in temperate and tropical regions," *Soil Science*, vol. 68, pp. 419–432, 1949.

[90] K. T. Steffen, A. Hatakka, and M. Hofrichter, "Degradation of humic acids by the litter-decomposing basidiomycete *Collybia dryophila*," *Applied and Environmental Microbiology*, vol. 68, no. 7, pp. 3442–3448, 2002.

[91] K. E. Eriksson, R. A. Blanchette, and P. Ander, *Microbial and Enzymatic Degradation of Wood and Wood Components*, Springer, Berlin, Germany, 1990.

[92] J. D. Ovington, "The circulation of minerals in plantations of *Pinus sylvestris* L.," *Annals of Botany*, vol. 23, no. 2, pp. 229–239, 1959.

[93] E. Holmsgaard and C. Bang, "Et traeartsforsøg med nåletraeer, bøg og eg. De Første 10 år," *Forstlig Forsøgsvaesen*, no. 35, pp. 159–196, 1977 (Danish).

[94] B. Berg, "Sequestration rates for C and N in humus at four N-polluted temperate forest stands," in *Biogeochemistry of Forested Catchments in a Changing Environment. A German Case Study*, E. Matzner, Ed., vol. 172 of *Ecological Studies*, pp. 361–376, Springer, Berlin, Germany, 2004.

[95] H. Meesenburg, K. J. Meiwes, and H. Bartens, "Veränderung der Elementvorräte im Boden von Buchen- und Fichtenökosystemen in Solling," *Berichte Freiburger Forstliche Forschung*, vol. 7, pp. 109–114, 1999 (German).

[96] K. J. Meiwes, H. Meesenburg, H. Bartens et al., "Accumulation of humus in the litter layer of forest stands at Solling. Possible causes and significance for the nutrient cycling," *Forst und Holz*, vol. 13-14, pp. 428–433, 2002 (German), English summary.

[97] C. Akselsson, B. Berg, V. Meentemeyer, and O. Westling, "Carbon sequestration rates in organic layers of boreal and temperate forest soils—Sweden as a case study," *Global Ecology and Biogeography*, vol. 14, no. 1, pp. 77–84, 2005.

[98] B. Berg, M. Johansson, Á. Nilsson, P. Gundersen, and L. Norell, "Sequestration of carbon in the humus layer of Swedish forests—direct measurements," *Canadian Journal of Forest Research*, vol. 39, no. 5, pp. 962–975, 2009.

[99] B. Berg, P. Gundersen, C. Akselsson, M. Johansson, Å. Nilsson, and L. Vesterdal, "Carbon sequestration rates in Swedish forest soils—a comparison of three approaches," *Silva Fennica*, vol. 41, no. 3, pp. 541–558, 2007.

Vegetation Response to Climate Change and Human Impacts in the Usambara Mountains

C. T. Mumbi,[1,2] R. Marchant,[1] and P. Lane[3]

[1] *Environment Department, York Institute for Ecosystem Dynamics (KITE), University of York, Heslington, York, YO10 5DD, UK*
[2] *Tanzania Wildlife Research Institute (TAWIRI), P.O. Box 661, Arusha, Tanzania*
[3] *Department of Archaeology, University of York, King's Manor, York YO 17 EP, UK*

Correspondence should be addressed to C. T. Mumbi; cassian.mumbi@tawiri.or.tz

Academic Editors: S. Davey, J. F. Mas, and M. Vitale

East and West Usambara Mountain blocks are unique based on three characteristics. Firstly, they are connected blocks; secondly, they have an oceanic-influenced climate; and thirdly, the rain seasons are not easily discernible due to their close proximity to the Indian Ocean and Equator. Sediment cores were collected from peat bogs in Derema (DRM) and Mbomole (MBML) in East Usambara and from Madumu (DUMU) in West Usambara. The multiproxy record provides an understanding on climate and vegetation changes during the last 5000 years. DRM and MBML cores result in radiocarbon ages and age-depth curve which showed hiatus at 20 cm and 61 cm and huge inversion for DUMU core at 57 cm. Period 5000–4000 ^{14}C yr BP for DUMU core revealed increased Montane forest indicative of relatively moist conditions. Periods 3000–2000 and 2000–1000 ^{14}C yr BP, DUMU core demonstrated increased submontane and lowland forests. Period 1000–200 ^{14}C yr BP, DUMU core signified increased coprophilous fungi while DRM and MBML cores signified fluctuating herbaceous pollen spectra (wet-dry episodes). Period 200 ^{14}C yr BP to present, all cores demonstrated stable recovery of forest types especially dominance of submontane forests. Abundant coprophilous fungi indicated increased human impacts including forest fires, cultivation, and grazing.

1. Introduction

The Eastern Arc Mountains comprise thirteen separate blocs with their location stretching from south-east Kenya through south-central Tanzania (Figure 1). They are situated between 3°20′ to 8°45′S latitude and 35°37′ to 38°48′E longitude covering an area of around 3300 km^2 of submontane, montane, and upper montane forests, which is less than 30%, or some 1440 km^2, of the estimated original forested area [1]. Their unique characteristic geological formation of isolation and connectivity played a crucial role in shaping the current distribution of species diversity within and between the mountain blocs. So to say, the Eastern Arc Mountains exhibit connectivity and isolation within blocks crucial to its existence. Connectivity is where blocs were formed as sister blocs were separated by a narrow gap without much difference in forest types. Connected mountain blocs include North and South Pare, West and East Usambara, North and South Uluguru, and the Udzungwa, all in Tanzania. The vice

versa is true for the isolated mountain blocs; these include Nguu, Nguru, Ukaguru, Rubeho, Mahenge, Malundwe, and Uvidunda in Tanzania; Taita hills are an isolated bloc, the only mountain bloc in Kenya with an estimated remnant forested area of 6 km^2 (Figure 1).

The Eastern Arc Mountains are evolutionary and ecologically quite distinct from adjacent highlands, grasslands, savannas, and woodlands in East Africa. They have more affinity to Guineo-Congolian lowland rainforests [2]. The Eastern Arc Mountains have been proposed as one of several refugia in Africa during geologic periods when the climate in tropics was generally unfavourable for forest development [3] (Hamilton [4]). This suggestion is based upon three lines of evidence: (1) the large numbers of species and endemics; (2) the centres of distribution for many disjunct species; and (3) declining species diversity with increasing distance from these regions. The three lines of evidence lack strong backing from the palaeoenvironmental point-of-view, that is, knowledge on past changes in the Eastern Arc Mountains.

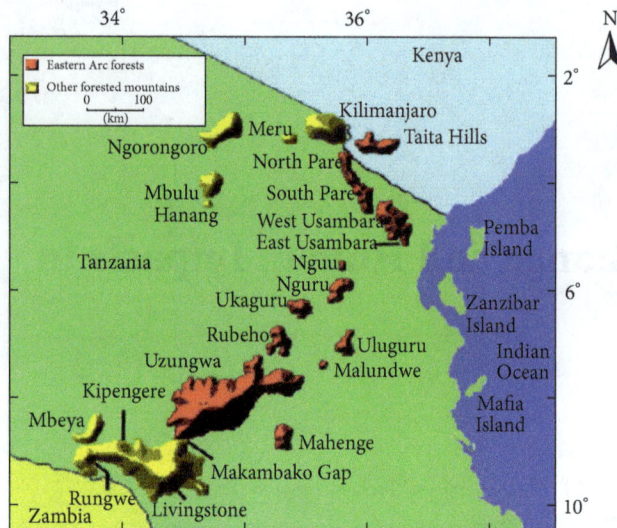

FIGURE 1: Map showing 12 Eastern Arc Mountain blocs in Tanzania and 1 in Kenya. Also shown are other tropical mountain blocs in Tanzania (adapted from [1]).

FIGURE 2: Amani pond showing the floating vegetation and the surrounding forest trees.

The first sedimentary record from this Mountain ecosystem was published in Mumbi et al., [5].

The climate of Eastern Arc Mountains is regulated by the topography of its blocs. Related to topography, two factors largely influence the weather and climate, namely, the Indian Ocean and Intertropical Convergence Zone (ITCZ). The influence of the Indian Ocean is determined by the distance from the mountain blocs, which brings about differences in the amount of precipitation and temperature variations between the blocs. They are located some 40 km from the Indian Ocean in East Usambara to 450 km in Udzungwa Mountains. The ITCZ influence determines seasonality, variability, and reliability in terms of annual rainfall and variability in terms of annual temperatures between the mountain blocs. The steep sloping eastern sides are generally wetter than the gentle sloping western sides of all the mountain ranges. The distance from the Indian Ocean determines the amount of annual precipitation, not temperature as the latter is influenced locally with the changing altitude. Within connected blocks, East Usambara is generally wetter than West Usambara; South Pare is wetter than North Pare; and South Uluguru is wetter than North Uluguru. Between isolated blocs, East and West Usambara are wetter than Nguru Mountains; North and South Uluguru are wetter than Udzungwa and Mahenge. The northern blocs have more oceanic-influenced climate having a few kilometers from the Indian Ocean compared to southern blocks which are both oceanic and continental influenced in daily, monthly, and annual rainfall and temperatures.

In East Usambara and West Usambara the rain seasons are not easily discernible partly because of its close proximity to the Indian Ocean and also to equator [7], while moving westwards and southwards, rain seasonality is clear. In some forest areas of East and West Usambara Mountains, there are three rain seasons. Short rainy season occurs from October to December based on Northeast trade winds. Long rains

occur from March to May mainly convectional and the third rain occurs from July to August based on Southeast trade winds [7–9]. In Udzungwa and other southern mountain ranges, there is one long rainy season relatively constant from November to May [10]. Average rainfall measures at 1800 mm yr^{-1}. Mean annual temperatures measure at 22°C (maximum) during December and 17°C (minimum) during July. Generally, in all mountain ranges, frosts occur during July, which is the driest month of the year.

The main objective of the study in the Eastern Arc Mountains in Tanzania is to reconstruct vegetation changes from a hitherto underresearched area using several proxies, to infer environmental and climate change. The current hypothesis on the long-term ecological functioning of forest ecosystems in the Eastern Arc Mountains will be evaluated, in particular the claim that the biodiversity relates to long-term ecosystem stability. Sediment cores were collected from peat bogs in Derema and Mbomole (East Usambara) and Madumu (West Usambara). The multiproxy record provides an understanding on climate and vegetation changes across the interglacial-glacial periods.

2. Materials and Methods

2.1. Study Sites: Description of Swamps

(a) Amani Pond (Altitude: 870 m a.s.l; 05°07′ S, 38°40′ E). This is a man-made pond constructed in 1902 during German colonial period. It is cut through by a stream that used to be a swimming pool (Figure 2).

(b) Derema Swamp (Altitude: 935 m a.s.l., 05°07′ S, 38°40′ E). This swamp (Figure 3) is located near the Derema Tea Estate in the Integrated Business Community (IBC) area of Msasa village. Derema corridor is a proposed IBC area in the Amani Nature Reserve (ANR).

Dominant Vegetation Zones of Derema Swamp

(i) Swamp (Bog) Vegetation. The main vegetation types on the swamp floors are *Cyperus alopecuroides*, *Typha latifolia*, *Paepalanthus* sp. (Papyrus). Ferns include *Polypodium* sp.,

FIGURE 3: Derema Swamp area showing part of an actively cultivated area proposed as an integrated business community area.

FIGURE 4: Madumu Swamp showing a far part of Vugiri plateau where Vurumi stream originates that flows into this swamp.

Pteridium sp. and *Thelypteris serrata* on the swamp floors. Another fern, *Cyathea manniana* is also found along the valley banks, roads and disturbed forest areas. Along the edge of the swamp are few stands of the African mountain bamboo (*Sinarundinaria alpina*) and *Dracaena demesne*.

(ii) Transition Zone. These are found between the swamps and the regional vegetation. Shrubs, herbs, and climbers include *Commelina benghalensis, Justicia insularis, Lantana camara, Ludwigia leptocarpa* and *Piper umbellatum*; and grasses such as *Andropogon* sp., and *Digitaria diagonalis*. Cultivated plants include *Mussa capensis, Trifolium odonata, Psidium guava, Camellia sinensis* (Tea) and *Cinnamomum* sp. (Cinnamon), and *Elettaria cardamomum* (Cardamon).

(iii) Forest Area Vegetation. The regional vegetation includes *Allanblackia stuhlmannii, Beilschmiedia* sp., *Cephaelasphaera usambarensis, Chrysophyllum gorungosanum, Credemia hirta, Ficus natalensis, Isoberlinia* sp., *Macaranga kilimandscharica, Maesa lanceolata, Maesopsis eminii, Melianthus holistii, Milicia excelsa, Myrianthus holistii, Newtonia buchananii, Ocotea usambarensis, Psidium africana, Strombosia scheffleri, Syzygium guineense, Tarbanaemontana pachysiphon,* and *Voacanga africana*.

(c) Mbomole Swamp (Altitude: 935 m a.s.l., 05°07′ S, 38°40′ E). This swamp is located just 1.5 km away from Derema Swamp. It is also located near Derema Tea Estate in the Integrated Business Community (IBC) area of Msasa village. For vegetation description, see Derema Swamp vegetation.

(d) Madumu Swamp (Altitude: 950 m a.s.l., 04°58′ S, 38°25′ E). A catchment swamp (Figure 4) fed by inflowing water and sediments via Vurumi River originating from West Usambara Mountains. This swamp forms part of the south-western edge of Vugiri plateau located at 950 m a.s.l. altitude. The diameter of the swamp is approximately 2 km. The swamp area is used by local people in the surrounding villages of Chekereni and Makuyuni for cultivating *Saccharum officinarum* (Sugar cane), *Oryza sativa* (Rice), and *Amaranthus chlorostachys* (Amaranth). It is frequently burnt for this purpose for the large part of it.

Vegetation Description. The swamp area is dominated by Cyperaceae, *Ludwigia leptocarpa,* and *Typha latifolia*. The area surrounding the swamp is cultivated mainly for *Agave sisalana* (Sisal), *Zea mays* (Maize), *Phaseolus vulgaris* (Kidney beans), *Oryza sativa* (Rice), and *Amaranthus chlorostachys* (Amaranth).

2.2. Sampling Procedure. Sediment cores were raised from three sedimentary basins in the Eastern Arc Mountain ranges; these include Derema and Mbomole (East Usambara) and Madumu (West Usambara) Catchment areas in Usambara Blocs. Sediments were collected using a 50 mm-diameter Russian sampler. Sediment cores were wrapped in plastic, fixed with PVC guttering, and transported to York for storage in the dark at 4°C. Lithological changes were described along the cores before sampling for pollen and radiocarbon dating. Sediment samples of 2 cm^3 were taken at 2 cm intervals along the profile for pollen analysis. For preparing the pollen samples, we used the standard pretreatment technique, including sodium pyrophosphate, acetolysis, and heavy liquid separation with bromoform. Exotic *Lycopodium* spores were added before treatment to each sample to calculate the pollen concentration [11–15].

Chronological control is provided by using Accelerator Mass Spectrometry (AMS) radiocarbon dating on samples. Ages were calibrated to calendar years using PC-Based software Radiocarbon Calibration Program Calib Rev 4.4.2 [6]. The cores applied are Derema, Mbomole, and Madumu.

For pollen identification, we used morphological descriptions published by the African Pollen Database (APD) [11], [12–15]. For identification of fungal spores, the morphological descriptions published by Van Geel et al. [16, 17] were used. We used a pollen sum of minimally 300 pollen grains from regional vegetation. Pollen grains from aquatic taxa and spores of ferns, mosses, fungi, and algae were excluded from the pollen sum. According to altitudinal and ecological preference, taxa were classified into three groups: upper montane herb and shrub, upper montane forest, and montane forest (Table 1). Clusters of similar pollen spectra were identified using CONISS; the results were used to delimit pollen zones. Results were plotted using PC-based software TILIA and TILIA GRAPH [18].

TABLE 1: Specific data of AMS ^{14}C samples from cores DRM, MBML and DUMU. Calibration of radiocarbon years is based on Calib Rev 4.4.2 [6].

Sample Depth (cm)	Sample Code	^{14}C yr BP	Interpolated age (cal yr BP)	δ^{14}C
Derema-20	GrA-33279	no age	100	−20.50
Derema-40	GrA-33071	195 ± 35	190	−23.81
Derema-61	GrA-33073	no age	225	−28.18
Derema-110	GrA-33075	265 ± 35	260	−24.92
Derema-160	GrA-33076	285 ± 35	290	−26.23
Mbomole-20	GrA-33279b	no age	100	−20.31
Mbomole-40	GrA-33071b	195 ± 35	190	−23.53
Mbomole-61	GrA-33073b	no age	230	−26.69
Dumu-25	GrA-33662	2680 ± 45	2600	−13.54
Dumu-57	GrA-33077	185 ± 35	3500	−15.43
Dumu-90	GrA-33663	4315 ± 45	4300	−13.81

FIGURE 5: Derema core showing lithological changes between 50–100 cm sediment profiles.

FIGURE 6: Dumu core showing lithological changes between 50 and 100 cm sediment profile.

3. Results

3.1. Lithology and Chronology

(a) Lithological Descriptions (Abbreviated Cores: Derema (DRM) (Figure 5), Mbomole (MBML), and Madumu (DUMU), (Figure 6))

(i) Amani Pond (was not analysed due to large hiatus/break in sediment core)

Grey in colour
0–40 cm: water
40–91 cm: sediments.

(ii) DRM core I

0–50 cm: decomposed peat/grey/fibrous material
50–100 cm: clay material/grey material
100–150 cm: dark lamination on top/sand at bottom/grey material.

(iii) DRM core II

0–50 cm: decomposed peat/fibrous/grey/very loose material
50–100 cm: decomposed peat/grey/fibrous/fibrous/loose material
100–150 cm: decomposed peat/light yellowish material
150–180 cm: fine sand/grayish material.

(iv) MBML core

MBML core I

0–30 cm: decomposed peat material
30–43 cm: coarse sand material
43–50 cm: decomposed peat material
50–100 cm: decomposed peat on top, coarse, sand material at the bottom;

MBML core II

0–10 cm: decomposed peat material
10–30 cm: fine sand material
30–50 cm: dense material/dark grey material

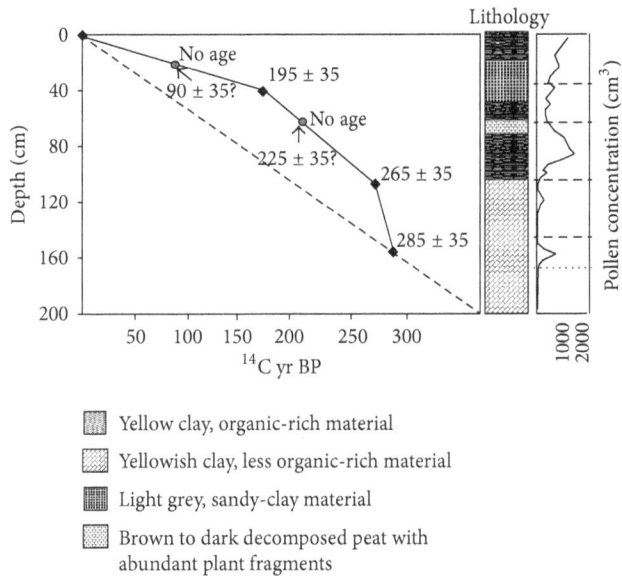

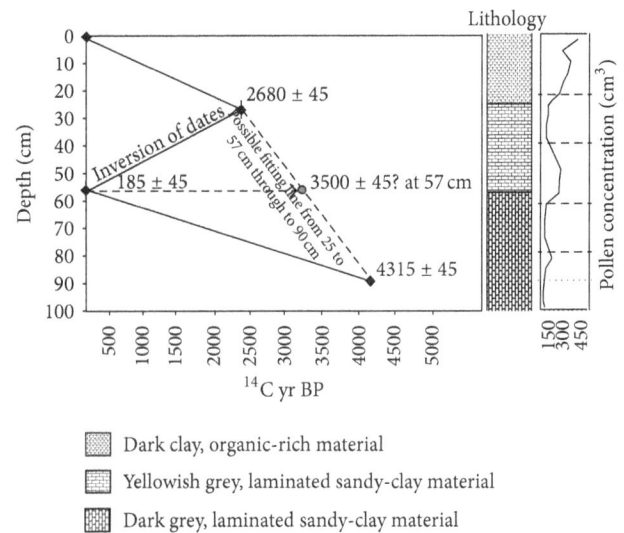

FIGURE 7: Construction of age-depth curves for DRM and MBML cores.

FIGURE 8: Construction of age-depth curves for DUMU core.

50–100 m: decomposed peat on top, coarse, sand particles at the bottom.

(v) DUMU core

DUMU I

0–20 cm: decomposed peat/dark in colour/fibrous material
20–50 cm: dark-yellowish/fine sand/calcareous material
50–90 cm: dark-yellowish/fine sand/calcareous material;

DUMU II

0–20 cm: decomposed peat/very dark/fibrous material
20–50 cm: dark-yellowish/fine sand/calcareous material
50–100 cm: dark/grey/fine sand material.

The sediment characteristics and radiocarbon ages are presented in Figures 7 and 9. Radiocarbon results show that the ages of the DUMU core are the longest dated 4315 ± 45 ^{14}C yr BP at 90 cm depth. This is the generally agreed onset of late Holocene period (4000 yr BP to present). Results are graphed in a depth versus time plot (Figures 7, 8, and 9). The DUMU core age at 57 cm shows an inversion of almost 1000 radiocarbon years. The younger age 185 ± 35 ^{14}C yr BP cm may be explained by two reasons. Firstly, a contamination by possibly Cyperaceous roots that penetrate deeper and cause older sediments to reflect younger age, and secondly an erosional effect that displaced bottom/older sediments and replaced it with top/younger sediments. In general, the depth versus time curves of all cores show relatively similar sediment accumulation rates. An interpolated time scale in

calendar age (cal yr BP) for each core was obtained by using the ages of core depth indicated in Table 1.

3.2. Chronologies

(a) DRM and MBML Cores. DRM and MBML cores results on radiocarbon ages and age-depth curve do not show an inversion common to other cores (Figure 7). However, they present a totally different problem; there is a loss of ages at depths 20 cm and 61 cm possibly due to contamination. The tricky situation here is even on the difficulty in providing reasons for contamination. The sediments at these depths are yellowish clay, rich in organic matter, which suggest a combination of many contaminating factors, bearing in mind the natural setting of an area. Hard water effect from parent rock, bioturbation, and human-related contamination (e.g., lipids) could be responsible for the loss of dates.

(b) DUMU Core. The results of radiocarbon ages from DUMU core showed that the middle ages hugely invert the depth versus time relationship at 57 cm (Figure 8). The sample was taken at the interval between yellowish gray, laminated sand-clay sediments (due to the contribution of parent rock due to carbon effect) and dark gray laminated sand-clay (due to the contribution of organic matter), at radiocarbon date (185 ± 45 ^{14}C yr BP). As for the above inversions, the relatively young age could be caused by contaminations of young samples through the occurrence of vertical migration of datable components in the sediment sequence. Lipids, humics and fluvics, and insoluble carbon, or the hard water effect could be responsible for DUMU core.

3.3. Pollen Zonation and Construction of Diagrams. Pollen zones and constructed diagrams for cores DRM, MBML, and

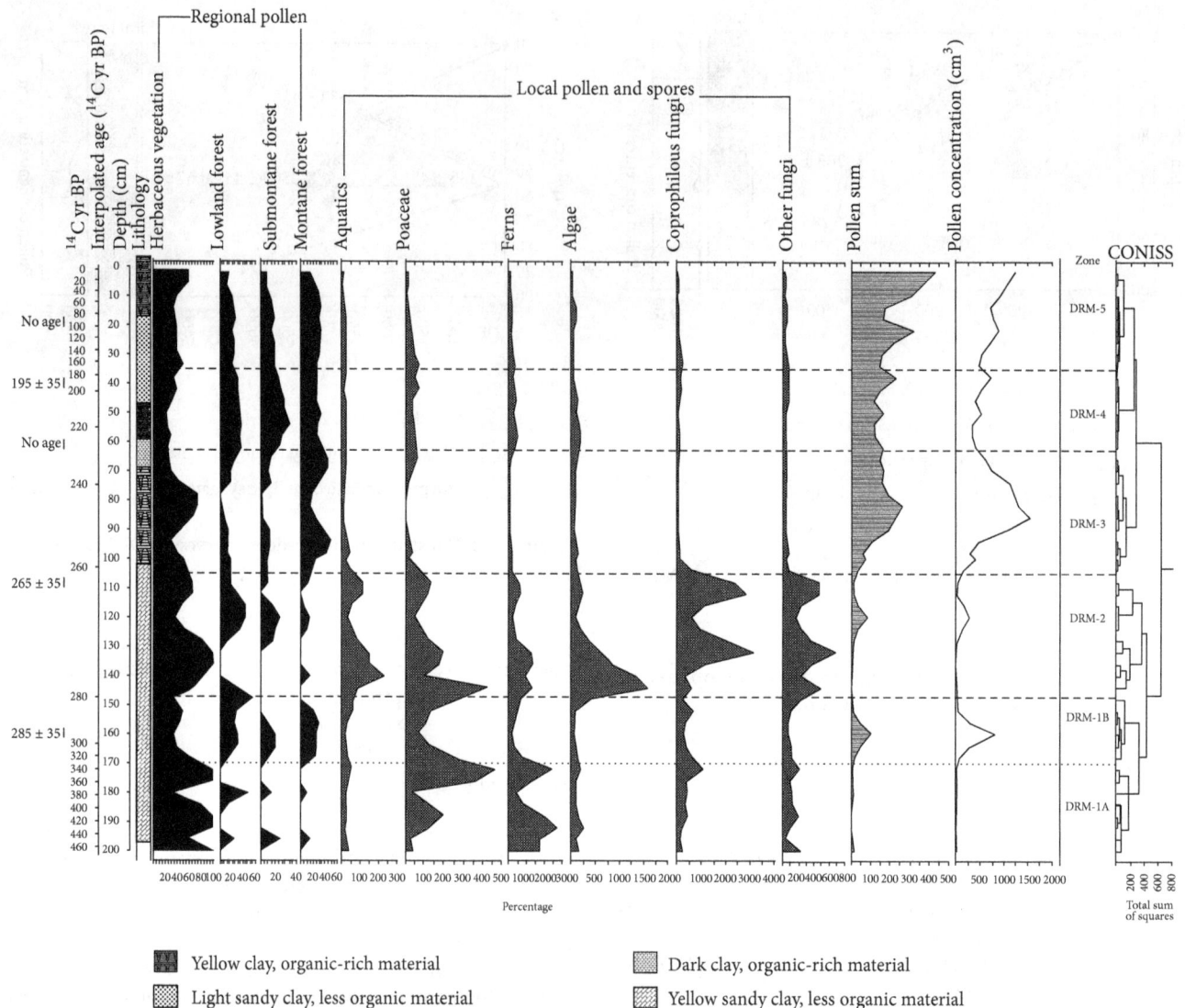

FIGURE 9: Summary diagram DRM core showing radiocarbon ages, interpolated scale of estimated ages, depth scale, down core changes of the main forest belts (regional pollen), local pollen and spores, pollen sum, pollen concentration, pollen zones, and CONISS cluster dendrogram.

DUMU are shown in Figures 9, 10, and 11. The interpretation of pollen diagrams follows in Section 4.

(a) DRM Core

Pollen Zone DRM-1A (200–170 cm, 460–340 ^{14}C yr BP). This zone is characterized by representation of high percentages of herbaceous vegetation, Poaceae, and ferns. The main forest types (regional pollen) that include lowland, submontane, and montane vegetation types are represented by low percentages, as do algal and fungal groups (Figure 9).

Pollen Zone DRM-1B (170–145 cm; 340–280 ^{14}C yr BP). This zone is characterized by representation of increased percentages of the main forest types (regional pollen) that include lowland, submontane, and montane vegetation types. Algal and fungal groups also showed a slight increase. Herbaceous vegetation, Poaceae, and ferns are represented by low percentages.

Pollen Zone DRM-2 (145–105 cm; 280–260 ^{14}C yr BP). This zone is characterized by higher proportions of all vegetation types than other pollen zones in this record. Herbaceous vegetation, Poaceae, ferns, algae, and fungal spores score higher percentages. Increased percentages also are shown by the main forest types (regional pollen) that include lowland, submontane, and montane vegetation types.

Pollen Zone DRM-3 (105–61 cm; 260–230 ^{14}C yr BP). This zone is characterized by higher proportions of Herbaceous and Montane vegetation types. This increase is reflected in the pollen sum and concentration as they show higher proportions in this pollen zone compared to other pollen zones. Poaceae, ferns, algae, and fungal spores show decreased percentages.

Pollen Zone DRM-4 (61–35 cm; 230–170 ^{14}C yr BP). This zone is characterized by high percentages of herbaceous, lowland, submontane, and montane vegetation types . Poaceae, ferns,

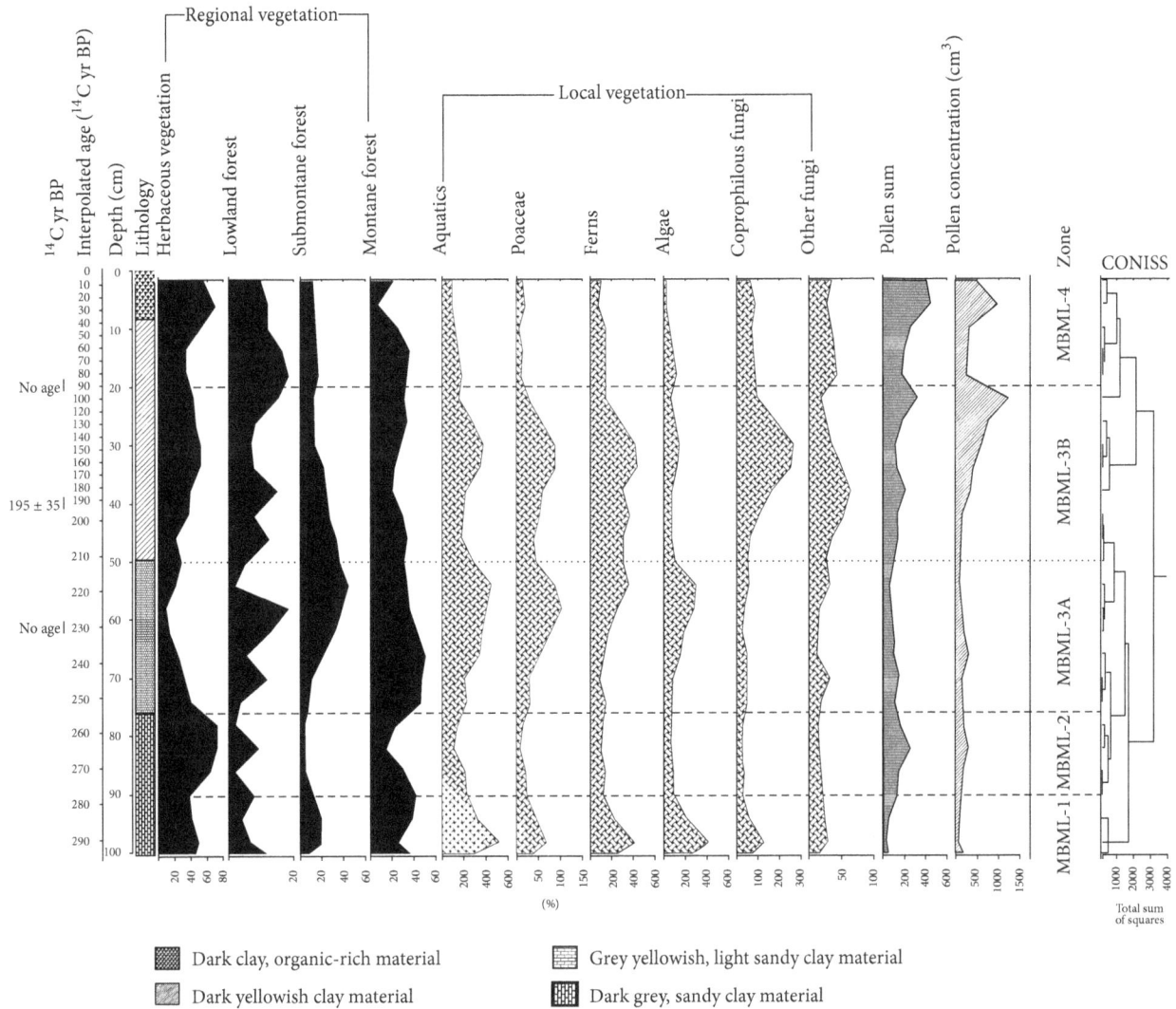

FIGURE 10: Summary diagram MBML core showing radiocarbon ages, interpolated scale of estimated ages, depth scale, down core changes of the main forest belts (regional pollen), local pollen and spores, pollen sum, pollen concentration, pollen zones, and CONISS cluster dendrogram.

algae, and fungal spores show low percentages. Pollen sum and pollen concentration showed relatively constant proportions.

Pollen Zone DRM-3 (35–0 cm; 170 ^{14}C yr BP to Present). This zone is characterized by high percentages of herbaceous, lowland, submontane, and montane vegetation types. Poaceae, ferns, algae, and fungal spores show low percentages. Pollen sum and pollen concentration showed increased proportions.

(b) MBML Core

Pollen Zone 1 (100–90 cm, 300–275 ^{14}C yr BP). This zone is characterized by high percentages of all vegetation types. High representation is particularly shown by montane vegetation, aquatic vegetation, ferns, algal, and fungal spores (Figure 10).

Pollen Zone 2 (90–78 cm, 275–265 ^{14}C yr BP). This zone is characterized by representation of highest percentages of

herbaceous vegetation than any other pollen zone in MBML record. Apart from lowland vegetation type which showed low-high percentage interchanges, all other vegetation types showed low percentages.

Pollen Zone 3A (78–50 cm, 265–215 ^{14}C yr BP). This zone is characterized by representation of highest percentages of montane forest with also increased submontane and lowland forest types. Aquatics, ferns, algae, and fungal spores showed increased percentages towards the top of the zone. Herbaceous vegetation showed the lowest percentages than any other pollen zone in MBML record.

Pollen Zone 3B (50–20 cm, 215–100 ^{14}C yr BP). Herbaceous vegetation, lowland, and montane forest types showed increasing percentages. Submontane forest type showed a gradual decrease. The high percentages of herbaceous vegetation and forest types are reflected in the pollen sum and pollen concentration. These vegetation types make up the pollen

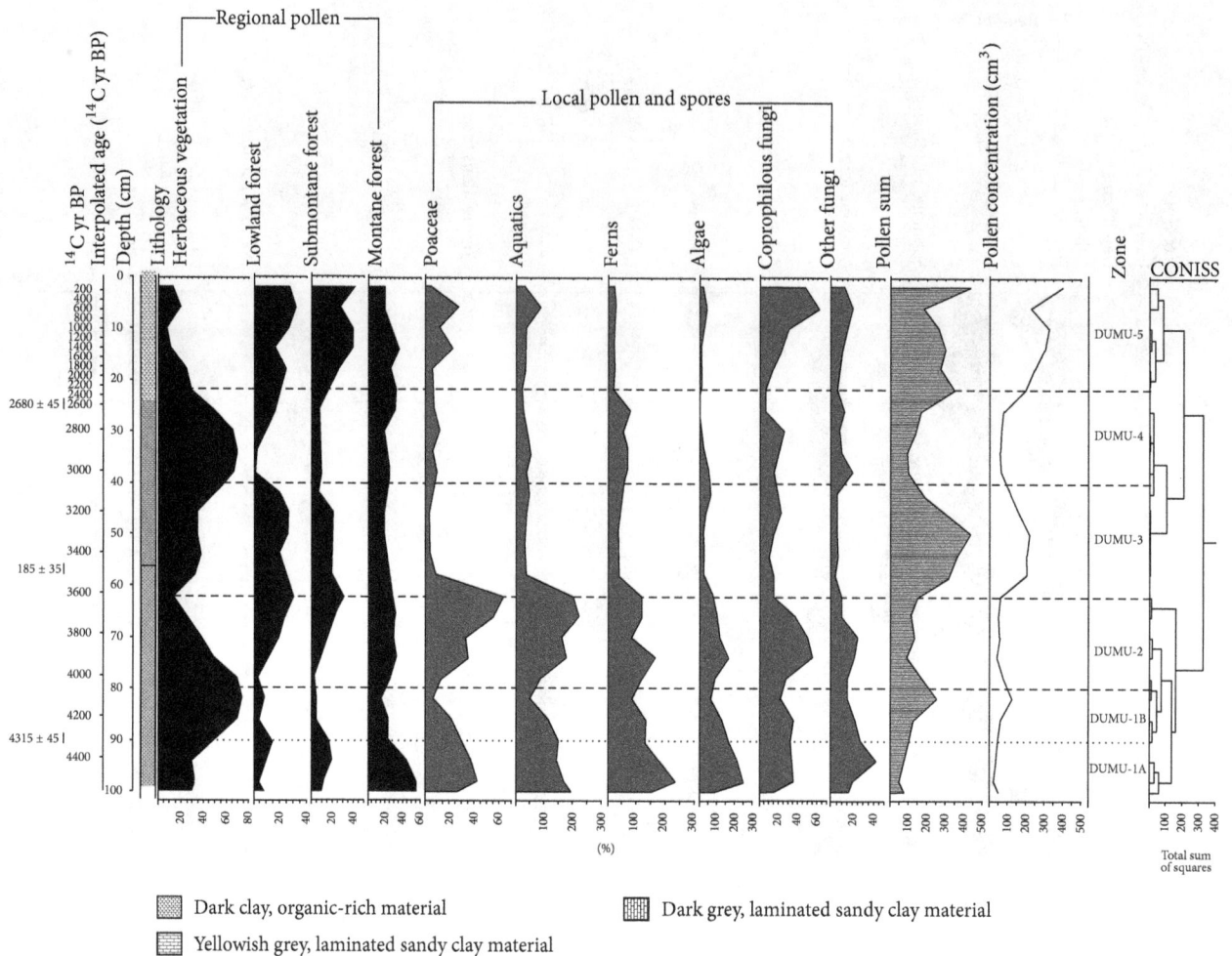

FIGURE 11: Summary diagram DUMU core showing radiocarbon ages, interpolated scale of estimated ages, depth scale, down core changes of the main forest belts (regional pollen), local pollen and spores, pollen sum, pollen concentration, pollen zones, and CONISS cluster dendrogram.

sum. Local vegetation (aquatics, ferns, algae, and fungal spores) showed higher percentages, with coprophilous fungi scoring the highest percentages than in any other zone.

Pollen Zone 4 (20–0 cm, 100–0 $^{14}C\,yr\,BP$). The high percentages of herbaceous vegetation and forest types are reflected in the pollen sum and pollen concentration. Local vegetation (aquatics, ferns, algae, and fungal spores) showed lower percentages.

(c) DUMU Core

Pollen Zone 1A (100–90 cm, 4600–4300 $^{14}C\,yr\,BP$). This zone is characterized by the highest percentages of montane forest types than in any other zone; herbaceous vegetation also showed high percentages. Lowland and submontane forest types showed low percentages. The local vegetation (Poaceae, aquatics, algae, and fungi) showed increased percentages; this is reflected in the pollen sum and pollen concentration.

Pollen Zone 1B (90–80 cm, 4300–4050 $^{14}C\,yr\,BP$). The herbaceous vegetation showed highest percentages, and because

it is included in the pollen sum, its dramatic increase is reflected. The main forest types showing decreased percentages and the local vegetation showed lower percentages (Figure 11).

Pollen Zone 2 (80–62 cm, 4050–3600 $^{14}C\,yr\,BP$). This zone is characterized by sharp decrease in percentages of herbaceous vegetation as opposed by sharp increase in percentages of local vegetation. Main forest types showed gradual increase in percentages. Pollen sum and concentration showed a gradual decrease in proportions.

Pollen Zone 3 (62–40 cm, 3600–3050 $^{14}C\,yr\,BP$). This zone is characterized by increased percentages of main forest types with gradual increase in herbaceous vegetation. In particular, lowland vegetation showed highest percentages than in any other zone. Local vegetation showed lowest percentages than in any other zone. This is supported by higher proportions of pollen sum and concentration.

Pollen Zone 4 (40–22 cm, 3050–2300 $^{14}C\,yr\,BP$). This zone showed similar characteristics as in pollen zone 1B. The herbaceous vegetation showed highest percentages, and because it is included in the pollen sum, its dramatic increase is reflected. The main forest types are showing decreased percentages and the local vegetation showed lower percentages.

Pollen Zone 5 (22–0 cm, 2300 ^{14}C yr BP to Present). This zone is characterized by remarkable increased percentages of main forest types and coprophilous fungi. The local vegetation groups of Poaceae and aquatics also showed gradual increase. Herbaceous vegetation showed a sharp increase towards the top of the zone. Pollen sum and concentration showed highest proportions in this zone reflective of increased percentages of main forest types.

4. Discussion

This section provides an interpretation of pollen spectral changes as explained in pollen zones for DRM, MBML, and DUMU cores. An approach will be to travel through time scales as provided by radiocarbon dating results encompassing all the cores where the period in interpretation is concerned. For the purpose of providing a summarized interpretation, the following time scales (periods) will be used: 5000–4000, 4000–3000, 3000–2000, 2000–1000, 1000–200, and 200–0 ^{14}C yr BP. The periods will be interpreted in light of the existing inferred palaeoenvironmental records in East Africa during the Late Holocene.

4.1. Period 5000–4000 ^{14}C yr BP. This period deals with DUMU core in which it is characterized with higher proportions of montane forest and local vegetation which can be indicative of relatively moist conditions at the higher altitude areas of East and West Usambara. Water table in swamps in the higher altitude areas could have been higher that favoured submerged plants. In East Africa, most records indicated the end of "African humid period" that brought about drier conditions [19–21]. This period is coincident with the "First Dark Age," the period of the greatest historically recorded drought in tropical Africa where some lakes (L. Victoria) dried up, which apparently extended to the Middle East and western Asia (cf. [21–25]). In some lake records, intervals of dry and wetter conditions were reported [21, 24, 26, 27].

4.2. Period 4000–3000 ^{14}C yr BP. The DUMU record shows conditions that were dryer than the previous period and all the proceeding periods. The period showed that montane forest type and herbaceous vegetation were higher in abundance, an indication that there were forest remnants at the high altitude areas probably related to moisture supply from the Indian Ocean. The inferred conditions indicate there was an abrupt aridity regarding the fact that all the vegetation types indicated low percentages. In East African region, the longer and more extensive drought event at c. 4000–3000 ^{14}C yr BP was registered in both the pollen and diatom records at different sites (Hamilton, [4]) [25, 28, 29]. However, Taylor [30] observed that expansions of dry forest taxa around Ahakagyezi and Muchoya occurred after ca 3900 and 3400 yr BP, respectively.

4.3. Period 3000–2000 ^{14}C yr BP. The period in DUMU core demonstrates the recovery of the forest vegetation types with sharp reduction in herbaceous vegetation indicative of dominance of submontane and lowland forests. This has been explained by a temperature rise which caused the clouds to be formed at a higher altitude than at present [31–33]. Hamilton et al. [34] noted increases in *Podocarpus, Olea,* and *Syzygium* pollen at Ahakagyezi Swamp around this period. In Kashiru Swamp, a significant extension of Poaceae and Ericaceae, accompanied by a decline of all forest elements except *Podocarpus, Maytenus,* and *Hypericum* occurred [35]. Roche and Bikwemu [35] noted a cold and dry period centred around 2500 yr BP in the Kashiru area.

4.4. Period 2000–1000 ^{14}C yr BP. The DUMU core showed a significant increase in coprophilous (dungi) fungi indicative of increased human impacts in the forest, including forest fires, cultivation, and grazing. In East Africa, a final dry interval on a millennia timescale occurred between 2000 and 1000 yr BP. Several lake sediment records show a dry period of about 300 year's length [36–40]. This brief summary of the paleoclimate record from eastern Africa reveals several regionally felt intervals, but with the length and character differing from one site to another. Differences between the records may illuminate the range of variability that is found in the lake proxy records and/or in the large uncertainties inherited in radiocarbon age models and/or in the ambiguity in the interpretation of the proxy data. In many records there is evidence of increased human impact on their environments. In this period many sites showed that the landscapes were subjected to frequent burning, with a period of increase in fire at ca. 1300–1600 AD [41], which coincides with an increased human population in the region. In more recent times, an increase in species typically found in human-induced environments is found. The near shore vegetation indicates lower lake levels, suggesting that drier conditions dominated.

4.5. Period 1000–200 ^{14}C yr BP. The DUMU core showed a significant increase in coprophilous (dungi) fungi indicative of increased human impacts in the forest, including forest fires, cultivation, and grazing. DRM core showed that the local vegetation was influenced by stagnation of water on swampy floor resulting from fluctuating wet-dry episodes as this is also supported by fluctuating herbaceous pollen spectra. In East Africa, a 1,100-year rainfall-drought history for Kenya's Lake Naivasha, based on sediments, fossil diatoms, and midge species and numbers was recorded [36]. The data indicate that equatorial East Africa, over the past millennium, has alternated between contrasting climate conditions. There has been a significantly drier climate than today during the "Medieval Warm Period" (1000–1270) and a relatively wet climate during the "Little Ice Age" (1270–1850) that was interrupted by three prolonged dry episodes. During the end of a dry period that has been identified from both lake records at around 1400 AD, a famine, known as the Wamara drought (1400–1412 AD) occurred [42]. The Bacwezi Empire in western Uganda collapsed shortly thereafter, at around 1450 AD, and settlements declined as a combined result of drier conditions, overgrazing, and diseases [41, 43]. Following more humid conditions, new settlements were established in northern Uganda and East Africa generally experiencing a time of political stability and agricultural success. For example, a highly sophisticated irrigation system was developed in Engaruka in northern Tanzania—which today is a very dry region that cannot sustain a large population [44]. The next prolonged dry period can be observed from the lake records (not a perfect age match though) and is centred at around 1600–1625 AD. This period is near in time with a period that is known from oral traditions to have been one of extensive political instability, migrations, and famines, including the famine known as the Nyarubanga, at 1587–1589 AD [42]. An extreme drought in the 1620s may have resulted in the fall of the capital at Munsa in Uganda at 1700 AD [41, 45]. During this period, the Engaruka system, an extensive fossil irrigation system located in northern Tanzania, was initiated in the early 15th century. It was in its most extensive use between 1620 AD and 1720 AD and it declined in the early 19th century. The period of humid conditions followed thereafter, the Engaruka system continued to expand during the wetter conditions that followed, and its maximum size most probably occurred at some time between 1680 AD and 1820 AD [44], where in east Ukamba, Kenya, large settlements developed, possibly encouraged by favourable environmental conditions, the Kikuyu people expanded, and maize cultivation was initiated [37]. The third dry period observed from the lake records is contemporary with a famine known as the Lapanarat-Mahlatule drought (1760–1840 AD) [42]. At this time, famine and short-term drought struck Ukamba and agropastoral production declined. One response to the harsh environmental conditions could have been the beginning of trade in food staples between the Kamba and the Kikuyu [46].

4.6. Period 200 ^{14}C yr BP to Present. In this period, all cores demonstrate the stable recovery of the forest vegetation types with sharp reduction herbaceous vegetation indicative of dominance of mid-latitude forest types (submontane and lowland forests). Abundance of coprophilous (dungi) fungi is indicative of increased human impacts in the forest, including forest fires, cultivation, and grazing. Detailed records of hydrological changes that are based on lake sediments are available from Lake Naivasha, Kenya [36, 37], and northern Lake Victoria, Uganda, [21, 38, 47]. Peaks in the records indicate wetter conditions and trough drier ones, as deduced from lake-level changes in metres (Lake Naivasha) and the abundance of shallow water diatoms in percentage (Lake Victoria). A shorter temperature record, spanning the past 200 years, is available from stable isotope studies on corals outside the Kenyan coast [48]. The system was still active into the early 19th century when it fell into decline, probably due to a number of catastrophic events, such as imported diseases (smallpox), environmental deterioration due to population pressure, and increased Masaai aggression. Later at the wake of these events, some people took up irrigation at Engaruka. A drastic climatic dislocation is indicated for the last two decades of the 19th century, this is manifested by the abrupt drop of lake levels [49–51] and in the onset of

glacier recession in East Africa [49]. Increasing atmospheric humidity accompanying the warming contributed to the accelerated ice wastage in the most recent decades of the 20th century [52]. This was forced by enhanced solar radiation due to diminished cloud cover accompanying the reduced precipitation; continuation of ice retreat beyond the early decades of the 20th century was favored by a warming trend [53, 54].

5. Conclusion

The climate system is a naturally complex dynamic that proved in the past and will continue to do so in the present and future cause effects that are both positive and negative to civilizations. For the negative part of global climate effects, understanding the past on time scales provides us with knowledge to forecast the magnitudes that do affect human life. At the present, anthropogenically-induced global climate changes, particularly, global warming will have major implications on future climate, but it is important to also bear in mind that climate will change even without anthropogenic impact. With the increased understanding of the regional and temporal patterns and causes behind climate processes, people have today the challenging possibility of applying this knowledge to plan for the future.

Conflict of Interests

The authors declare that there is no conflict of interests regarding the publication of this paper.

Acknowledgments

This research was partly funded by the Critical Ecosystem Partnership Fund (CEPF) Small Grant Programme for Building Research Capacity among Tanzanian and Kenyan students. This research has also been partly funded by the Marie-Curie Excellence programme of the European 6th Framework under Contract MEXT-CT-2004-517098 to Dr. Robert Marchant carried out under the York Institute for Ecosystem Dynamics (KITE), Environment department, University of York, United Kingdom. The following organisations in Tanzania are highly acknowledged for their support: Wildlife Conservation Society of Tanzania (WCST) (the Birdlife Partner in Tanzania) as part of Critical Ecosystem Partnership Fund (CEPF), Tanzania Wildlife Research Institute (TAWIRI), Amani Nature Reserve and Forestry and Beekeeping Division all under Ministry of Natural Resources and Tourism for logistical support.

References

[1] N. D. Burgess, T. M. Butynski, N. J. Cordeiro et al., "The biological importance of the Eastern Arc Mountains of Tanzania and Kenya," *Biological Conservation*, vol. 134, no. 2, pp. 209–231, 2007.

[2] J. C. Lovett, "Endemism and affinities of the Tanzanian montane forest flora," in *Proceedings of the Eleventh Plenary Meeting of the Association for the Taxonomic Study of Tropical Africa*, P. Goldblatt and P. P. Lowry, Eds., vol. 25 of *Monographs in*

Systematic Botany from the Missouri Botanical Garden, pp. 591–598, Missouri Botanical Garden Press, St Louis, Mo, USA, 1988.

[3] A. W. Diamond and A. C. Hamilton, "The distribution of forest passerine birds and Quaternary climatic change in tropical Africa," *Journal of Zoology*, vol. 191, no. 3, pp. 379–402, 1980.

[4] A. C. Hamilton, *Environmental History of East Africa: A Study of the Quaternary*, Academic Press, London, UK, 1982.

[5] C. T. Mumbi, R. Marchant, H. Hooghiemstra, and M. J. Wooller, "Late Quaternary vegetation reconstruction from the Eastern Arc Mountains, Tanzania," *Quaternary Research*, vol. 69, no. 2, pp. 326–341, 2008.

[6] M. Stuiver, P. J. Reimer, E. Bard et al., "INTCAL98 radiocarbon age calibration, 24,000-0 cal BP," *Radiocarbon*, vol. 40, no. 3, pp. 1041–1083, 1998.

[7] S. T. Iversen, *The Usambara Mountains, North East Tanzania: Phytogeography of the Vascular Plant Flora*, Almqvist & Wiksell, 1991.

[8] J. Brunnthaler, "Vegetationsbilder aus Deutsch-Ostafrika: Regenwald von Usambara," *Vegetations-Bilder*, vol. 11, no. 8, pp. 43–48, 1914.

[9] M. Attems, "Permanent cropping in the Usambara Mountains," in *Smallholder Farming and Smallholder Development in Tanzania*, H. Ruthenberg, Ed., pp. 137–174, Weltforum, München, Germany, 1968.

[10] J. C. Lovett, "Tanzanian forest tree plot diversity and elevation," *Journal of Tropical Ecology*, vol. 15, no. 5, pp. 689–694, 1999.

[11] African Pollen Database (APD), 2004, http://www.ncdc.noaa.gov/paleo/apd.html.

[12] C. Caratini and P. Guinet, Eds., *Pollen et Spores d'Afrique Tropicale*, Center National de la Recherche Scientifique (CNRS), Center d'Etudes de Geographie Tropicale, Domaine Universitaire de Bordeaux, Talence, France, 1974.

[13] R. Bonnefille, "Atlas des pollens de Éthiopie. Principales espéces des forêts de montagne," *Pollen et Spores*, vol. 13, no. 1, pp. 15–72, 1971.

[14] R. Bonnefille and G. Riollet, *Pollens des Savanes d'Afrique Orientale*, Éditions du Center National de la Recherche Scientifique (CNRS), Paris, France, 1980.

[15] G. E. B. El Ghazali, "A study on the pollen flora of Sudan," *Review of Palaeobotany and Palynology*, vol. 76, no. 2–4, pp. 99–345, 1993.

[16] B. van Geel, "A palaeoecological study of holocene peat bog sections in Germany and The Netherlands, based on the analysis of pollen, spores and macro- and microscopic remains of fungi, algae, cormophytes and animals," *Review of Palaeobotany and Palynology*, vol. 25, no. 1, pp. 1–120, 1978.

[17] B. van Geel, J. Buurman, O. Brinkkemper et al., "Environmental reconstruction of a Roman period settlement site in Uitgeest (The Netherlands), with special reference to coprophilous fungi," *Journal of Archaeological Science*, vol. 30, no. 7, pp. 873–883, 2003.

[18] E. C. Grimm, "CONISS: a FORTRAN 77 program for stratigraphically constrained cluster analysis by the method of incremental sum of squares," *Computers & Geosciences*, vol. 13, no. 1, pp. 13–35, 1987.

[19] F. Gasse, "Diatom-inferred salinity and carbonate oxygen isotopes in Holocene waterbodies of the western Sahara and Sahel (Africa)," *Quaternary Science Reviews*, vol. 21, no. 7, pp. 737–767, 2002.

[20] P. Barker and F. Gasse, "New evidence for a reduced water balance in East Africa during the Last Glacial Maximum: implication for model-data comparison," *Quaternary Science Reviews*, vol. 22, no. 8-9, pp. 823–837, 2003.

[21] J. C. Stager, B. F. Cumming, and L. D. Meeker, "A 10,000-year high-resolution diatom record from Pilkington Bay, Lake Victoria, East Africa," *Quaternary Research*, vol. 59, no. 2, pp. 172–181, 2003.

[22] R. A. Marchant and H. Hooghiemstra, "Rapid environmental change in African and South American tropics around 4000 years before present: a review," *Earth-Science Reviews*, vol. 66, no. 3-4, pp. 217–260, 2004.

[23] P. A. Barker, F. A. Street-Perrott, M. J. Leng et al., "A 14,000-year oxygen isotope record from diatom silica in two alpine lakes on Mt. Kenya," *Science*, vol. 292, no. 5525, pp. 2307–2310, 2001.

[24] P. A. Barker, R. Telford, F. Gasse, and F. Thevenon, "Late pleistocene and holocene palaeohydrology of Lake Rukwa, Tanzania, inferred from diatom analysis," *Palaeogeography, Palaeoclimatology, Palaeoecology*, vol. 187, no. 3-4, pp. 295–305, 2002.

[25] L. G. Thompson, E. Mosley-Thompson, M. E. Davis et al., "Kilimanjaro ice core records: evidence of holocene climate change in tropical Africa," *Science*, vol. 298, no. 5593, pp. 589–593, 2002.

[26] M. R. Talbot and T. Lærdal, "The Late Pleistocene—Holocene palaeolimnology of Lake Victoria, East Africa, based upon elemental and isotopic analyses of sedimentary organic matter," *Journal of Paleolimnology*, vol. 23, no. 2, pp. 141–164, 2000.

[27] F. Chalié and F. Gasse, "Late Glacial-Holocene diatom record of water chemistry and lake level change from the tropical East African Rift Lake Abiyata (Ethiopia)," *Palaeogeography, Palaeoclimatology, Palaeoecology*, vol. 187, no. 3-4, pp. 259–283, 2002.

[28] J. Mworia-Maitima, "Vegetation response to climatic change in central Rift Valley, Kenya," *Quaternary Research*, vol. 35, no. 2, pp. 234–245, 1991.

[29] F. A. Street-Perrott and R. A. Perrott, "Holocene vegetation, lake levels and climate of Africa," in *Global Climates since the Last Global Maximum*, H. E. Wright Jr., J. E. Kutzbach, T. Webb III, W. F. Ruddiman, F. A. Street-Perrott, and P. J. Bartlein, Eds., pp. 318–356, University of Minnesota Press, London, UK, 1993.

[30] D. M. Taylor, "Late quaternary pollen records from two Ugandan mires: evidence for environmental changes in the Rukiga highlands of southwest Uganda," *Palaeogeography, Palaeoclimatology, Palaeoecology*, vol. 80, no. 3-4, pp. 283–300, 1990.

[31] J. A. Coetzee, "Evidence for a considerable depression of the vegetation belts during the upper pleistocene on the East African mountains," *Nature*, vol. 204, no. 4958, pp. 564–566, 1964.

[32] J. A. Coetzee, "Pollen analytical studies in East and Southern Africa," *Palaeoecology of Africa*, vol. 3, pp. 1–46, 1967.

[33] J. A. Coetzee, "Palynological intimations on the East African mountains," *Palaeoecology of Africa*, vol. 18, pp. 231–244, 1987.

[34] A. C. Hamilton, D. Taylor, and J. C. Vogel, "Early forest clearance and environmental degradation in south-west Uganda," *Nature*, vol. 320, no. 6058, pp. 164–167, 1986.

[35] E. Roche and G. Bikwemu, "Paleoenvironmental change on the Zaire-Nile ridge in Burundi, the last 2000 years: an interpretation of palynological data from the Kashiru Core, Ijenda, Burundi," in *Quaternary Environmental Research on East African Mountains*, W. C. Mahaney, Ed., pp. 231–244, Balkema, Rotterdam, The Netherlands, 1989.

[36] D. Verschuren, K. R. Lalrd, and B. F. Cumming, "Rainfall and drought in equatorial east Africa during the past 1,100 years," *Nature*, vol. 403, no. 6768, pp. 410–414, 2000.

[37] H. Lamb, I. Darbyshire, and D. Verschuren, "Vegetation response to rainfall variation and human impact in central Kenya during the past 1100 years," *The Holocene*, vol. 13, no. 2, pp. 285–292, 2003.

[38] J. C. Stager, D. Ryves, B. F. Cumming, L. D. Meeker, and J. Beer, "Solar variability and the levels of Lake Victoria, East Africa, during the last millenium," *Journal of Paleolimnology*, vol. 33, no. 2, pp. 243–251, 2005.

[39] J. M. Russell, T. C. Johnson, and M. R. Talbot, "A 725 yr cycle in the climate of central Africa during the late Holocene," *Geology*, vol. 31, no. 8, pp. 677–680, 2003.

[40] M. L. Filippi and M. R. Talbot, "The palaeolimnology of northern Lake Malawi over the last 25 ka based upon the elemental and stable isotopic composition of sedimentary organic matter," *Quaternary Science Reviews*, vol. 24, no. 10-11, pp. 1303–1328, 2005.

[41] P. Robertshaw and D. Taylor, "Climate change and the rise of political complexity in western Uganda," *Journal of African History*, vol. 41, no. 1, pp. 1–28, 2000.

[42] J. B. Webster, "Drought, migration and chronology in the Lake Malawi littoral," *Transafrican Journal of History*, vol. 9, pp. 70–90, 1980.

[43] D. Taylor, P. Robertshaw, and R. A. Marchant, "Environmental change and political-economic upheaval in precolonial western Uganda," *The Holocene*, vol. 10, no. 4, pp. 527–536, 2000.

[44] J. Sutton, "Engaruka: the success & abandonment of an integrated irrigation system in an arid part of the Rift Valley, c 15th to 17th centuries," in *Islands of Intensive Agriculture in the Eastern Africa*, M. Widgren and J. E. G. Sutton, Eds., Eastern African Studies, pp. 114–132, James Currey, 2004.

[45] B. J. Lejju, D. Taylor, and P. Robertshaw, "Late-Holocene environmental variability at Munsa archaeological site, Uganda: a multicore, multiproxy approach," *The Holocene*, vol. 15, no. 7, pp. 1044–1061, 2005.

[46] K. Jackson, "The dimensions of Kamba pre-colonial history," in *Kenya before 1900: Eight Regional Studies*, B. A. Ogot, Ed., pp. 174–261, East African Publishing House, Nairobi, Kenya, 1976.

[47] J. C. Stager, B. Cumming, and L. Meeker, "A high-resolution 11,400-yr diatom record from Lake Victoria, East Africa," *Quaternary Research*, vol. 47, no. 1, pp. 81–89, 1997.

[48] J. E. Cole, R. B. Dunbar, T. R. McClanahan, and N. A. Muthiga, "Tropical pacific forcing of decadal SST variability in the western Indian ocean over the past two centuries," *Science*, vol. 287, no. 5453, pp. 617–619, 2000.

[49] S. Hastenrath, *The Glaciers of Equatorial East Africa*, Reidel, Dordrecht, The Netherlands, 1984.

[50] S. E. Nicholson, "The nature of rainfall variability over Africa on time scales of decades to millenia," *Global and Planetary Change*, vol. 26, no. 1-3, pp. 137–158, 2000.

[51] S. E. Nicholson and X. Yin, "Rainfall conditions in equatorial East Africa during the nineteenth century as inferred from the record of Lake Victoria," *Climatic Change*, vol. 48, no. 2-3, pp. 387–398, 2001.

[52] S. Hastenrath and P. D. Kruss, "The dramatic retreat of Mount Kenya's glaciers between 1963 and 1987," *Annals of Glaciology*, vol. 16, pp. 127–133, 1992.

[53] G. Kaser and B. Noggler, "Observations on Speke Glacier, Ruwenzori Range, Uganda," *Journal of Glaciology*, vol. 37, no. 127, pp. 315–318, 1991.

[54] S. Hastenrath and L. Greischar, "Glacier recession on Kilimanjaro, East Africa, 1912–89," *Journal of Glaciology*, vol. 43, no. 145, pp. 455–459, 1997.

3

Modeling Develops to Estimate Leaf Area and Leaf Biomass of *Lagerstroemia speciosa* in West Vanugach Reserve Forest of Bangladesh

Niamjit Das

Department of Forestry and Environmental Science, Shahjalal University of Science and Technology, Sylhet-3114, Bangladesh

Correspondence should be addressed to Niamjit Das; niamjit.forestry@gmail.com

Academic Editors: J. Kaitera and G. Martinez Pastur

Leaf area and leaf biomass have an important influence on the exchange of energy, light interception, carbon cycling, plant growth, and forest productivity. This study showed development and comparison of models for predicting leaf area and leaf biomass of *Lagerstroemia speciosa* on the basis of diameter at breast height and tree height as predictors. Data on tree parameters were collected randomly from 312 healthy, well-formed tree species that were considered specifically for full tree crowns. Twenty-four different forms of linear and power models were compared in this study to select the best model. Two models (M_{10} and M_{22}) for the estimation of leaf area and leaf biomass were selected based on R^2, adjusted R^2, root mean squared error, corrected akaike information criterion, Bayesian information criterion and Furnival's index, and the three assumptions of linear regression. The models were validated with a test data set having the same range of DBH and tree height of the sampled data set on the basis of linear regression Morisita's similarity index. So, the robustness of the models suggests their further application for leaf area and biomass estimation of *L. speciosa* in West Vanugach reserve forest of Bangladesh.

1. Introduction

Leaf area (A_1) and leaf biomass (B_1) estimation are significant basics of studying gas-exchange processes and modeling ecosystems. It is key traits in ecophysiological studies that determine assessing photosynthetic efficiency, evapotranspiration, atmospheric deposition, biogenic volatile organic emissions, light interception, and other ecosystem processes. Leaf area is valuable for the evaluation and understanding of individual tree growth models [1], biogeochemical models [2], and gap models [3]. It is defined as the one-sided projected surface area which is an important consequence for the interception of radiant energy, the absorption of carbon dioxide, and the circulation of water between the foliage and the atmosphere [4]. Leaf biomass constitutes one of the most important pools of essential nutrients, which is vital for forest nutrient cycling [5], including carbon cycling in a forest ecosystem. Leaf biomass estimates were considerably improved when additional biometric information relating to crown structure was added, whereas canopy B_1 is the product of leaf dry matter content and leaf area index [6]. Interspecific variation in A_1 and B_1 has been connected with climatic variation, geology, altitude, or latitude, where heat stress, cold stress, drought stress, and high-radiation stress all tend to select relatively small leaves. Within climatic zones, leaf-size variation can also be linked to allometric factors (plant size, twig size, anatomy, and architecture) and ecological strategy [7].

Measurement of the destructive (direct) method of A_1 and B_1 is very time consuming, labour intensive, and ecounfriendly and depends on very small samples. But, nondestructive (indirect) methods were found user friendly and less expensive and can give accurate A_1 and B_1 estimates [8, 9]. To quantify the A_1 and B_1 of individual trees, the relationship between A_1 and B_1 with DBH (diameter at breast height) and height are widely used [10]. This relationship differs between tree species and to some degree also within a tree species [11, 12] and needs to be established for each species in question. For the development of models, the power function ($y = ax^b$) form is widely used in biological sciences particularly

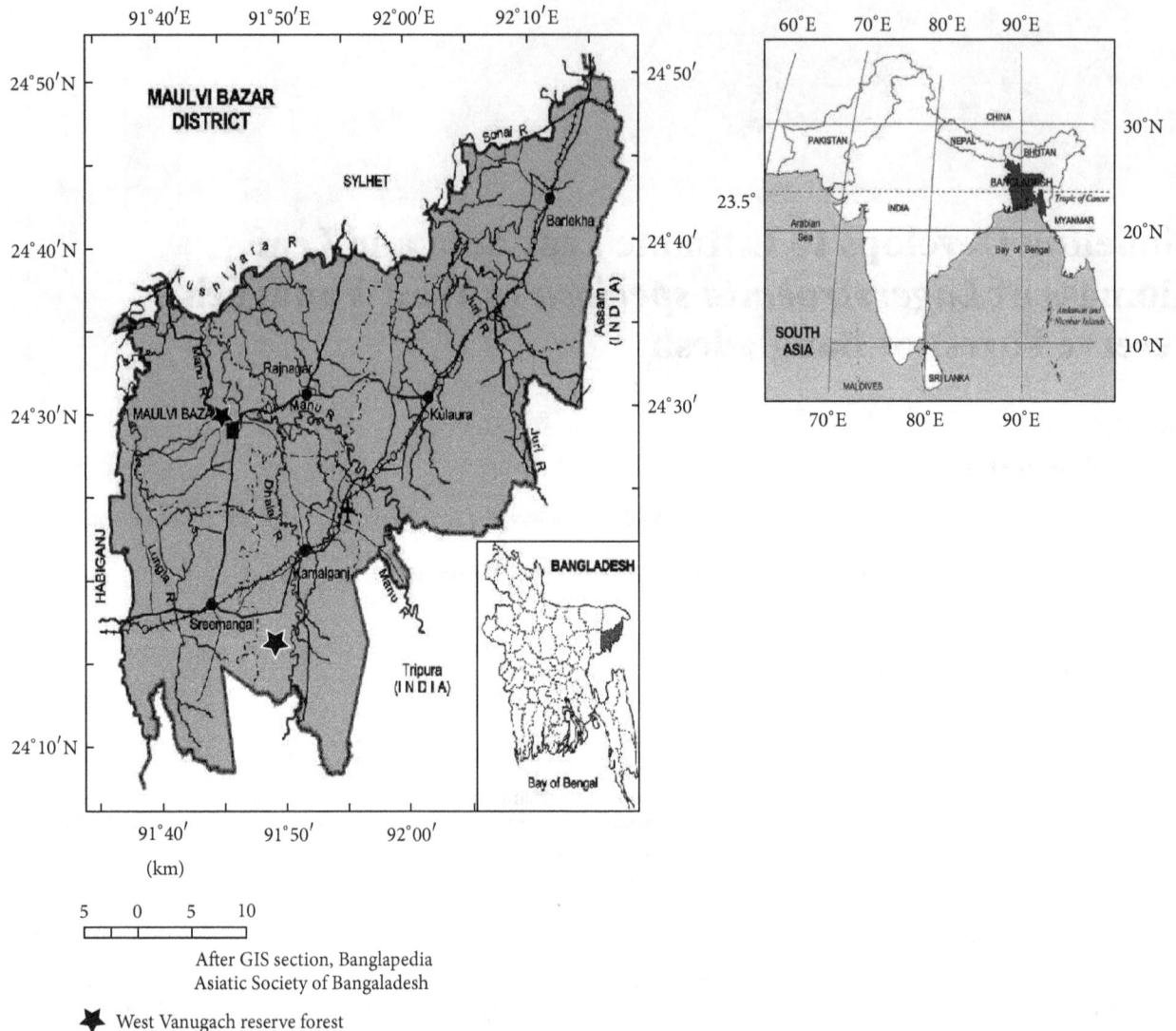

FIGURE 1: Study site in West Vanugach reserve forest of Bangladesh.

for these relationships [13]. But, in practice, the default nonlinear technique assumes homogeneity of errors that cannot be safely assumed with most model data. At present, models are developed by fitting a linear relationship between log-transformed diameter and leaf data. Although the log-transformed linear equation is mathematically equivalent to the power equation, they are not identical in the statistical sense [14]. However, many model characteristics of organisms are multiplicative by nature and thus fitting models to log-transformed data is perfectly acceptable because accounting for proportional rather than absolute variation is most important [15].

Lagerstroemia speciosa (L.) is the dominant tree both spatially and vertically in the forests of Bangladesh. It is well distributed in natural forests as well as in plantation forests [16]. It is also found at low to medium altitudes in comparatively open habitats, in disturbed or secondary forest and grassland. The habitat may vary from well drained to occasionally flooded but not peat soil and it is resistant to fire. *L. speciosa* is the most important wood species in parts of South and Southeast Asia [16], but to estimate A_1 and B_1 relationships of this species is not available. So, the main goal of my study was to develop and compare models of *L. speciosa* that will help to estimate A_1 and B_1 of the species in reserve forest of Bangladesh.

2. Materials and Methods

2.1. Site Description. The field work was conducted at the West Bhanugach Reserved Forest (2,740 ha) on the Moulvibazar Forest Range (24°19′11″ N, 91°47′1″ E, altitude of 45 m) in tropical semievergreen forest, Bangladesh (Figure 1). The soil of this forest is alluvial brown sandy clay loam to clay loam. Average annual rainfall is 4,332 mm and January is the coldest month (minimum temperature around 11°C), while May and October are the hottest months (average maximum temperature around 33°C). The relative humidity is about 75% during December and over 92% during July-August [17].

Modeling Develops to Estimate Leaf Area and Leaf Biomass of Lagerstroemia speciosa in West Vanugach Reserve
Forest of Bangladesh

37

This forest stand is a mixed forest stand and its vegetation is recognized as old growth plantation forest. It includes 78 species of trees, 14 species of shrubs, 42 species of herbs, and 25 species of climbers [18].

2.2. Species Description. *Lagerstroemia speciosa* (L.) is a medium-sized tree growing to 20 meters tall, much branched deciduous tree, with smooth and flaky bark [16]. Its leaves are opposite, elliptic, or oblong-lanceolate, 8.5–20.0 cm long and 3.0–7.5 cm broad, acuminate at the apex, acute to rounded at the base, and glabrous and finely reticulate on both surfaces, have lateral nerves 8–12 on either half, prominent beneath, and petioles 5–9 mm long, and are stout. Its wood is strong and fairly durable and is used for carts, furniture, and house posts. The root is considered as an astringent, stimulant, and febrifuge and due to the tree's dense and wide spreading root system, it is used in erosion control. Its extracts have been used as traditional medicines and are effective in controlling diabetes and obesity. A leaf poultice is used to relief malarial fever and is applied on cracked feet. It is planted in forests, edges of forest streams, swamps, ditches, and river banks. It is growing in South East Asia, Bangladesh, India, and Philippines and it is also widely cultivated as an ornamental plant in tropical and subtropical areas [19].

2.3. Field Measurements. Leaf, DBH, and tree height data were collected for three hundred twelve sampled healthy well-formed *L. speciosa* tree species with full crown by maintaining the protocols for standardized and easy measurement of plant functional traits worldwide developed by Cornelissen et al. 2003. Field sampling occurred in the rainy season during which A_1 and B_1 were maximum. Diameter at breast height (DBH; at 1.3 m height from ground level) and tree height (*H*) was measured with Suunto clinometers (Suunto PM5-360, Vantaa, Finland) and tree calipers (Haglöf Graduated Aluminum Tree Caliper, Ben Meadows, Janesville, Wisconsin, USA). Data on tree parameters are usually measured and collected after harvesting for improving data accuracy [6, 20]. Though using nondestructive methods is a highly reliable option [21] where whole tree removal is not possible, as is the case in Bangladeshi reserve forests [22], leaf data were collected manually (by climbing the tree). To retain high data accuracy, stratified random sampling was used for calculation of the leaf number per individual tree. Thereafter, the following steps were performed to count the leaves of each sampled tree: (I) the diameter of all of the main branches was measured; (II) the diameter of the main branch closest to the mean diameter was selected as the model main branch (MMB); (III) subbranches of all main branches were counted; (IV) three subbranches were selected randomly from the MMB; (V) twigs of each subbranch were counted and the mean twig number per subbranch was calculated; (VI) three twigs were selected randomly from each selected subbranch; (VII) the average number of leaves per twig was calculated; (VIII) the leaf number per subbranch was calculated by multiplying the mean twig number per subbranch and mean leaf number per twig; and (IX) the total leaf number per tree was calculated by multiplying the total number of subbranches and estimated leaf number per subbranch [9]. The leaves

were categorized into small, medium, and large size and fifteen leaves of each category were collected according to different directions and heights of a tree. A total of forty-five samples of leaves were taken from each sampled tree and packed in a plastic bag. The leaf area of each leaf sample was measured with a leaf area meter (CI-202, CID, Inc., Vancouver, Washington, USA) for small and medium leaves and Adobe Photoshop CS5 software for large leaves. Data of DBH, height, leaf samples, and total leaf number per tree from additional 40 trees were collected and used as a test data set for validating the models. The projected leaf area for each sample tree (A_1, m^2) was calculated by multiplying the average leaf area and total leaf number. Thickness of each sample leaf was estimated with digital caliper (Absolute Digimatic CD-6″ CS, Mitutoyo Corporation, Kanagawa, Japan) and the fresh mass of each leaf was measured with a balance meter. Measured leaves were oven-dried at 65°C for 72 h and weighed to determine fresh mass: dry mass ratios. Total leaf volume per tree was estimated by multiplying total leaf area and leaf thickness. Leaf density was calculated by dividing by the average leaf fresh mass and the leaf volume of forty-five sample leaf in a tree. Total leaf biomass (B_1, kg) per tree was estimated by multiplying the total leaf volume and leaf density.

2.4. Statistical Analysis. Regression analysis was used to evaluate the appropriateness of each of the two independent variables (*x*): height (m) and DBH (cm), as estimators of total A_1 and B_1 (dependent variables = *y*) for *L. speciosa* by using *R* statistical software version 2.15.2 and SPSS 17 software (SPSS, ver. 17, Chicago, IL, USA). The data with two regression models (linear and power) was analyzed. At first, scatter plots were used to see whether the relationship between independent and dependent variables was linear. Secondly, given the presence of heteroscedasticity, power models transformed the data for linear regression using natural logarithmic. A total of twenty-four models were developed using DBH and height for the best model selection (Table 2). The performance of the developed models was evaluated by examining the goodness of fit (R^2 and adjusted R^2), root mean squared error (RMSE), corrected akaike information criterion (AICc), Bayesian information criterion (BIC), and Furnival's index (FI). Thus, to select between models, the corrected akaike information criterion (AICc) for each model was then computed as

$$\text{AICc} = 2k - 2\ln(L) + \frac{2k(k+1)}{n-k-1}, \qquad (1)$$

where *n* is the sample size, *k* is the number of parameters in the statistical model, and *L* is the corresponding likelihood [23]. The formula for the BIC (Bayesian information criterion) is

$$\text{BIC} = n \cdot \ln\left(\widehat{\sigma_e^2}\right) + k \cdot \ln(n), \qquad (2)$$

where $\widehat{\sigma_e^2}$ is the error variance [23]. For models with different dependent variables using R^2 to compare them could sometimes produce false results [24]. So, Furnival's index

[25] which is able to compare models of different dependent variables or weights was used to compare the logarithmic and nonlogarithmic models. The index is calculated as follows:

$$\text{FI} = \frac{1}{[f'(Y)]} \sqrt{\text{MSE}}, \quad (3)$$

where $f'(Y)$ is the derivative of the dependent variable with respect to A_1 and B_1, MSE is the mean square error of the fitted model, and the square bracket ([]) is the geometric mean. After that, the three regression assumptions were used to judge the consistency of the models for the residues: (I) no outlier, (II) the distribution of these residuals should be normal with mean = 0 and a constant variance and (III) the data points must be independent [26]. The Durbin-Watson test was used to check for autocorrelation. Though logarithmic transformation is reported to increase the statistical validity of regression analysis by homogenising variance, it introduces a slight downward bias when data are back-transformed to arithmetic units [27]. To account for the bias, the back-transformed results from logarithmic unit are usually multiplied by a correction factor. Consequently, a correction factor (CF) was calculated for all logarithmic models. The CF is given by the following [28]:

$$\text{CF} = \exp^{((\text{SEE}*2.303)^2/2)}, \quad (4)$$

where SEE is the standard error of the estimate.

For validation of the models, the observed A_1 and B_1 data regressed from a test data set ($n = 40$) against the predicted A_1 and B_1 using linear regression. As a measure of distance between observed and predicted values, Morisita's [29] dissimilarity index (D_M) was calculated as follows:

$$D_M = \frac{(1/n)\sum_{i=1}^{n}\left(y_i - y_i^p\right)^2}{(1/n)\left(\sum_{i=1}^{n} y_1^2 + \left(\sum_{i=1}^{n} y_i^p\right)^2\right)}, \quad (5)$$

where y_i = observed A_1 and B_1 of the sampled data and y_i^p = predicted A_1 and B_1. Note that 1-D_M is essentially Morisita's measure of niche overlap or similarity [30].

3. Results

3.1. Model Development. DBH and tree height of sampled individuals of *L. speciosa* ranged from 7.52 to 65.16 cm and 3.27 to 19.01 m, respectively, and the A_1 and B_1 of sampled trees ranged from 36.20 to 141.31 cm^2 and 0.35 to 1.26 g, respectively. In addition, the range of DBH and H data in the model test data set fell within the boundary of the sampled data set (Table 1). Table 3 summarizes the different models derived to predict A_1 and B_1 with DBH and H as the independent variable. For the development of the best model, 24 different model forms were determined (Table 2). Model M_{10} and M_{22} were found to be the best predictor for A_1 and B_1 estimation, respectively. Whereas the RMSE value was the lowest and the goodness of fit (R^2 and adjusted R^2) was highly significant and explained more than 96% variation for the selected models (M_{10} and M_{22}). To compare the models,

TABLE 1: General characteristics of *L. speciosa* species.

Variables	Mean	SD	Range
DBH (cm)	27.6069	14.4302	7.52–65.16
Height (m)	12.4225	2.8879	3.27–19.01
Validation DBH (cm)	27.0027	12.9432	7.65–64.85
Validation height (m)	11.9863	2.6372	3.24–18.79
Individual leaf area (cm^2)	88.6291	9.3614	36.20–141.31
Individual dry leaf biomass (g)	0.8534	0.0526	0.35–1.26

TABLE 2: Twenty-four models used in this study.

Model number	Models
M_1	$A_1 = x_o + x_1\text{DBH}$
M_2	$A_1 = x_o + x_1 H$
M_3	$A_1 = x_o + x_1(\text{DBH} \times H)$
M_4	$A_1 = x_o + x_1\text{DBH} + x_2 H$
M_5	$A_1 = x_o + x_1\text{DBH} + x_2\text{DBH}^2$
M_6	$A_1 = x_o + x_1 H + x_2 H^2$
M_7	$\ln(A_1) = x_o + x_1\ln(\text{DBH})$
M_8	$\ln(A_1) = x_o + x_1\ln(H)$
M_9	$\ln(A_1) = x_o + x_1\ln(\text{DBH} \times H)$
M_{10}	$\ln(A_1) = x_o + x_1\ln(\text{DBH}) + x_2\ln(H)$
M_{11}	$\ln(A_1) = x_o + x_1\ln(\text{DBH}) + x_2\ln(\text{DBH}^2)$
M_{12}	$\ln(A_1) = x_o + x_1\ln(H) + x_2\ln(H^2)$
M_{13}	$B_1 = x_o + x_1\text{DBH}$
M_{14}	$B_1 = x_o + x_1 H$
M_{15}	$B_1 = x_o + x_1(\text{DBH} \times H)$
M_{16}	$B_1 = x_o + x_1\text{DBH} + x_2 H$
M_{17}	$B_1 = x_o + x_1\text{DBH} + x_2\text{DBH}^2$
M_{18}	$B_1 = x_o + x_1 H + x_2 H^2$
M_{19}	$\ln(B_1) = x_o + x_1\ln(\text{DBH})$
M_{20}	$\ln(B_1) = x_o + x_1\ln(H)$
M_{21}	$\ln(B_1) = x_o + x_1\ln(\text{DBH} \times H)$
M_{22}	$\ln(B_1) = x_o + x_1\ln(\text{DBH}) + x_2\ln(H)$
M_{23}	$\ln(B_1) = x_o + x_1\ln(\text{DBH}) + x_2\ln(\text{DBH}^2)$
M_{24}	$\ln(B_1) = x_o + x_1\ln(H) + x_2\ln(H^2)$

the AICc and BIC values were calculated and the results are presented in Table 3. The AICc and BIC values for models M_{10} and M_{22} were lower than that of other tested models indicating statistical robustness of the selected models. Based on Furnival's index (FI) of fit, models M_{10} and M_{22} performed better in both A_1 and B_1 estimation, whereas the FI values for both models were lower (0.36 and 0.47, respectively) than that of other tested models.

To avoid any confusion, the three regression assumptions were checked to select the best model. A good model should conform to all valid statistical assumptions [26]. First assumption, residual analysis of models M_{10} and M_2, showed that the values of standard residual were 2.93 and 2.48, respectively. A model to be outlier free requires that the value of standard residual must be below 3 [26]. In both models this requirement was fulfilled. The normality of residuals was measured for selected models by histogram and normal

Modeling Develops to Estimate Leaf Area and Leaf Biomass of Lagerstroemia speciosa in West Vanugach Reserve Forest of Bangladesh

39

TABLE 3: Different models tested for the estimation of leaf area (A_1, m²) and leaf biomass (B_1, kg) with diameter at breast height (DBH, cm) and tree height (H, m) of *L. speciosa* species.

Model	Estimated coefficients $x_0 \pm$ SE	$x_1 \pm$ SE	$x_2 \pm$ SE	F	RMSE	R^2	Adjusted R^2	FI	AICc	BIC	CF	Durbin Watson
M$_1$	2.36383 ± 2.76814	5.33557 ± 0.08891		3602	18.59	0.9452	0.9449	0.64	1836.22	1846.22	—	1.807
M$_2$	−132.85 ± 13.52	22.74 ± 1.06		460	44.36	0.6877	0.6862	3.21	2203.20	2213.21	—	0.753
M$_3$	32.165522 ± 1.982558	0.313651 ± 0.004419		5037	15.84	0.9602	0.9600	0.59	1768.75	1778.75	—	1.411
M$_4$	−44.6109 ± 5.1513	4.4608 ± 0.1128	5.7255 ± 0.5637	2733	15.24	0.9633	0.9630	0.52	1753.26	1766.59	—	1.736
M$_5$	−11.235989 ± 5.19719	6.465311 ± 0.378365	−0.018144 ± 0.005913	1878	18.23	0.9475	0.9470	0.74	1828.91	1842.23	—	1.767
M$_6$	−9.5289 ± 34.0715	0.2453 ± 5.8305	0.9602 ± 0.2450	253	42.91	0.7092	0.7064	2.95	2190.20	2203.53	—	0.813
M$_7$	1.57067 ± 0.05028	1.03390 ± 0.01563		4378	0.1301	0.9544	0.9542	0.53	−257.78	−247.77	1.0084	2.014
M$_8$	0.03719 ± 0.18967	1.93313 ± 0.07582		650	0.3006	0.7567	0.7555	2.62	95.66	105.67	1.0462	0.854
M$_9$	0.681579 ± 0.052864	0.736374 ± 0.009257		6328	0.1090	0.9680	0.9679	0.43	−332.53	−322.52	1.0059	1.520
M$_{10}$	0.93516 ± 0.06836	0.83960 ± 0.02101	0.50296 ± 0.04413	3604	0.1023	0.9720	0.9717	0.36	−358.13	−344.80	1.0052	2.003
M$_{11}$	1.57067 ± 0.05028	1.03390 ± 0.01563	−0.002271 ± 0.0062	4378	0.1301	0.9544	0.9542	0.68	−257.75	−247.77	1.0084	2.014
M$_{12}$	0.03719 ± 0.18967	1.93313 ± 0.07582	0.00373 ± 0.05283	650	0.3006	0.7567	0.7555	2.41	95.70	105.67	1.0462	0.854
M$_{13}$	0.389894 ± 0.240441	0.495224 ± 0.007723		4112	1.615	0.9516	0.9514	0.64	805.08	815.08	—	1.933
M$_{14}$	−11.0649 ± 1.3503	2.0227 ± 0.1059		364	4.431	0.6358	0.6341	3.33	1231.05	1241.06	—	0.875
M$_{15}$	3.3471251 ± 0.237553	0.0286013 ± 0.00052		2917	1.899	0.9332	0.9328	0.86	873.37	883.37	—	1.227
M$_{16}$	−2.23654 ± 0.50829	0.44631 ± 0.01113	0.32012 ± 0.05562	2389	1.503	0.9583	0.9579	0.59	775.93	789.26	—	1.914
M$_{17}$	−0.411165 ± 0.456914	0.5617675 ± 0.03326	−0.0010687 ± 0.00051	2090	1.603	0.9526	0.9521	0.62	802.87	816.19	—	1.934
M$_{18}$	−4.26959 ± 3.48979	0.78302 ± 0.59719	0.05291 ± 0.02509	187	4.395	0.6435	0.6400	3.51	1228.62	1241.95	—	0.903
M$_{19}$	−0.73397 ± 0.05330	1.01654 ± 0.01656		3766	0.1379	0.9474	0.9472	0.54	−233.16	−223.15	1.0095	1.626
M$_{20}$	−2.22910 ± 0.19082	1.89560 ± 0.07628		617	0.3025	0.7472	0.7459	2.69	98.21	108.21	1.0468	1.062
M$_{21}$	−1.6049 ± 0.0588	0.7234 ± 0.0103		4938	0.1212	0.9594	0.9592	0.56	−287.66	−277.65	1.0073	1.708
M$_{22}$	−1.34008 ± 0.07665	0.83123 ± 0.02357	0.47969 ± 0.04948	2768	0.1147	0.9638	0.9634	0.47	−309.79	−296.46	1.0066	1.971
M$_{23}$	−0.73397 ± 0.05330	1.01654 ± 0.01656	−0.00421 ± 0.00284	3766	0.1379	0.9474	0.9472	0.52	−233.13	−223.15	1.0095	1.626
M$_{24}$	−2.22910 ± 0.19082	1.89560 ± 0.07628	0.04858 ± 0.03842	617	0.3025	0.7472	0.7459	2.24	98.24	108.21	1.0468	1.062

RMSE: root mean squared error, FI: Furnival's index, AICc: corrected akaike information criterion, BIC: Bayesian information criterion, and CF: correction factor.

(a)

(b)

(c)

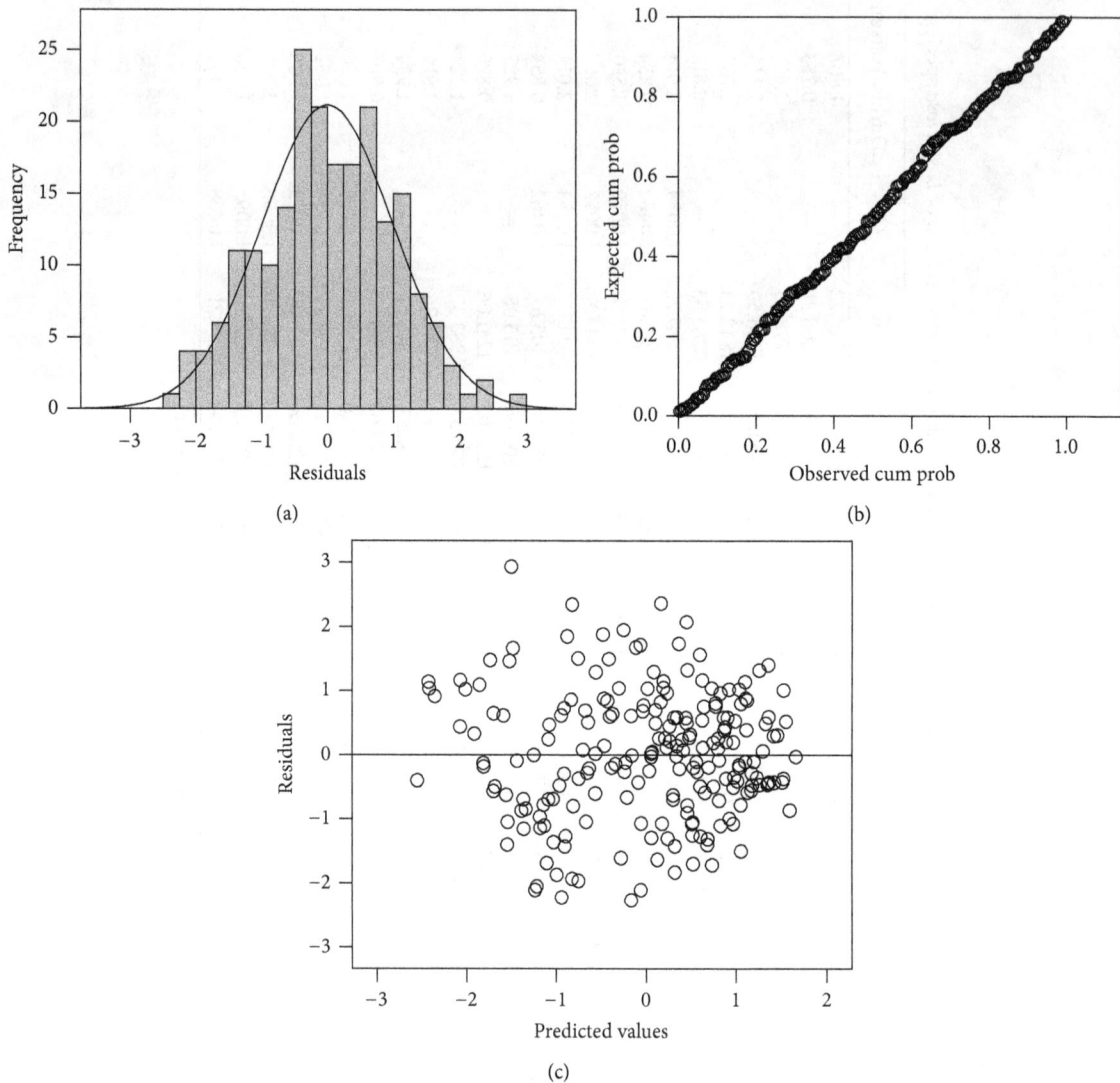

FIGURE 2: Residual distribution (a), normal P-P plot of regression standardized residuals (b), and scatter plot of residuals (c) for model M_{10}.

probability plot. The distribution of the residual satisfies the normality assumptions for both models M_{10} (Figures 2(a) and 2(b)) and M_{22} (Figures 3(a) and 3(b)). Second assumption, diagnosis for constant variance, was done by plotting the regression standardized residuals and predicted values in a scatter plot. It was shown that the scatter of the residual points was randomly distributed (Figures 2(c) and 3(c)). Thus it can be concluded that the variance was constant for both models. Third assumption, the Durbin-Watson statistic, lies in the range 0.0–4.0. A value of 2 or nearly 2 indicates that there is independence of the data points [31]. This range of the selected models was 2.003 for A_1 and 1.971 for B_1. Thus, the independence assumption was satisfied for both models. However, due to the systematic bias introduced by logarithmic transformation of data, correction factor is normally calculated for logarithmic transformed models so as to account for the bias. Correction factors showed a rather narrow variation for A_1 and B_1 (Table 3).

As a result, the final model of A_1 is

$$A_1 = \exp(0.93516 + 0.83960 \times \ln(\text{DBH}) + 0.50296 \times \ln(H)), \tag{6}$$

$$\text{or } A_1 = 2.547621 \times (\text{DBH})^{0.83960} \times (H)^{0.50296}.$$

The bias corrected model is $A_1 = 2.547621 \times (\text{DBH})^{0.83960} \times (H)^{0.50296} \times 1.005246 = 2.560986 \times (\text{DBH})^{0.83960} \times (H)^{0.50296}$.

The final model of B_1 is

$$B_1 = \exp(-1.34008 + 0.83123 \times \ln(\text{DBH}) + 0.47969 \times \ln(H)), \tag{7}$$

$$\text{or } B_1 = 0.261825 \times (\text{DBH})^{0.83123} \times (H)^{0.47969}.$$

The bias corrected model is $B_1 = 0.261825 \times (\text{DBH})^{0.83123} \times (H)^{0.47969} \times 1.0066 = 0.263553 \times (\text{DBH})^{0.83123} \times (H)^{0.47969}$.

Modeling Develops to Estimate Leaf Area and Leaf Biomass of Lagerstroemia speciosa in West Vanugach Reserve
Forest of Bangladesh

41

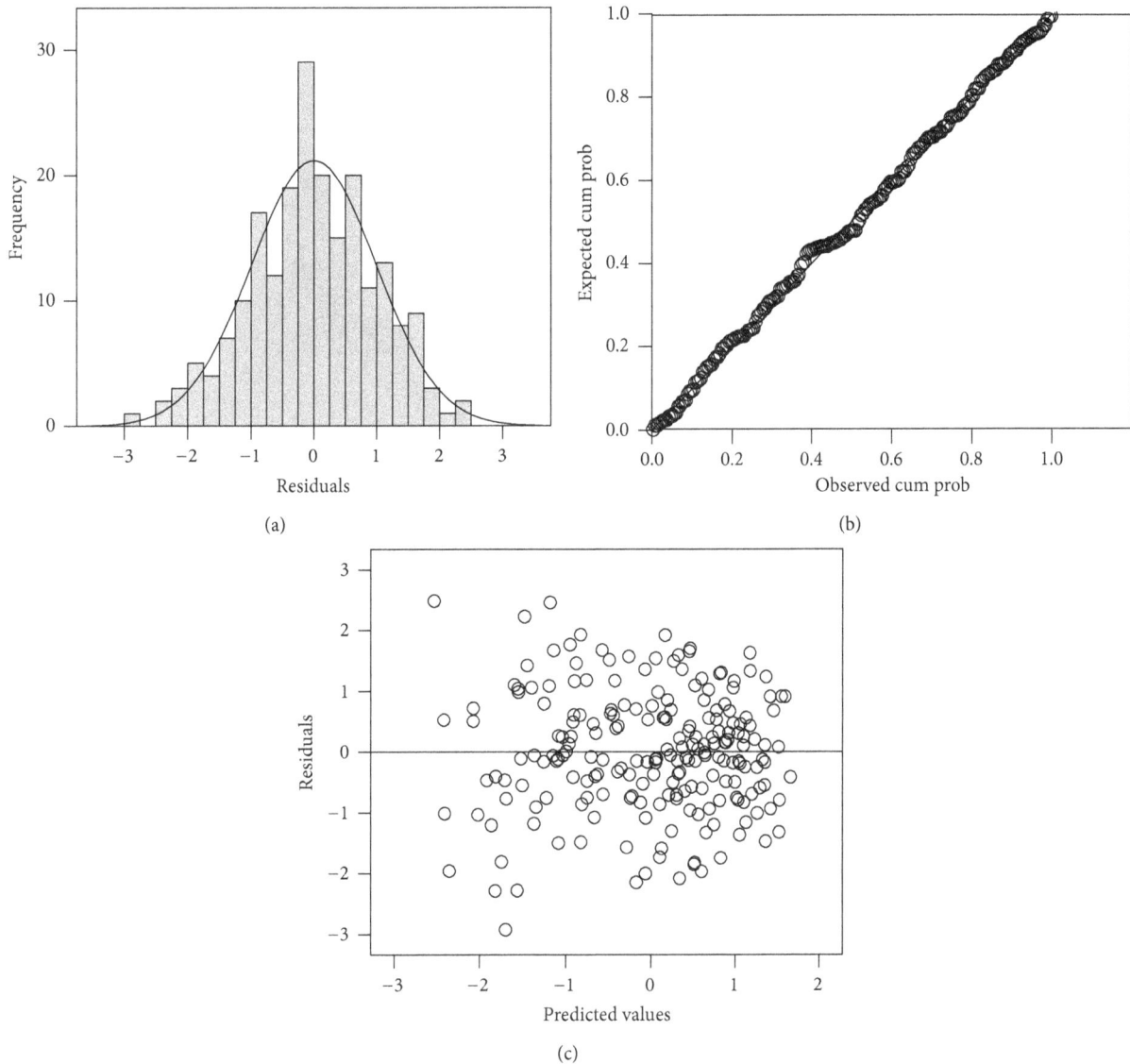

FIGURE 3: Residual distribution (a), normal P-P plot of regression standardized residuals (b), and scatter plot of residuals (c) for model M_{22}.

3.2. Model Validation. In the model validation, the goodness of fit (R^2) showed that there was a highly reliable relationship between predicted and observed data for A_1 and B_1. The R^2 values between them for models using A_1 or B_1 were 0.9716 and 0.9736, respectively (Figure 4). Morisita's [29] index of similarity ($1 - D_M$) was 0.994836 for A_1 and 0.993567 for B_1 (almost 1), which indicated that the models can predict A_1 and B_1, outside the sampled data set, effectively.

4. Discussion

Different combinations and forms of model of diameter and tree height were used to select the best models (M_{10} and M_{22}) that are more than 96% variation and lower RMSE, AICc, BIC, FI value in measuring A_1 and B_1 of *L. speciosa*. After that, logarithmic transformation induces a systematic bias in the estimation, which was corrected using a CF in the final

model [27]. Given the paucity of models for this forest, I expect these models to prove valuable in future research that requires estimates of A_1 and B_1 of this species. Burton et al. [32] found that DBH is the best predictor for estimating A_1 and B_1 ($R^2 > 0.90$). Using a nondestructive sampling technique for A_1 estimation, Grace [33] found that DBH could explain 91% variation in A_1 of *Acacia Koa*. Although sapwood area (AS) or sapwood volume has proved a good predictor of tree canopy properties in different temperate in the past [34, 35], it has not been measured in this study or any national forest inventories in Bangladesh. However, Turner et al. [36] found that, in *Pseudotsuga menziesii*, the A_1 estimates based on DBH were more accurate than those on AS. So, I excluded AS from this study. Other studies found higher [37–41] correlations with DBH. Then it provides support for estimation of A_1 and B_1 which are positively correlated to DBH and H [20]. Sarker et al. [9]

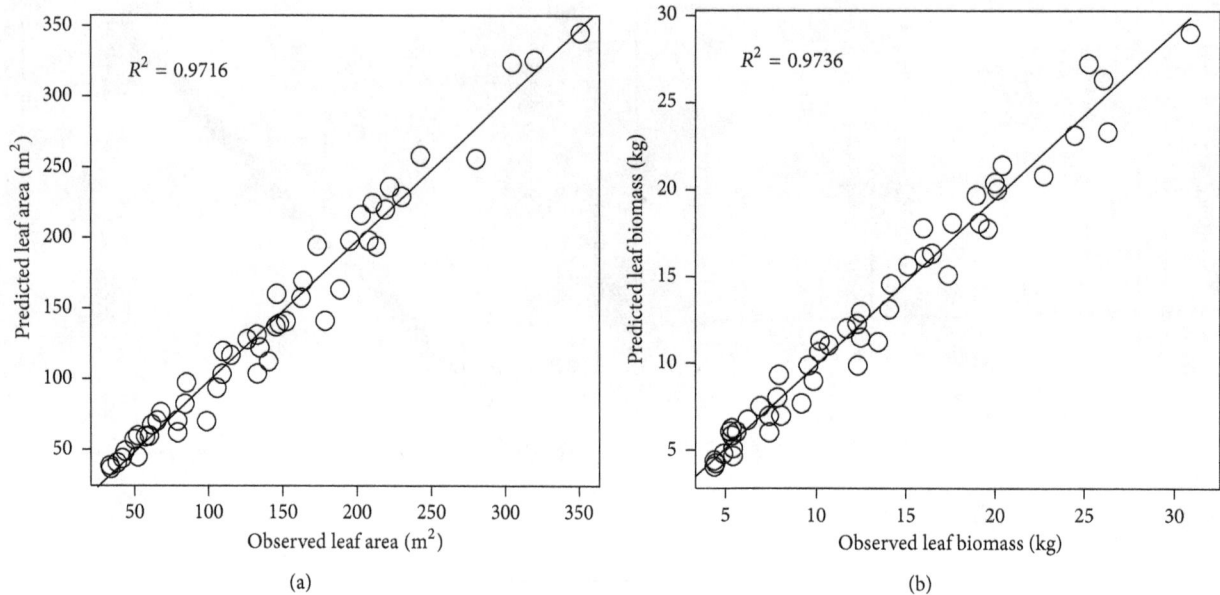

FIGURE 4: Linear regression between observed and predicted leaf area (a); between observed and predicted leaf biomass (b).

also show that allometric models are explained for more than 95% of the variation based on DBH with A_1 and B_1. *L. speciosa* is the dominant and medium sized deciduous tree species in tropical semievergreen forest of Bangladesh. Thus, this study is important to tree growth models of this species and ecological purposes, including light interception, transpiration rate, biomass estimation, and carbon storage. As a result, I suggest estimating A_1 and B_1 of this species based on available standard inventory measurements such as DBH and tree height in Bangladesh.

5. Conclusion

The consequences of this result showed a strong statistical dependence and best fitted models (M_{10} and M_{22}) between A_1 and B_1 with DBH and H. These models can be used effectively at the stand level for estimation of A_1 and B_1 of *L. speciosa* quickly, accurately, and nondestructively. Therefore, I propose that the models obtained and validated in this study could be employed with confidence by foresters, forest ecologists, and other scientists for estimating leaf area and leaf biomass in Bangladeshi reserve forests.

Conflict of Interests

The author declares that there is no conflict of interests regarding the publication of this paper.

Acknowledgment

The author wish to thank field staffs of the Divisional Forest Office, Sylhet, for their assistance during the field works.

References

[1] H. Pretzsch, P. Biber, and J. Ďurský, "The single tree-based stand simulator SILVA: construction, application and evaluation," *Forest Ecology and Management*, vol. 162, no. 1, pp. 3–21, 2002.

[2] S. A. Pietsch, H. Hasenauer, and P. E. Thornton, "BGC-model parameters for tree species growing in central European forests," *Forest Ecology and Management*, vol. 211, no. 3, pp. 264–295, 2005.

[3] M. J. Lexer, K. Hönninger, H. Scheifinger et al., "The sensitivity of Austrian forests to scenarios of climatic change: a large-scale risk assessment based on a modified gap model and forest inventory data," *Forest Ecology and Management*, vol. 162, no. 1, pp. 53–72, 2002.

[4] H. Margolis, R. Oren, D. Whitehead, and M. R. Kaufmann, "Leaf area dynamics ofconifer forests," in *Ecophysiology of Coniferous Forests*, W. K. Smith and T. M. Hinckley, Eds., pp. 255–308, Academic Press, San Diego, Calif, USA, 1995.

[5] R. H. Waring and W. H. Schlesinger, *Forest Ecosystems: Concepts and Management*, Academic Press, London, UK, 1985.

[6] B. Tobin, K. Black, B. Osborne, B. Reidy, T. Bolger, and M. Nieuwenhuis, "Assessment of allometric algorithms for estimating leaf biomass, leaf area index and litter fall in different-aged Sitka spruce forests," *Forestry*, vol. 79, no. 4, pp. 453–464, 2006.

[7] J. H. C. Cornelissen, S. Lavorel, E. Garnier et al., "A handbook of protocols for standardised and easy measurement of plant functional traits worldwide," *Australian Journal of Botany*, vol. 51, no. 4, pp. 335–380, 2003.

[8] J. M. Norman and G. S. Campbell, "Canopy structure," in *Plant Physiologicalecology: Field Methods and Instrumentation*, R. W. Pearcy, J. Ehleringer, H. A. Mooney, and P. W. Rundel, Eds., pp. 301–325, Chapman and Hall, New York, NY, USA, 1989.

[9] S. K. Sarker, N. Das, M. Q. Chowdhury, and M. M. Haque, "Developing allometric equations for estimating leaf area and leaf biomass of *Artocarpus chaplasha* in Raghunandan Hill Reserve, Bangladesh," *Southern Forests*, vol. 75, pp. 51–57, 2013.

Modeling Develops to Estimate Leaf Area and Leaf Biomass of Lagerstroemia speciosa in West Vanugach Reserve
Forest of Bangladesh

43

[10] P. A. Gajardo-Caviedes, M. A. Espinosa, U. D. T. Gonzalez, and D. G. Ríos, "The influence of thinning and tree size on the sapwood area/leaf area ratio in coigue," *Canadian Journal of Forest Research*, vol. 35, no. 7, pp. 1679–1685, 2005.

[11] N. M. Good, M. Paterson, C. Brack, and K. Mengersen, "Estimating tree component biomass using variable probability sampling methods," *Journal of Agricultural, Biological, and Environmental Statistics*, vol. 6, no. 2, pp. 258–267, 2001.

[12] M. Mencuccini and L. Bonosi, "Leaf/sapwood area ratios in Scots pine show acclimation across Europe," *Canadian Journal of Forest Research*, vol. 31, no. 3, pp. 442–456, 2001.

[13] K. J. Niklas, "A phyletic perspective on the allometry of plant biomass-partitioning patterns and functionally equivalent organ-categories," *New Phytologist*, vol. 171, no. 1, pp. 27–40, 2006.

[14] J. H. Zar, "Calculation and miscalculation of the allometric equation as a model inbiological data," *BioScience*, vol. 18, pp. 1118–1120, 1968.

[15] A. J. Kerkhoff and B. J. Enquist, "Multiplicative by nature: why logarithmic transformation is necessary in allometry," *Journal of Theoretical Biology*, vol. 257, no. 3, pp. 519–521, 2009.

[16] D. K. Das and M. K. Alam, *Trees of Bangladesh*, Bangladesh Forest Research Institute, Chittagong, Bangladesh, 2001.

[17] BBS/UNDP (Bangladesh Bureau of Statistics/United Nations DevelopmentProgramme), *Compendium of Environment Statistics of Bangladesh*, Ministry of Planning, Dhaka, Bangladesh, 2005.

[18] J. C. Malaker, M. M. Rahman, A. K. M. Azad-ud-doula, S. K. Malaker, and M. A. H. Khan, "Floristic composition of Lawachara Forest in Bangladesh," *International Journal of Experimental Agriculture*, vol. 1, pp. 1–9, 2010.

[19] Asiatic Society of Bangladesh (A S B), *Encyclopedia of Flora and Fauna of Bangladesh*, Asiatic Society of Bangladesh, Dhaka, Bangladesh, 2007.

[20] J. C. Calvo-Alvarado, N. G. McDowell, and R. H. Waring, "Allometric relationships predicting foliar biomass and leaf area:sapwood area ratio from tree height in five Costa Rican rain forest species," *Tree Physiology*, vol. 28, no. 11, pp. 1601–1608, 2008.

[21] C. Dobbs, J. Hernández, and F. Escobedo, "Above ground biomass and leaf area models based on a non destructive method for urban trees of two communes in central chile," *Bosque*, vol. 32, no. 3, pp. 287–296, 2011.

[22] S. K. Sarker, J. C. Deb, and M. A. Halim, "A diagnosis of existing logging bans in Bangladesh," *International Forestry Review*, vol. 13, no. 4, pp. 461–475, 2011.

[23] K. P. Burnham and D. R. Anderson, *Model Selection and Inference: A PracticalInformation-Theoretic Approach*, Springer, New York, NY, USA, 2002.

[24] B. R. Parresol, "Assessing tree and stand biomass: a review with examples and critical comparisons," *Forest Science*, vol. 45, no. 4, pp. 573–593, 1999.

[25] G. M. Furnival, "An index for comparing equations used in constructing volume tables," *Forest Science*, vol. 7, pp. 337–341, 1961.

[26] S. Chatterjee and A. S. Hadi, *Regression Analysis by Example*, JohnWiley & Sons, New Jersey, NJ, USA, 2006.

[27] G. L. Baskerville, "Use of Logarithmic regression in the estimation of plant biomass," *Canadian Journal of Forest Research*, vol. 2, no. 1, pp. 49–53, 1972.

[28] D. G. Sprugel, "Correcting for bias in log-transformed allometric equations," *Ecology*, vol. 64, no. 1, pp. 209–210, 1983.

[29] M. Morisita, "Measuring of interspecific association and similarity between communities," *Memoirs of the Faculty of Science of Kyushu University E: Biology*, vol. 3, pp. 65–80, 1959.

[30] E. P. Smith and K. A. Rose, "Model goodness-of-fit analysis using regression and related techniques," *Ecological Modelling*, vol. 77, no. 1, pp. 49–64, 1995.

[31] D. I. Warton, D. S. Falster, and M. Westoby, *Bivariate Line Fitting Methods Forallometry*, Department of Biological Sciences, Macquarie University, Sydney, Australia, 2005.

[32] A. J. Burton, K. S. Pregitzer, and D. D. Reed, "Leaf area and foliar biomass relationships in northern hardwood forests located along an 800 km acid deposition gradient," *Forest Science*, vol. 37, pp. 1041–1059, 1991.

[33] K. T. Grace, "Leaf area allometry and evaluation of non-destructive estimates of total leaf area and loss by browsing in a silvopastoral system," *Agroforestry Systems*, vol. 40, no. 2, pp. 139–147, 1998.

[34] D. Whitehead, W. R. N. Edwards, and P. G. Jarvis, "Conducting sapwood area, foliage area, and permeability in mature trees of Picea sitchensis and Pinus contorta," *Canadian Journal of Forest Research*, vol. 14, no. 6, pp. 940–947, 1984.

[35] M. G. Keane and G. F. Weetman, "Leaf area-sapwood cross-sectional arearelationships in repressed stands of lodgepole pine," *Canadian Journal of Forest Research*, vol. 17, pp. 205–209, 1987.

[36] D. P. Turner, S. A. Acker, J. E. Means, and S. L. Garman, "Assessing alternative allometric algorithms for estimating leaf area of Douglas-fir trees and stands," *Forest Ecology and Management*, vol. 126, no. 1, pp. 61–76, 2000.

[37] H. Peter, E. Otto, and S. Hubert, "Leaf area of beech (*Fagus sylvatica* L.) from different stands in eastern Austria studied by randomized branch sampling," *European Journal of Forest Research*, vol. 129, no. 3, pp. 401–408, 2010.

[38] R. H. Waring, P. E. Schroeder, and R. Oren, "Application of the pipe model theory to predict canopy leaf area," *Canadian Journal of Forest Research*, vol. 12, no. 3, pp. 556–560, 1982.

[39] J. N. Long and F. W. Smith, "Leaf area-sapwood area relations of lodgepole pine as influenced by stand density and site index," *Canadian Journal of Forest Research*, vol. 18, no. 2, pp. 247–250, 1988.

[40] J. H. Fownes and R. A. Harrington, "Allometry of woody biomass and leaf area infive tropical multipurpose trees," *Journal of Tropical Forest Science*, vol. 4, pp. 317–330, 1991.

[41] D. Zianis and M. Mencuccini, "Aboveground biomass relationships for beech (*Fagus moesiaca* Cz.) trees in Vermio Mountain, Northern Greece, and generalised equations for *Fagus* sp," *Forest Science*, vol. 60, no. 5, pp. 439–448, 2003.

4

The Influence of Landscape and Microhabitat on the Diversity of Large- and Medium-Sized Mammals in Atlantic Forest Remnants in a Matrix of Agroecosystem and Silviculture

Juliano André Bogoni,[1,2] **Talita Carina Bogoni,**[3] **Maurício Eduardo Graipel,**[2] **and Jorge Reppold Marinho**[4]

[1] *Instituto de Ciências Biológicas, Universidade de Passo Fundo (UPF), Campus de Passo Fundo, Rodovia BR 285, Bairro São José, Caxia Postal 611, 99052-900 Passo Fundo, RS, Brazil*

[2] *Departamento de Ecologia e Zoologia, Universidade Federal de Santa Catarina (UFSC), Campus de Florianópolis, Campus Universitário Reitor João David Ferreira Lima-Trindade, 88040-970 Florianópolis, SC, Brazil*

[3] *Instituto Federal Catarinense (IFET), Campus de Concórdia-SC, Curso de Medicina Veterinária, Rodovia SC 283, Distrito de Santo Antônio, 89900-000 Concórdia, SC, Brazil*

[4] *Programa de Pós Graduação em Ecologia, Universidade Regional Integrada do Alto Uruguai e das Missões (URI), Campus de Erechim-RS, Avenida Sete de Setembro, 1621, Caxia Postal 743, 99700-000 Erechim, RS, Brazil*

Correspondence should be addressed to Juliano André Bogoni; bogoni.colorado@gmail.com

Academic Editors: K. Kielland, M. Kitahara, F. Le Tacon, G. Martinez Pastur, S. F. Shamoun, and S. Turton

Fragmentation and destruction of a habitat are strongly relevant aspects to determine the richness and the dynamics of the mammals in ecosystems. This study, developed from October, 2010 to July, 2011 in three Atlantic Forest remnants in Ipumirim, SC, Brazil, aims at identifying the diversity of large- and medium-sized nonflying mammals and verifying associations of the patterns obtained with features of the researched areas. The approximate measurement of the inventoried areas is 51 ha. The data collection of the mammal fauna was obtained through direct registers, with the use of a photographical trap, and indirect records through the search of material that indicated the presence of species. The total amount of species studied was 13, pertaining to nine families: Canidae (1), Cebidae (1), Dasyproctidae (1), Dasypodidae (2), Didelphidae (2), Felidae (2), Mustelidae (2), and Procyonidae (2). In addition, landscape data was obtained through the development of a chorological matrix of the areas and the data about the microhabitats. From these data, 20 models for analysis were stipulated and this selection was determined with the corrected Akaike Information Criterion (AICc). The aspect of greater influence on the magnitude of the obtained data was the degree of human occupation in the landscape.

1. Introduction

Several human activities are eroding ecosystems, species, and biological features in an alarming pace and such loss will certainly alter the way the ecosystems and their goods and services operate. This alteration, to different degrees, forces the ecosystems to critical thresholds tending to approximate to a problematic planetary scale. Besides that, there are very scarce data about the geographical and taxonomical distribution for most species, which have been called Wallacean and Linnean deficiencies, respectively. This perspective invariably shows that plenty of information on ecology is being lost, mainly in less known groups in tropical environments, previously to its understanding [1–4].

The fragmentation of tropical forests has a strong impact on biodiversity [5], with more than one-third of the species disappearing when the habitats are fragmented [6, 7]. By that means, the conservationist biology in fragmented tropical ecosystems has to concentrate not only on preserved areas, but also on managed ecosystems [8]. Throughout time, diversity tends to decrease and eventually reaches a less diverse steady state [9]. Several authors consider that habitat loss and fragmentation are the main factors for the decrease of diversity. For instance, Chiarello [10] studied the effects of fragmentation on mammals in the Atlantic Forest and it has shown that fragments smaller than 200 ha are too small and perturbed to keep the assemblies intact and, in addition to the impoverished group of species, the size of the population is reduced.

The future of species in tropical forests thus depends partially on their ability to survive in landscapes altered by human activity [11]. Historically, plenty of research has concentrated on the fragmentation of habitats, the importance of the area, the degree of isolation, and the persistence of species. Although the decrease in biodiversity in altered landscapes is well documented, Perfecto and Vandermeer [8] argue that the conservationist paradigm generally concentrates on forest reserves ignoring the agricultural landscape where they are placed due to the fragmented nature of the majority of the tropical ecosystems.

The Atlantic Forest is considered one of the five most important biodiversity hotspots and one of the most endangered ecosystems in the world [12]. The expansion of the agricultural frontier, the construction of infrastructure, the expansion of the cities, and the nonsustainable exploration of forests are the main causes of the deforestation process [13]. Approximately 88% of the original forest cover of the Atlantic forest biome has been lost and, in a landscape scale, the remaining cover is incorporated in the dynamics of agromosaics and the monoculture of exotic species [14]. In the state of Santa Catarina, there are only 23.37% of the original Atlantic Forest cover remaining [15], and nineteen species of large- and medium-sized nonflying terrestrial mammals (approximately 1/4) are in any of the categories of risk of extinction [16].

The Atlantic Forest mammal fauna comprises 298 species, with 90 endemisms [17] and several studies are noteworthy in Brazil regarding the mammal fauna. These studies make use of several research methodologies and of important results. Studies on the mammal fauna in Santa Catarina between the decade of 1990 and the beginning of the years 2000 focused on the shore regions and islands (e.g., [18, 19]). In the last decade, however, this framework has been altered due to some studies on species register and distribution (e.g., [20–22]). A number of studies has been developed in the western mesoregion of Santa Catarina in recent years, and indicated that the region has shown an important richness of mammals (e.g., [23, 24]).

For this reason, the aim of this study is assessing the richness of large- and medium-sized mammals in three areas in Ipumirim, a municipality in the middle-west region of Santa Catarina, associating the magnitude of the data to local factors which potentially influence on the richness of mammals in the sampled areas, assuming as a hypothesis that richness and diversity (presence/absence) suffer the influence of the predominance of human occupation in the areas, the microhabitats, and the species registers in the meteorological variables.

2. Materials and Methods

2.1. Area of Study.
Located in the Atlantic forest biome, the region is characterized for the large number of small rural

properties and has been effectively colonized from the 1940s and its main activity is logging (Figure 1) [25]. This activity has decreased in present days due to the efficacy of legislation though there are many reforesting areas with exotic species. The area presents phytophysionomic characteristics of the Mixed Ombrophilous and the Stational Semidecidual forest, hydrologically the main basins are the ones of Jacutinga, Irani, and Engano rivers, the pedological formation follows the characteristics of Serra Geral and the climate is humid mesothermal with average temperature between 17°C and 28°C [26]. Three fragments with average size of 51 ha were studied featuring low direct anthropic activity though areas B and C suffered the logging process (larger trees) approximately 40 years ago. The relief of the area is uneven and water is provided from springs and drainage troughs. The general climatic conditions during the research period presented large temperature amplitude with an average of 21.7°C (maximum 33.5°C and minimum −1.9°C), average relative humidity 77.32%, and precipitation accumulation of 1,795 mm.

2.2. Sampling.
There is a great variation in body size, habits, and preferred habitats among mammals, and these factors force the use of several specific methodologies in order to inventory the mammal fauna [27]. As they are discreet animals and have nocturnal habits, techniques of direct and indirect registers were employed [28]. The set of methods employed to obtain the information concerning the richness of mammals in the areas included the use of photographic trap, visualization, and observation of traces.

The collections of direct registers of the mammals were carried out between October 2010 and July 2011 with a photographic trap model *Tigrinus* 6.0D and the direct visualization of the animal. The trap was exposed in each area for 2,160 hours, being weekly checked and equally switched from one area to another every 360 hours and, in order to maximize the possibility to register the species, fresh attractive baits, such as fruit, birds entrails and blood, sardines and coarse salt, we use between 100 and 200 grams each bait type and all were used together. The trap was set between 35 and 45 cm from the ground. The visual registers were obtained through 12 expeditions with random walks of three hours each per area, in a total of 36 hours per area.

Indirect registers indicating the presence of species were obtained during the same period and through the methodology of random walks employed by Rabinowitz [29], Carrillo et al. [30], Gibbs [31], and White and Edwards [32]. The search for traces (footprints, feces, and other signs) was carried out through 18 expeditions of three hours each per area, in a total of 36 hours to obtain the indirect registers. Each register was photographed and measured to be used as evidence. All the registers comprised a data matrix of the presence or absence of species. Dubious cases were discarded, especially the ones related to indirect registers. Registers of the genus *Leopardus* (both direct and indirect) were included as *Leopardus* spp..

The meteorological variables during the sampling period (temperature, humidity, and pluviosity) were provided by the meteorological station of the Brazilian Agricultural Research Corporation (Embrapa) Swines and Birds (Concórdia, SC).

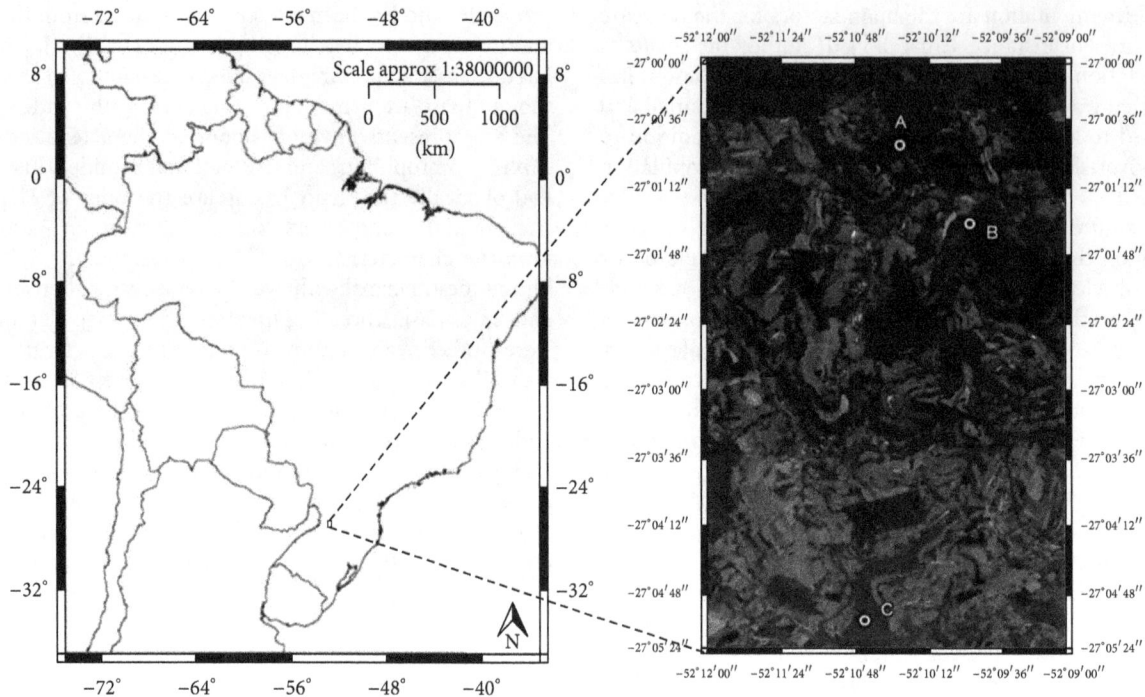

FIGURE 1: Area of study in the municipality of Ipumirim-SC utilized for the sampling of medium and large mammals through camera trapping between October 2012 and July 2011. The upper left area A (27°01′39″S, 52°09′58″O, 893 m, 55 ha), upper right area B (27°02′20″S, 52°09′43″O, 899 m, 47 ha), below the area C (27°05′10″S, 52°10′30″O, 664 m, 51 ha). Source: National Geophysical Data Center (NOAA) and *Google Earth*.

Concomitantly with it, the *Software R* was employed in order to elaborate a chorological matrix of an area of 12.5 km^2 (area average of the life of the species registered (or genus, as in the case of *Galictis*) according to Sunquist et al. [33], Gentile et al. [34], Parera [35], Cheida et al. [36], and Saab et al. [37]) surrounding the central coordinates of each area, subdivided into quadrants of 100 m^2, through a data matrix obtained with the spatial analysis of the satellite image of *Google Earth*. Conflict values were attributed, being 0 (preserved forest area), 1 (pasture/farming/field area), 2 (silviculture), and 3 (area with human occupation and/or roads) in order to obtain visual information that evidences potential conflicts with the local fauna for subsequent pattern selection.

In addition to the data of mammal fauna and landscape conditions, the environmental characterization of the sampling sites (microhabitats) was made in each of the fragments, following a protocol of a series of measurements in each sampling site, through the adapted quadrant-point method [38], being a cross allocated in the collection sites demarcated into four quadrants (N, S, E, W) and vegetation and environment measured. In each quadrant, trees with diameter at breast height (DBH) greater than 16 cm and shrubs with DHB smaller than 16 cm and height smaller than 1 m closer to the center of the cross, had their distances from the cross measured as well as their height, the diameter of their canopies and trunks. The diameter was measured at breast height (1.3 m) for the trees and at ankle height (0.1 m) (DAH) for the shrubs. Furthermore, the height of the undergrowth vegetation and the terrain inclination were measured in each quadrant, at 1 m^2, as well as the percentage of undergrowth vegetation cover, green cover, exposed soil, and canopy cover in the four directions through visual estimation. The latter measurement was carried out with the support of a square of 10 cm^2, set at a distance of 40 cm from the observer's view, at an inclination of 20° in relation to the zenith [39]. In addition to these measurements, the distance of each site in relation to the closest watercourse was also measured. The averages of the four quadrants were considered for each site, and for each area the average of all the sites.

2.3. Data Analysis. The statistical analyses employed were carried out with the *Software R*. The normality and homocedasticity tests were used for the analyses.

The species accumulation curve [40] was carried out to indicate the sampling sufficiency. The species richness, which makes no distinction among species and treats equally the species that are exceptionally abundant or rare [41], was estimated through the Chao2 index [40, 42]. Classical diversity measurements were developed with the Shannon-Wiener index [43] and Pielou's equitability index [44]. In order to ascertain the existence of difference among the diversity obtained in the three areas, the nonparametric test of Kruskall-Wallis [43], followed by the Wilcoxon test with Bonferroni correction [43] were carried out in the case of statistically significant results. The dissimilarity between the mammal fauna composition and the environmental characteristics among the areas was estimated with Jaccard's index [45]. Due to the nonnormality of the data, the correlation analyses of Spearman [43] was employed in order to ascertain

The Influence of Landscape and Microhabitat on the Diversity of Large- and Medium-Sized Mammals in Atlantic Forest Remnants in a Matrix of Agroecosystem and Silviculture

47

if the meteorological variables had any relation with mammals temporal registers.

Twenty candidatemodels were established *a priori* through the data of landscape and environmental characterization, and the corrected Akaike Information Criterion (AICc) was employed in the selection of the most parsimonious model regarding these conditions and the richness of species, and between these data and the presence or absence matrix. The AIC is one of the methods of model selection that determines which better minimizes the expected discrepancy, once it is an unbiased estimator which assumes that all model candidates contain the real model [46]. For this context, it was assumed that there are data and a set of models and that statistical inference is based on the models. Thus, based on the classical idea, it was presumed that there is a single model or at least a better model though the identity of the model is unknown, and it was also presumed that this identity may be estimated [47]. In order to avoid the increase in the probability of model selection with too many parameters (overfitting), the AICc was employed [48–50].

For the direct register of species in 25% of the collections or more in one area, the individual model selection with AICc was carried out following the same conditions previously described and the same models. However, one model was added for each microhabitat conditions, comprising 35 candidate models, in order to establish the most preponderant variables in relation to these species register.

3. Results

Three hundred and forty-three direct registers, including one visualization, and 28 indirect registers were obtained. Thirteen mammal species were registered altogether. Only *Procyon cancrivorus* (G.[Baron] Cuvier, 1798) and *Puma concolor* (Linnaeus, 1771) out of the 13 species were registered in altitudes higher than 830 meters, and only 46.15% below 830 m, especially in area C. Some species recorded are shown in Figure 2.

The species accumulation curve for A, B, and C and for the total set of species in the mammals' community of the three areas showed that the sampling effort was effective, the stabilization of the curve taking place in the third collection for areas A and B, and in the fourth collection in area C. When considering the three areas together the stabilization occurred from the fourth collection (Figure 3).

Area A showed diversity estimated in 22,5 by Chao2 index; for area B, it was 7.1 and 8.2 for area C. The diversity index of Shannon-Wiener was, respectively, 2.117, 1.795, and 1.696 (Table 1). Through Kruskall-Wallis there was no significant difference among the sampled areas [$x^2 = 2.315$; df = 2; $P = 0.314$].

The dissimilarity index of Jaccard was smaller between A and B (10%), intermediate between B and C (38%), and greater between A and C (45%). For the environmental characterization of the microhabitats the dissimilarity was smaller between areas B and C (15%), intermediate between A and C (44%), and greater between A and B (48%) (Figure 4).

(a)

(b)

(c)

(d)

FIGURE 2: Some recorded species of medium and large mammals of three Atlantic Forest fragments located in the western of Santa Catarina state. (a) *Eira barbara* (Linnaeus, 1758); (b) *Leopardus* sp.; (c) *Dasypus novemcinctus* (Linnaeus, 1758) and (d) *Procyon cancrivorus* (G.[Baron] Cuvier, 1798).

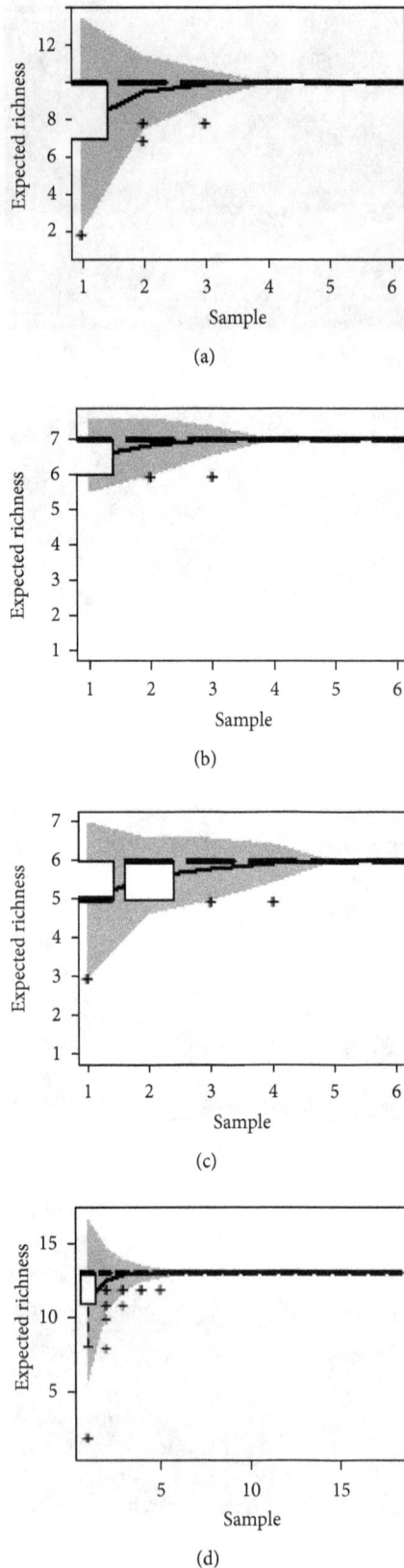

FIGURE 3: Accumulation curve with a confidence interval of 95% (gray), standard and error deviation (bars) and cumulative number of species per collection (+) of medium and large mammals of three Atlantic Forest fragments located in the western of Santa Catarina state. (a) Area A; (b) Area B; (c) Area C and (d) total of all areas.

TABLE 1: Records of medium and large mammal's species per sampling points between October 2010 and July 2011 in three Atlantic Forest remnants located in the municipality of Ipumirim-SC, Brazil. With its richness, diversity index of Shannon-Wiener, evenness of Pielou, and estimated diversity by Chao2 index.

Specie	Area A	Area B	Area C
Sapajus nigritus (Goldfuss, 1809)	0	2	1
Cerdocyon thous (Linnaeus, 1766)	2	6	2
Dasyprocta azarae (Lichtenstein, 1823)	1	0	0
Dasypus novemcinctus (Linnaeus, 1758)	0	1	1
Didelphis aurita (Wied-Neuwied, 1826)	3	0	0
Eira barbara (Linnaeus, 1758)	3	2	3
Euphractus sexcinctus (Linnaeus, 1758)	1	0	0
Galictis cuja (Molina, 1782)	0	2	2
Leopardus spp.	3	4	0
Nasua nasua (Linnaeus, 1766)	1	2	1
Philander frenatus (Olfers, 1818)	1	0	0
Procyon cancrivorus (G.[Baron] Cuvier, 1798)	3	0	0
Puma concolor (Linnaeus, 1771)	1	0	0
Richness	10	7	6
Shannon-Wiener	2.117	1.795	1.696
Equitability	0.946	0.922	0.946
Chao2	22.500	7.125	8.250

The correlation analyses of Spearman between the abiotic factors and the temporal richness of species demonstrated that there is significant negative correlation for relative humidity and precipitation ($S = 1720.298$, rho $= -0.775$, $P < 0.01$; $S = 1500.406$, rho $= -0.548$, $P < 0.05$) and nonsignificant for temperature ($S = 701.260$, rho $= 0.276$, $P = 0.267$).

Through the chorological matrix (Figure 5), it was observed that 35.5% of area A and surroundings are occupied by silviculture, 26.4% by open fields, farming or pastures, 9.1% by human occupation (buildings and roads), and 29.0% of remaining native forest. In area B and surroundings 31.6% is occupied by fields, farming or pastures, 22.7% by silviculture, 8.7% by human occupation and 37% of native forest. In area C and surroundings, the farming, fields, and pastures are predominant (38.4%) followed by human occupation (15.3%) and silviculture (14.7%), remaining 31.6% of native forest (Figure 4).

The results of the measurements of environmental characterization (microhabitat) showed that area A has larger trees with lower density, besides being closer to watercourse and presenting lower canopy cover and undergrowth vegetation and, also, lower inclination (Table 2).

The Influence of Landscape and Microhabitat on the Diversity of Large- and Medium-Sized Mammals in Atlantic Forest
Remnants in a Matrix of Agroecosystem and Silviculture

49

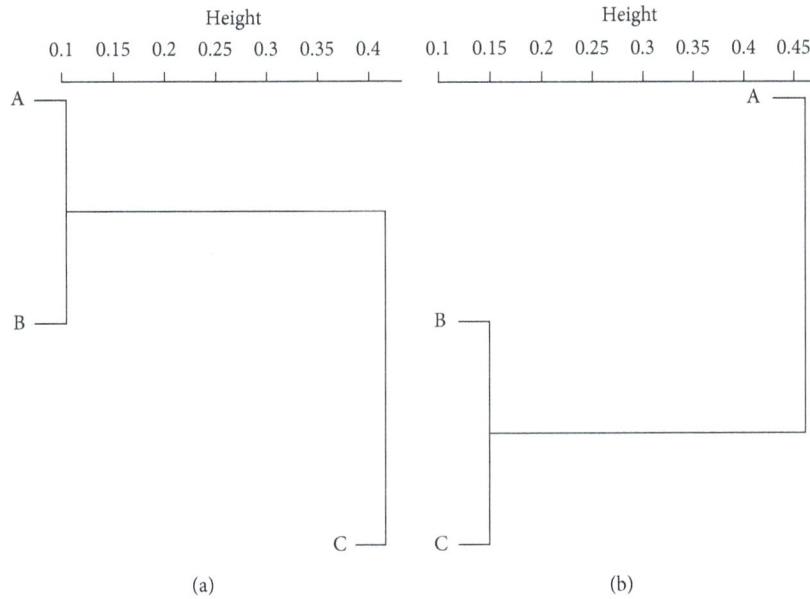

FIGURE 4: Dissimilarity dendrogram obtained by the Jaccard index for (a) medium and large mammals records of three Atlantic Forest fragments located in the municipality of Ipumirim-SC, Brazil obtained from October 2010 to July 2011 and (b) between the microhabitat features presented in the same areas. Agglomerative coefficient of 0.5 (a) and 0.45 (b). Where: A: Area A; B: Area B and C: Area C.

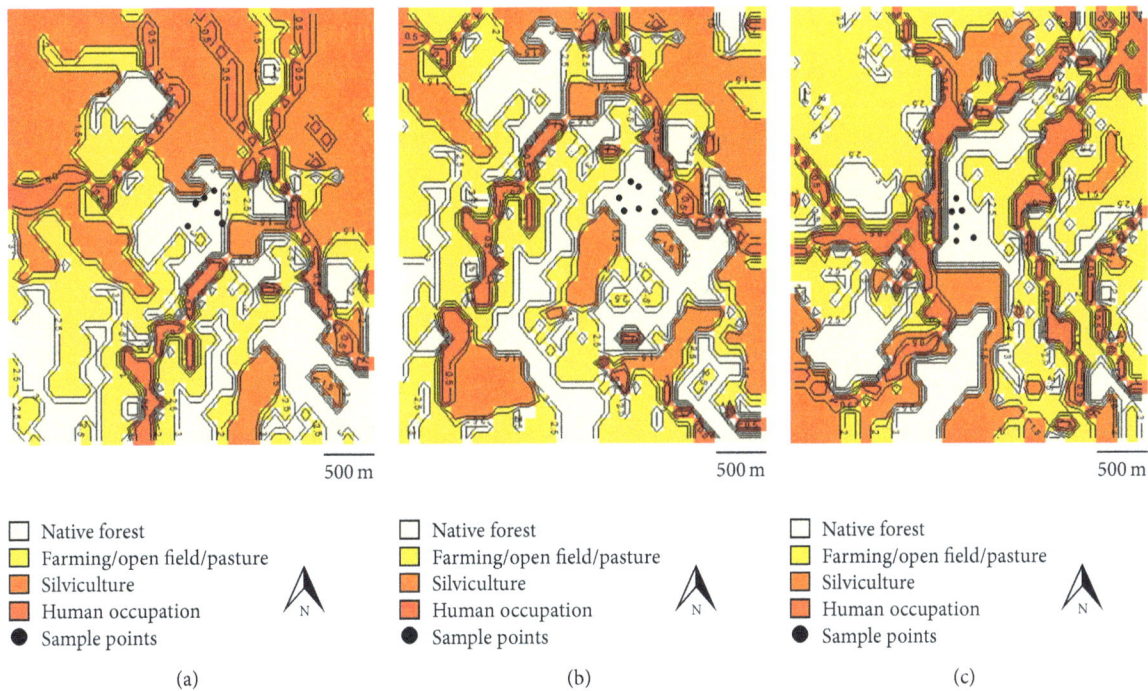

FIGURE 5: Chorological matrix of three Atlantic Forest fragments in the municipality of Ipumirim-SC, Brazil where the mammals were sampled from October 2010 to July 2011, showing the types of occupation of the landscape and enumerating (dark colors) for possible conflicts of medium and large mammalian fauna. (a) Area A, (b) Area B, and (c) Area C.

The model calculations of the parameters of landscape and environmental characterization made through the AICc show that the three most parsimonious models for the data of richness and qualitative matrix (presence/absence) in the assessed areas are the ones which include landscape criteria such as the proportion of human occupation (AICc = 70.78;

AICc = 140.61) followed by farming, field, and pasture areas for species richness (AICc = 71.61) and altitude for species presence/absence (AICc = 140.82).

Amongst the microhabitat conditions, the best model of richness prediction was the canopy cover (AICc = 74.88) although a model containing landscape and microhabitat

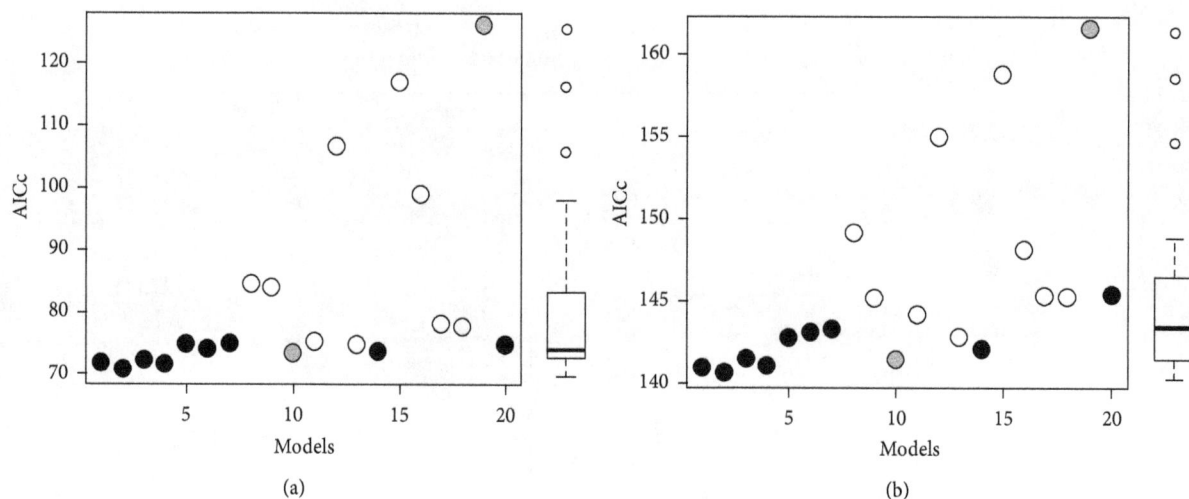

FIGURE 6: Scatter the result of the selection of 20 candidate models based on AICc for (a) species richness of mammals and (b) matrix of presence/absence in time. In black predictors models of landscape conditions, in white microhabitat models and gray mixed models (see Table 2).

TABLE 2: Mean (±SD) of environmental characterization of three Atlantic Forest areas sampled in the municipality of Ipumirim-SC, Brazil.

Parameters	A (Mean (±SD))	B (Mean (±SD))	C (Mean (±SD))
Distance tree (m)	2.97 (0.29)	1.98 (0.22)	3.53 (1.01)
DAP tree (cm)	86.83 (31.25)	39.48 (11.45)	56.83 (28.2)
Height tree (m)	10.02 (1.91)	7.66 (2.27)	9.76 (3.69)
Diameter tree (m)	6.02 (2.13)	3.73 (1.17)	4.61 (2.43)
Distance shrub (m)	1.84 (0.68)	0.95 (0.32)	1.34 (0.29)
DAT shrub (cm)	7.67 (0.81)	9.36 (3.1)	7.1 (1.81)
Height shrub (m)	2.16 (0.24)	2.44 (0.54)	2.29 (0.5)
Diameter shrub (cm)	97.2 (13.76)	111.2 (24.25)	70.9 (16.77)
Litterfall height (cm)	2.6 (0.96)	3.8 (0.42)	2.6 (0.87)
Litterfall coverage (%)	78 (9.59)	99.67 (0.52)	90.49 (3.99)
Green coverage (%)	48.83 (16.91)	12.92 (3.32)	30.2 (7.2)
Exposed soil (%)	22 (9.59)	0.33 (0.52)	9.51 (3.99)
Rocky out distance (m)	1.44 (0.24)	2.02 (0.56)	1.2 (0.12)
Canopy coverage (%)	77.63 (9.29)	89.75 (5.34)	80.21 (12.68)
Inclination (°)	16.17 (1.65)	22.83 (5.36)	22.88 (6.92)
Distance of stream (m)	25.64 (15.88)	329.5 (36.33)	336.17 (37.56)

Where: A: Area A, B: Area B, and C: Area C.

parameters was ranked in the fifth place. For the presence and/or absence of species, the most parsimonious model of microhabitat predictor data was also the canopy cover (AICc = 142.82) also showing a mixed model as the fourth best (Table 3).

The microhabitat conditions as predictive factors for species richness presented as the most parsimonious model the conditions that included parameters of the canopy cover, followed by soil cover, undergrowth vegetation, green cover, exposed soil, and rocky outcrop. Amongst the microhabitat explanatory variables listed and their respective model candidates for the prediction of the patterns of presence and

absence, the same models previously described were also the most parsimonious (Figure 6).

The result of the model selection through AICc for species which presented detectability ≥25% in all samples showed that the factor with greater importance for predictors of landscape was the proportion of native forests in the sampled areas and surroundings for all species, except for D. aurita. As a microhabitat criterion there was a great variation among species although conditions of rocky outcrop, undergrowth vegetation and green cover have been relevant for the majority of species (Table 4).

The Influence of Landscape and Microhabitat on the Diversity of Large- and Medium-Sized Mammals in Atlantic Forest
Remnants in a Matrix of Agroecosystem and Silviculture

51

TABLE 3: Results of the model selection based on AICc of a predefined set of candidate models for predicting richness and presence/absence of nonflying mammals of three Atlantic Forest remnants in the municipality of Ipumirim-SC, Brazil.

RV	Model description[a]	Criterion[b]	K	AICc	ΔAICc	WAIC	LL
Richness	HM OCCP	Landscape	3	70.78	0.00	0.53	−24.59
	Farm/OPN field/pasture	Landscape	3	71.61	0.83	0.24	−23.38
	ALT	Landscape	3	71.78	1.00	0.17	−19.76
	SILV	Landscape	3	72.41	1.63	0.01	−31.53
	ALT + INCL	Landscape + microhabitat	6	73.48	2.70	0.01	−31.56
	CAN COV	Microhabitat	6	74.88	4.10	0.00	−31.53
	LIT HGT + LIT COV + GRN COV + EXP SOIL + RCK OUT	Microhabitat	6	75.18	4.40	0.00	−33.02
	LIT HGT + CAN COV	Microhabitat	8	77.70	6.92	0.00	−25.37
	GRN COV + CAN COV	Microhabitat	11	78.18	7.40	0.00	−33.50
	DIST TREE + DAP TREE + HGT TREE + DMT TREE + DIST SHR + DAT SHR + HGT SHR + DMT SHR + CAN COV + FRG SIZE	Landscape + microhabitat	4	125.91	55.13	0.00	−32.35
Presence/ absence	HM OCCP	Landscape	3	140.61	0.00	0.12	−66.96
	ALT	Landscape	3	140.82	0.21	0.12	−66.98
	Farm/OPN field/pasture	Landscape	3	141.03	0.42	0.11	−67.07
	ALT + INCL	Landscape + microhabitat	6	141.42	0.81	0.10	−66.12
	SILV	Landscape	3	141.44	0.84	0.10	−63.16
	CAN COV	Microhabitat	7	142.82	2.21	0.06	−61.72
	LIT HGT + LIT COV + GRN COV + EXP SOIL + RCK OUT	Microhabitat	6	144.19	3.58	0.04	−67.03
	DIST TREE + DAP TREE + HGT TREE + DMT TREE	Microhabitat	4	145.17	4.56	0.04	−68.07
	LIT HGT + CAN COV	Microhabitat	9	145.18	4.57	0.03	−65.27
	DIST TREE + DAP TREE + HGT TREE + DMT TREE + DIST SHR + DAT SHR + HGT SHR + DMT SHR + CAN COV + FRG SIZE	Landscape + microhabitat	4	161.54	20.93	0.00	−62.77

[a] Abbreviations used: RV: response variable; M: model; K: parameters number of model; AIC: AIC value; ΔAIC: AIC delta; WAIC: AIC weight; LL: Likelihood; LIT HGT: litterfall height; LIT COV: litterfall covering; GRN COV: green covering; EXP SOIL: exposed soil; RCK OUT: rocky out; INCL: inclination of terrain; CAN COV: canopy covering; HM OCCP: human occupation; SILV: silviculture; FARM: farming; OPN FLD: open field; FRG SIZE: fragment size; ALT: altitude; DIST STR: distance of stream; DIST TREE: distance of tree; HGT TREE: height tree; DMT TREE: diameter tree; DIST SHR: distance of shrub; HGT SHR: height shrub; DMT SHR: diameter shrub.
[b] We present only the four best models each criterion, except Landscape + Microhabitat that only two models were defined.

TABLE 4: Result of the model selection-based AICc of a predefined set of candidate models for mammalian species with greater detectability in three remnants in the Atlantic Forest located in the municipality of Ipumirim-SC, Brazil.

Specie	Landscape[a(1)]	AICc	Major[1]	Minor[2]	Microhabitat[a(1)]	AICc	Major[1]	Minor[2]	Microhabitat[b(2)]
Cerdocyon thous	FRAG SIZE	25.6	34.30%	30.30%	RCK OUT	24.71	1.64 m	1.28 m	GRN COV
Eira barbara	FRAG SIZE	33.21	32.00%	33.00%	RCK OUT	25.96	1.34 m	1.64 m	CAN COV
Galictis cuja	FRAG SIZE	20.79	35.20%	32.00%	HGT TREE	16.95	5.84 m	9.8 m	DMT TREE
Sapajus nigritus	FRAG SIZE	20.79	35.20%	32.00%	LIT HGT	17.78	4.07 cm	2.82 cm	DIST SHR
Nasua nasua	FRAG SIZE	26.61	34.00%	32.20%	RCK OUT	21.83	1.76 m	1.42 m	LIT HGT
Didelphis aurita	DIST STR	13.97	20.80 m	272.3 m	GRN COV	17.49	51.30%	26.50%	DAP TREE

[a] Abbreviations used: AICc: AICc value; FRG SIZE: fragment size; DIST STR: distance of stream; RCK OUT: rocky out; HGT TREE: height tree; LIT HGT: litterfall height; GRN COV: green covering; DMT TREE: diameter tree; DIST SHR: distance of shrub; DAP TREE: diameter at breast height of tree.
[1] Conditions of greatest occurrence.
[2] Conditions of minor occurrence or absence.
[(1)] Best model.
[(2)] Second best model.

4. Discussion

4.1. Registered Species. The review of the list of Brazilian mammals indicates the occurrence of 701 species, classified into 243 genera, 12 orders, and 50 families [17]. In Santa Catarina, whose territory is completely inside the Atlantic Forest biome, 178 species have confirmed and/or potential occurrence and more than 64% of the total number of species belong to the orders Chiroptera and Rodentia [51]. Therefore, 64 are species of orders likely to be sampled with the methodology adopted in this research, and based on these numbers, 20.3% of the species of mammals of this category, for this biome, were registered. Although, we infer that the region in particular, the number of species reported by us was higher than 65% can still occur. According to Caro et al. [52] and Coelho et al. [53], we highlight that medium and large mammals do not usually have seasonal variations in subtropical environments.

Mammals of the order Carnivora compose the main guild of predators of vertebrates in the terrestrial ecosystems [54]. It was observed that they corresponded to more than half of the mammals registered, indirectly indicating the existence of resources for the maintenance of these predators and/or the use of the researched sites as foraging and/or breeding areas. *C. thous* and *E. barbara* were the most frequently registered species. Once the first is an opportunistic and generalist species, it is potentially less susceptible to environmental changes having a wide geographical distribution in America [35, 55]. *E. barbara* is featured by the occupation of more densely wooded areas [35] which allows to infer that there is heterogeneity of environments. The description of habitat occupation for other species is considerably variable, from the occupation of large areas (e.g., *P. concolor* and *Leopardus* spp.), to species with great ecological flexibility (e.g., *P. concolor*, *N. nasua*, *D. aurita*, and *S. nigritus*), edge environments (e.g., *Euphractus sexcinctus* (Linnaeus, 1758) and *Dasyprocta azarae* (Lichtenstein, 1823), environments whose important condition is the presence of waterbody (*P. frenatus* and *P. cancrivorus*) and animals whose occurrence takes place in forests and occasionally in open areas (e.g., *Dasypus novemcinctus* (Linnaeus, 1758) and *G. cuja*) [35, 56].

Thus, even the studied fragments with considerably small areas for species that generally live in large areas, and being a fragmented region with strong evidences of predatory hunting, in addition to the lack of studies in the field, the results obtained were of great importance, once endangered species of great ecological importance such as *P. concolor*, were registered [16]. The richness observed in the three areas was similar to the one found in studies carried out in protected natural areas in the same biome, as in Kuhnen's study [57], which registered 16 species in the Sea ridge (Serra do Mar) in Santa Catarina. The study conducted by Luiz [58] in Aguaí's Ecological Station (Estação Ecológica do Aguaí) registered 10 species. Cherem [20], using a similar methodology, including an interview, studied seven areas in southeast Santa Catarina and along Uruguai River and registered 46 species. In the western region of Santa Catarina, Cherem et al. [23] registered 27 species (employing several methodologies) in the areas that suffer influence of the Quebra Queixo Hydroelectric Power Plant, located between Ipuaçu and São Domingos, two towns in Santa Catarina, approximately 90 km from the areas studied in this research. In the basin of the Irani River, Santa Catarina, Cherem et al. [24] registered 26 species in a sampling period of four years, including several different methodologies.

4.2. Comparison among the Fragments. Among the studied areas, where the richness observed varied between six and ten species and the estimated richness by the Chao2 index showed amplitude between 22.5 and 7.13, the greatest variation happened due to the characteristics of the registers, which in area A, six (46.1%) of the thirteen species registered were exclusive. Although the richness estimated presented great variation among the areas, the result of the Kruskall-Wallis test showed that the areas are statistically similar in relation to the composition of the mammal fauna. The equitability index of Pielou also showed very similar results, with a variation of 0.044 among the areas, what does not indicate the increase in dominance, and the same also happened with the classical index of Shannon-Wiener, which did not show great variations. Even though area A presented greater richness and exclusive species, the similarity index of Jaccard demonstrated that areas A and B are similar, as well as B and C. The environmental characterization presented greater similarity between areas B and C, which had previously been lodging areas. Thus, even with the differences between the adopted indexes, the results lead to the conclusion that the three areas presented a great similarity.

4.3. The Influence of Biotic and Abiotic Factors. Through the model selection based on the AICc for the variables response richness and presence/absence, in the magnitude of occupation of the areas, the factors that stood out were, in order, conditions of human occupation, open field, farming and/or pastures and silviculture (richness), and altitude (presence/absence). The landscape conditions, despite the amount of ongoing studies, do not allow many generalizations regarding the consequences for the communities [59] and the conditions, which include the spatial structure originated from processes of temporal and spatial scales, involve several parameters interfering in a single diversity measurement [60, 61]. Even considering these difficulties, it is perceptible that the three main factors are conditions that suffer strong and direct human interference (except for altitude), either decreasing or altering the native areas thus confirming the aspect of conflicts because of human occupation and the fauna. The altitude also presented a clear effect because 13 species were registered in altitudes higher than 830 meters, even though part of them were also found in lower altitudes, what confirms the smaller indirect anthropic impact due to conditions of access and relief.

Regarding the conditions of microhabitats listed by us, especially the case of canopy cover, Jennings et al. [62] describe that the cover promoted controls the quantity, quality and spatial and temporal distribution of luminosity, determining differentiated levels of humidity, temperature, and conditions of soil humidity. Thus, the canopy cover is the greatest determinant of the internal microhabitat of the forest once it directly affects the growth and development of seedlings determining the floristic composition of the community [63]. Several authors reinforce this condition assuming the intensity of incident light as the main determinant of these systems, and the microclimatic factors dependent on this variable [64, 65]. In the event of the undergrowth vegetation, Portes et al. [66] indicate that its deposition is influenced by the type of vegetation, successional stages, altitude, latitude, wind, precipitation, temperature, herbivory, hydric availability, and the storage of nutrients in the soil. This way, it works as a bioindicator responding with alterations of deposition processes because of alterations in the environment [67]. These conditions therefore provide a contribution regarding the resulting conditions.

Biological communities undergo changes as there is variation among the abiotic factors along a spatial gradient. Species richness tends to increase towards lower latitudes [68–70]. Nevertheless, the effects on the latitudinal change have rarely been considered distinct from this factor [71]. Furthermore, specific microclimatic changes such as temperature and precipitation interact in order to ascertain conditions and resources available for plant growth, which in turn influence the local fauna. However, both the intensity and the frequency of the events along time are relative measures which depend on the organism that suffers them [72]. Thus, the conditions of great amplitude for obtaining registers temporally may be directly or indirectly associated to meteorological factors. It was confirmed through the correlation analysis between the outcome of the species registers and the abiotic factors obtained in the period, with emphasis to relative humidity and precipitation.

The faunal responses to fragmentation depend on a series of factors, especially if the species will use the fragment edges, if there is fidelity to the interior sections of the fragment [73] or the interior sections present single attributes [74]. The generality of the edge effect, however, may be limited as well as the results vary among different systems [75]. Fragmentation is an important problem for biodiversity preservation, which has received special attention from the scientific community. Consequently, the perception and evaluation of landscape fragmentation are recommended and presumably necessary in any assessment, assistance, or decisive process which involve landscape alteration [76].

A good example to be considered is the study by Lantschner et al. [77] in the Argentinian Patagonia where they assessed the habitat and landscape characteristics and the presence of species of wild carnivores, registering four species and two of them (*Lycalopex culpaeus* (Molina, 1782) and *Conepatus chinga* (Molina, 1782)) used continuous native vegetation with greater frequency but also used dense coniferous plantations, while *Leopardus geoffroyi* (d'Orbigny and Gervais, 1844) was almost totally restricted to native forest. On the other hand, individuals of *P. concolor* did not show any preferences, being detected in all types of habitats.

Most studies regarding microhabitat selection carried out in the Atlantic Forest are developed with small mammals (e.g., [78–80]), while investigating microhabitat use by large- and medium-sized mammals in Atlantic Forest environments revealed tendencies to differential uses by the mammal community. In relation to microhabitat from the perspective of foraging, a fragment which is used for this purpose may occasionally be used as a refuge once it presents richer food resources, even though it is structurally different or distant from central areas.

Under such circumstances, the distance and the structure of the microhabitat may have effects on the risk of predation and in the animals' decision making [81]. All the organisms may potentially alter their environments. This ecological trait leads to spatial self-organized heterogeneity of the environments and, consequently, to the limits of species distribution. Within concurrent systems, the capacity of alteration leads to competitive alternative consequences, resulting in trade-offs among the competition abilities, colonization and niche construction, becoming important for the competitive coexistence [82]. The habitat heterogeneity hypothesis [83] states that structurally more complex and thus more diverse habitats tend to an increase in the diversity of species in the landscape [84].

The individualization in models selection through species with greater numbers of registers has provided distinct results. Conditions of quantity of native areas in the landscape were predominant, although generalist species such as *C. thous* presented a greater number of registers in larger though less preserved areas, based on microhabitat criteria. In an opposite way, *E. barbara*, more sensitive to alterations, was registered in greater numbers in smaller areas, though with better conditions in local scale. This condition of quantity of native areas was an exception solely for *D. aurita* whose condition of importance was closeness to water (although the analysis has not been carried out for *P. cancrivorus*, this condition has also been important once the species was registered solely in the area closest to water).

When species were assessed individually, the main microhabitat condition for half of the species was the density of rocky outcrop, although in opposite conditions (smaller or greater quantity) among some species. This condition influences on vegetation dynamics and composition, creating insular environments in the landscape and refuges for plants sensitive to grazing [85, 86]. Other important criteria were the green cover of the soil, height, DBH and diameter of the trees, height of undergrowth vegetation, canopy cover, and shrub density. These conditions provide an overview of the local quality. They determine the success of a population and in heterogeneous sites the variation of these conditions is common, leading to the formation of microhabitats of higher or lower quality in a reduced scale [72, 87]. It is thus argued that the landscape conditions are relevant associated with the quality conditions in local scale and they exert influences of different magnitude on different species.

5. Conclusions

The use of associated methodologies and the climatic conditions must be taken into consideration in the faunal inventory of mammals, as well as aspects and paradigms of landscape alteration. Notwithstanding, a great number of mammals likely to be inventoried was registered with the methods employed. Yet considering that several species of large- and medium-sized mammals are no longer found, the areas presented, in fact, a considerable richness and it was estimated that the study has registered more than 65% of species that are potential dwellers of the region. The ecological aspects of the areas match with the results, attesting the ecological importance of the Atlantic Forest remnants for the maintenance of the biological diversity.

The richness and presence/absence of mammals had influence on the landscape, with emphasis on the proportion of human occupation and open areas for farming, fields and pastures, potential foraging areas as well as the altitude of the areas, due to the lower anthropic disturbance. In the microhabitat scale, the most noteworthy aspects were canopy cover, undergrowth vegetation cover and height, percentage of green cover, exposed soil and rocky outcrop, which are usually indicators of environmental conditions in spatial-temporal scales allowing contributions regarding the resulting conditions.

Several of the environmental and climatic conditions here emphasized may be included in the development of maneuver plans as well as in the establishment of new areas of environmental preservation in Santa Catarina aiming at preserving the diversity of large- and medium-sized mammals and their structural and ecological ecosystemic relationships.

Conflict of Interests

The authors reported no conflict of interests.

Acknowledgments

The authors thank owners of the research areas Mr. Paulo Senger and Mr. Urbano Wildner and their families and the newspaper "O Jornal" of the city of Concórdia SC. They also thank Clodoaldo J. Pozzebon for providing them baits through slaughterhouse in the region.

References

[1] A. D. Barnosky, E. A. Hadly, J. Bascompte et al., "Approaching a state shift in Earth's biosphere," *Nature*, vol. 486, no. 7401, pp. 52–58, 2012.

[2] B. J. Cardinale, J. E. Duffy, A. Gonzalez et al., "Biodiversity loss and its impact on humanity," *Nature*, vol. 486, no. 7401, pp. 59–67, 2012.

[3] J. A. F. Diniz-Filho, P. de Marco Jr., and B. A. Hawkins, "Defying the curse of ignorance: perspectives in insect macroecology and conservation biogeography," *Insect Conservation and Diversity*, vol. 3, no. 3, pp. 172–179, 2010.

[4] R. J. Whittaker, M. B. Araújo, P. Jepson, R. J. Ladle, J. E. M. Watson, and K. J. Willis, "Conservation biogeography: assessment and prospect," *Diversity and Distributions*, vol. 11, no. 1, pp. 3–23, 2005.

[5] T. G. Wade, K. H. Riitters, J. D. Wickham, and K. B. Jones, "Distribution and causes of global forest fragmentation," *Conservation Ecology*, vol. 7, no. 2, 2003.

[6] B. C. Klein, "Effects of forest fragmentation on dung and carrion beetle communities in central Amazonia," *Ecology*, vol. 70, no. 6, pp. 1715–1725, 1989.

[7] D. A. Driscoll, "Extinction and outbreaks accompany fragmentation of a reptile community," *Ecological Applications*, vol. 14, no. 1, pp. 220–240, 2004.

[8] I. Perfecto and J. Vandermeer, "Biodiversity conservation in tropical agroecosystems: a new conservation paradigm," *Annals of the New York Academy of Sciences*, vol. 1134, pp. 173–200, 2008.

[9] R. H. MacArthur and E. O. Wilson, *The Theory of Island Biogeography*, Princeton University Press, Princeton, NJ, USA, 1967.

[10] A. G. Chiarello, "Effects of fragmentation of the Atlantic forest on mammal communities in south-eastern Brazil," *Biological Conservation*, vol. 89, no. 1, pp. 71–82, 1999.

[11] J. Barlow, J. Louzada, L. Parry et al., "Improving the design and management of forest strips in human-dominated tropical landscapes: a field test on Amazonian dung beetles," *Journal of Applied Ecology*, vol. 47, no. 4, pp. 779–788, 2010.

[12] N. Myers, R. A. Mittermeler, C. G. Mittermeler, G. A. B. Da Fonseca, and J. Kent, "Biodiversity hotspots for conservation priorities," *Nature*, vol. 403, no. 6772, pp. 853–858, 2000.

[13] WWF, "Mata Atlântica, herança em perigo," 2009, http://www.wwf.org.br/?24780/Mata-Atlantica-heranca-em-perigo.

[14] M. Tabarelli, A. V. Aguiar, M. C. Ribeiro, J. P. Metzger, and C. A. Peres, "Prospects for biodiversity conservation in the Atlantic Forest: lessons from aging human-modified landscapes," *Biological Conservation*, vol. 143, no. 10, pp. 2328–2340, 2010.

[15] INPE and SOS Mata Atlântica, "Atlas dos remanescentes florestais da Mata Atlântica," 6th edition, 2010, http://www.inpe.br/noticias/noticia.php?Cod_Noticia=2199.

[16] CONSEMA, *Lista Oficial de Espécies da Fauna Ameaçadas de Extinção no Estado de Santa Catarina*, Secretaria de Estado do Desenvolvimento Econômico Sustentável, Conselho Estadual do Meio Ambiente de Santa Catarina, Florianópolis, Brazil, 2011.

[17] A. P. Paglia, G. A. B. Da Fonseca, A. B. Rylands et al., *Lista Anotada dos Mamíferos do Brasil*, Occasional papers in conservation biology, no.60, Conservation International, Washington, DC, USA, 2nd edition, 2012.

[18] M. E. Graipel, J. J. Cherem, D. A. Machado, P. C. Garcia, M. E. Menezes, and M. Soldateli, "Vertebrados da Ilha de Ratones Grande, Santa Catarina, Brasil," *Biotemas*, vol. 10, no. 2, pp. 105–122, 1997.

[19] M. E. Graipel, J. J. Cherem, and A. Ximienez, "Mamíferos terrestres não voadores da Ilha de Santa Catarina, sul do Brasil," *Biotemas*, vol. 14, pp. 109–140, 2001.

[20] J. J. Cherem, "Registro de mamíferos não voadores em estudos de avaliação ambiental no sul do Brasil," *Biotemas*, vol. 18, no. 2, pp. 169–202, 2005.

[21] N. C. Cáceres, J. J. Cherem, and M. E. Graipel, "Distribuição geográfica de mamíferos terrestres na região sul do Brasil," *Ciência & Ambiente*, vol. 35, pp. 167–180, 2007.

[22] J. J. Chrem, M. Kammers, I. R. Ghizoni Jr., and A. Martins, "Mamíferos de médio e grande porte atropelados em rodovias do estado de Santa Catarina, sul do Brasil," *Biotemas*, vol. 20, pp. 81–96, 2007.

[23] J. J. Cherem, S. L. Althoff, and R. C. Reinicke, "Mamíferos," in *A fauna das áreas de Influência da Usina Hidrelétrica Quebra Queixo*, J. J. Cherem and M. Kammers, Eds., p. 192, Habilis, Erechim, Brazil, 2008.

[24] J. J. Cherem, S. L. Althoff, and A. F. Testoni, "Mamíferos," in *Fisiografia, Flora E Fauna Do Rio Irani*, J. J. Cherem and V. Salmoria, Eds., p. 159, ETS, Florianópolis, Brazil, 1st edition, 2012.

[25] C. Locatelli, *O Município de Lpumirim: Estudo Histórico e Político*, Prefeitura Municipal de Ipumirim, Ipumirim, Brazil, 1985.

[26] IBGE, *Mapa Do Clima Do Brasil*, Instituto Brasileiro de Geografia e Estatística, Brasília, Brazil, 2002.

[27] R. S. Voss and L. H. Emmons, "Mammalian diversity in neotropical lowland rainforests: a preliminary assessment," *Bulletin of the American Museum of Natural History*, no. 230, pp. 1–86, 1996.

[28] M. Becker and J. C. Dalponte, *Rastros De Mamíferos Silvestres Brasileiros*, Universidade de Brasília, Brasília, Brazil, 1991.

[29] A. Rabinowitz, *Wildlife Field Research and Conservation Training Manual*, Wildlife Conservation Society, New York, NY, USA, 1997.

[30] E. Carrillo, G. Wong, and A. D. Cuarón, "Monitoring mammal populations in Costa Rican protected areas under different hunting restrictions," *Conservation Biology*, vol. 14, no. 6, pp. 1580–1591, 2000.

[31] J. P. Gibbs, "Monitoring populations," in *Research Techniques in Animal Ecology: Controversies and Consequences*, L. Boitani and T. K. Fuller, Eds., p. 442, Columbia University Press, New York, NY, USA, 2000.

[32] L. White and A. Edwards, *Conservation Research in the African Rainforests: A technical handbook*, Wildlife Conservation Society, New York, NY, USA, 2000.

[33] M. E. Sunquist, F. Sunquist, and D. E. Daneke, "Ecological separation in a Venezuelan llanos carnivore community," in *Advances in Neotropical Mammalogy*, K. H. Redford and J. F. Eisenberg, Eds., Sandhil Crane Press, Gainesville, Fla, USA, 1989.

[34] R. Gentile, P. S. D'Andrea, and R. Cerqueira, "Home ranges of Philander frenata and Akodon cursor in Brazilian restinga (Coastal shrubland)," *Mastozoología Neotropical*, vol. 4, no. 4, pp. 105–112, 1997.

[35] A. Parera, *Los Mamíferos de la Argentina y la Región Austral de Sudamérica*, El Ateneo, Buenos Aires, Argentina, 1st edition, 2002.

[36] C. C. Cheida, F. H. G. Rodrigues, and G. M. Mourão, "Ecologia espaço-temporal de guaxinins Procyon cancrivorus (Carnivora, Procyonidae) no Pantanal central," in *Anais do 6th Congresso Brasileiro de Mastozoologia*, 2012.

[37] J. L. Saab, L. G. R. Oliveira-Santos, and G. M. Mourão, "Período de atividade e área de vida de quati (Procyonidae: Nasua nasua) mediante o uso de colar-GPS no Pantanal da Nhecolândia," in *Anais do 6th Congresso Brasileiro de Mastozoologia*, 2012.

[38] J. E. Brower, J. H. Zar, and C. N. Von Ende, *Field and Laboratory Methods For General Ecology*, McGraw-Hill, Boston, Mass, USA, 4th edition, 1998.

[39] F. A. Ramos, "Nymphalid butterfly communities in an Amazonian forest fragment," *Journal of Research on the Lepidoptera*, vol. 35, pp. 29–41, 2000.

[40] R. K. Colwell and J. A. Coddington, "Estimating terrestrial biodiversity through extrapolation," *Philosophical transactions of the Royal Society of London B*, vol. 345, no. 1311, pp. 101–118, 1994.

[41] A. E. Magurran, *Measuring Biological Diversity*, Blackwell Publishing, Malden, Mass, USA, 2004.

[42] A. Chao, "Nonparametric estimation of the number of classes in a population," *Scandinavian Journal of Statistics*, vol. 11, pp. 265–270, 1984.

[43] J. H. Zar, *Biostatistical Analysis*, Prentice Hall, New Jersey, NJ, USA, 4 edition, 1999.

[44] J. Daget, *Les Modeles Mathèmatiques En Écologie*, Masson, Paris, France, 1976.

[45] D. Muller-Dombois and H. Ellemberg, *Aims and Methods of Vegetation Ecology*, Blackburn Press, New York, NY, USA, 1974.

[46] H. Yanagihara and C. Ohmoto, "On distribution of AIC in linear regression models," *Journal of Statistical Planning and Inference*, vol. 133, no. 2, pp. 417–433, 2005.

[47] K. P. Burnham and D. R. Anderson, "Multimodel inference: understanding AIC and BIC in model selection," *Sociological Methods and Research*, vol. 33, no. 2, pp. 261–304, 2004.

[48] H. Akaike, "Information theory and the maximum likelihood principle," in *Proceedings of the 2nd International Symposium on Information Theory*, B. N. Petrov and F. Csaki, Eds., Akademiai Kiado, Budapest, Hungary, 1973.

[49] H. Bozdogan, "Model selection and Akaike's Information Criterion (AIC): the general theory and its analytical extensions," *Psychometrika*, vol. 52, no. 3, pp. 345–370, 1987.

[50] K. P. Burnham and D. R. Anderson, *Model Selection and Multi-Model Inference. A Practical Information-Theoretic Approach*, Springer, New York, NY, USA, 2nd edition, 2002.

[51] J. J. Cherem, P. C. Simões-Lopes, S. Althoff, and M. E. Graipel, "Lista dos mamíferos do estado de Santa Catarina, sul do Brasil," *Mastozoología Neotropical*, vol. 11, no. 2, pp. 151–184, 2004.

[52] T. M. Caro, J. A. Shargel, and C. J. Stoner, "Frequency of medium-sized mammal road kills in an agricultural landscape in California," *The American Midland Naturalist*, vol. 144, no. 2, pp. 362–369, 2000.

[53] I. P. Coelho, A. Kindel, and A. V. P. Coelho, "Roadkills of vertebrate species on two highways through the Atlantic Forest Biosphere Reserve, southern Brazil," *European Journal of Wildlife Research*, vol. 54, no. 4, pp. 689–699, 2008.

[54] Pitman, M. R. P. L, T. G. Oliveira, R. C. Paula, and C. Indrusiak, *Manual de identificação, Prevenção e Controle de Predação por Carnívoros*, Instituto Brasileiro do Meio Ambiente e Recursos Naturais Renováveis (IBAMA), Brasília, Brazil, 2002.

[55] L. H. Emmons and F. Feer, *Neotropical Rainforest Mammals. A Field Guide*, University of Chicago Press, Chicago, Ill, USA, 2nd edition, 1997.

[56] N. R. Reis, A. L. Peracchi, W. A. Pedro, and I. P. Lima, *Mamíferos do Brasil*, Universidade Estadual de Londrina-PR (UEL), Londrina, Brazil, 2006.

[57] V. V. Kuhnen, *Diversidade de Mamíferos e a Estrutura do Hábitat. Estudo da Composição da Mastofauna Terrestre em Diferentes Estágios Sucessionais de Regeneração da Floresta Ombrófila Densa, Santa Catarina, Brasil*, Dissertação de mestrado do programa de pós-graduação em Ecologia, Universidade Federal de Santa Catarina (UFSC), Florianópolis, Brazil, 2010.

[58] M. R. Luiz, *Ecologia e Conservação de Mamíferos de Médio e Grande Porte na Reserva Biológica Estadual do Aguaí. Dissertação de Especialização em Gestão de Recursos Naturais*, Universidade do Extremo Sul Catarinense, Criciúma, Brazil, 2008.

[59] N. Olfiers and R. Cerqueira, "Fragmentação de habitat: efeitos históricos e ecológicos," C. F. D. Rocha, H. G. Bergallo, M. V. Sluys, and M. A. S. Alves, Eds., p. 582, Biologia da conservação. Essências, São Carlos, 2006.

[60] D. L. Urban, R. V. O'Neill, and H. H. Shugart Jr., "Landscape ecology: a hierarchical perspective can help scientists to understand spatial patterns," *Bioscience*, vol. 37, no. 2, pp. 119–127, 1987.

[61] J. P. Metzger, "O que é ecologia de paisagens?" *Biota Neotropica*, vol. 1, pp. 2–9, 2004.

[62] S. B. Jennings, N. D. Brown, and D. Sheil, "Assessing forest canopies and understorey illumination: canopy closure, canopy cover and other measures," *Forestry*, vol. 72, no. 1, pp. 59–73, 1999.

[63] A. C. G. De Melo, D. L. C. De Miranda, and G. Durigan, "Cobertura de copas como indicador de desenvolvimeto estrutural de reflorestamentos de restauração de matas ciliares no Médio Vale do Paranapanema, SP, Brasil," *Revista Árvore*, vol. 31, no. 2, pp. 321–328, 2007.

[64] T. C. Whitmore, "Canopy gaps and the two major groups of forest trees," *Ecology*, vol. 70, pp. 536–538, 1989.

[65] J. A. A. Meira-Neto, F. R. Martins, and A. L. Souza, "Influência da cobertura e do solo na composição florística do sub-bosque em uma floresta estacional semidecidual em Viçosa, MG, Brasil," *Acta Botania Brasilica*, vol. 19, no. 3, pp. 473–486, 2005.

[66] M. C. G. O. Portes, A. Koehler, and F. Galvão, "Variação sazonal de deposição de serapilheira em uma Floresta Ombrófila Densa Altomontana no morro do Anhangava-PR," *Floresta*, vol. 26, no. 2, pp. 3–10, 1996.

[67] A. Klumpp, " Utilização de bioindicadores de poluição em condições temperadas e tropicais," in *Indicadores Ambientais: Conceitos e Aplicações*, N. B. Maia, H. L. Martos, and W. Barrella, Eds., p. 226, ECUC/COMPED/INEP, São Paulo, Brazil, 2001.

[68] E. R. Pianka, "Latitudinal gradients in species diversity: a review of the concepts," *The American Naturalist*, vol. 100, pp. 33–46, 1966.

[69] G. C. Stevens, "The latitudinal gradient in geographical range: how so many species coexist in the tropics," *The American Naturalist*, vol. 33, no. 2, pp. 240–256, 1989.

[70] K. J. Gaston, "Global patterns in biodiversity," *Nature*, vol. 405, no. 6783, pp. 220–227, 2000.

[71] G. C. Stevens, "The elevational gradient in altitudinal range: an extension of Rapoport's latitudinal rule to altitude," *The American Naturalist*, vol. 140, no. 6, pp. 893–911, 1992.

[72] R. E. Ricklefs, *A Economia Da Natureza*, Guanabara Koogan, Rio de Janeiro, Brazil, 5th edition, 2009.

[73] D. J. Bender, T. A. Contreras, and L. Fahrig, "Habitat loss and population decline: a meta-analysis of the patch size effect," *Ecology*, vol. 79, no. 2, pp. 517–533, 1998.

[74] H. D. Harwell, M. H. Posey, and T. D. Alphin, "Landscape aspects of oyster reefs: effects of fragmentation on habitat utilization," *Journal of Experimental Marine Biology and Ecology*, vol. 409, no. 1-2, pp. 30–41, 2011.

[75] T. M. Donovan, P. W. Jones, E. M. Annand, and F. R. Thompson III, "Variation in local-scale edge effects: mechanisms and landscape context," *Ecology*, vol. 78, no. 7, pp. 2064–2075, 1997.

[76] A. Llausàs and J. Nogué, "Indicators of landscape fragmentation: the case for combining ecological indices and the perceptive approach," *Ecological Indicators*, vol. 15, no. 1, pp. 85–91, 2012.

[77] M. V. Lantschner, V. Rusch, and J. P. Hayes, "Habitat use by carnivores at different spatial scales in a plantation forest landscape in Patagonia, Argentina," *Forest Ecology and Management*, vol. 269, pp. 271–278, 2012.

[78] R. Gentile and F. A. S. Fernandez, "Influence of habitat structure on a streamside small mammal community in a Brazilian rural area," *Mammalia*, vol. 63, no. 1, pp. 29–40, 1999.

[79] A. D. Dalmagro and E. M. Vieira, "Patterns of habitat utilization of small rodents in an area of Araucaria forest in Southern Brazil," *Austral Ecology*, vol. 30, no. 4, pp. 353–362, 2005.

[80] F. V. B. Goulart, N. C. Cáceres, M. E. Graipel, M. A. Tortato, I. R. Ghizoni Jr., and L. G. R. Oliveira-Santos, "Habitat selection by large mammals in a southern Brazilian Atlantic Forest," *Mammalian Biology*, vol. 74, no. 3, pp. 182–190, 2009.

[81] D. J. Druce, J. S. Brown, J. G. Castley et al., "Scale-dependent foraging costs: habitat use by rock hyraxes (*Procavia capensis*) determined using giving-up densities," *Oikos*, vol. 115, no. 3, pp. 513–525, 2006.

[82] C. Hui, Z. Li, and D. X. Yue, "Metapopulation dynamics and distribution, and environmental heterogeneity induced by niche construction," *Ecological Modelling*, vol. 177, no. 1-2, pp. 107–118, 2004.

[83] E. H. Simpson, "Measurement of diversity," *Nature*, vol. 163, no. 4148, p. 688, 1949.

[84] R. H. MacArthur and J. W. MacArthur, "On bird species diversity," *Ecology*, no. 594, p. 598, 1961.

[85] P. J. Clarke, K. J. E. Knox, K. E. Wills, and M. Campbell, "Landscape patterns of woody plant response to crown fire: disturbance and productivity influence sprouting ability," *Journal of Ecology*, vol. 93, no. 3, pp. 544–555, 2005.

[86] C. Smit, D. Béguin, A. Buttler, and H. Müller-Schärer, "Safe sites of tree regeneration in wooded pastures: a cade of associational resistance?" *Journal of Vegetation Science*, vol. 16, pp. 209–214, 2005.

[87] J. L. Harper, *Population Biology of Plants*, Academic Press, New York, NY, USA, 1977.

5

An Overview of Indian Forestry Sector with REDD+ Approach

Vandana Sharma and Smita Chaudhry

Institute of Environmental Studies, Kurukshetra University, Kurukshetra 136119, Haryana, India

Correspondence should be addressed to Smita Chaudhry; smitachaudhry11@gmail.com

Academic Editors: K. Kielland and G. Martinez Pastur

Forest ecosystems cover large parts of the terrestrial land surface and are major components of the terrestrial carbon (C) cycle. The primary objective of REDD+ is to minimize the carbon emissions from deforestation in developing countries and enhance their carbon storage capacities through sustainable management programme. The recognition of REDD+ throughout the international community, its support by donors and promotion in the perspectives of the UNFCCC negotiations are mainly due to vital functions of forests in regulating the world's climate. This paper gives an overview of REDD+ approach and its methodological guidance in context of Indian forestry sector. The strengthening of governance arrangements and institutions in India needs to integrate learning through piloting, adaptive management, and knowledge transfer. A phased approach for India for REDD+ implementation having safeguards for local communities and biodiversity along with a system of their reporting and capacity building has to be developed. Successfully designed REDD+ implementation in India entirely depends on a rigid, scalable, and reliable finance mechanism, technological assistance, and effective forest-related legislation along with transparent and equitable political momentum which has support of core stakeholder groups.

1. Introduction

Forests, like other ecosystems, are affected by climate change. Forests also influence climate, absorbing CO_2 from the atmosphere and storing carbon in wood, leaves, litter, roots, and soil. The carbon is released back into the atmosphere when forests are cleared or burnt. By acting as sinks, forests are considered to moderate global climate change [1]. Climate change is one of the most significant global challenges of our time, and addressing it requires the urgent formulation of comprehensive and effective policy responses [2]. Natural forests are more resilient to climate change and disturbances than plantations because of their genetic, taxonomic, and functional biodiversity. This resilience includes regeneration after fire, resistance to and recovery from pests and diseases, and adaptation to changes in radiation, temperature, and water availability (including those resulting from global climate change). While the genetic and taxonomic composition of forest ecosystems changes over time, natural forests will continue to take up and store carbon as long as there is adequate water and solar radiation for photosynthesis [3]. Forests play a major role in the global carbon (C) cycle because they store 80% of the global aboveground C of the vegetation and about 40% of the soil C and interact with atmospheric processes through the absorption and respiration of CO_2 [4–8]. Human activities such as fuel consumption and land-use change are the main causes of an increase in the atmospheric carbon dioxide concentration, which is generally recognized as a factor of climate change and global warming [9]. The Stern Review [10] highlighted that forest conservation, afforestation, reforestation, and sustainable forest management can provide up to 25% of emission reductions needed to effectively combat climate change and that curbing deforestation has the potential to offer significant emission reductions fairly quickly in a highly cost-effective manner. Reducing emissions from deforestation and forest degradation (REDD+) is a mechanism for providing financial rewards to countries that reduce carbon emissions caused by the loss and degradation of their forests. In concept, REDD resembles other Payment for Environmental Services (PES) programs; however, REDD emphasizes a reduction in deforestation and degradation rates from expected levels, also known as avoided deforestation and degradation [11].

India has a total land area of 329 million hectares [12] of which around 23.4%, that is, 76.87 million hectares (Mha), is classified as the forestland with tree cover. More than 40% of the country's forests are degraded and understocked [13, 14]; subsequently there is a large potential of REDD+ activities in the country. Hence an overview of REDD+ approach along with its methodological guidance is discussed in this article in context of Indian forestry sector.

2. RED to REDD+: Evolution Since Its Conceptualization

Fuelled by the continuing destruction of forests in developing countries, the aspired development and implementation of a REDD+ mechanism under the United Nations Framework Convention on Climate Change (UNFCCC) evolved into one of the major issues in the negotiations on a post-Kyoto agreement [15]. The concept of REDD was first introduced in its preliminary form at the climate change negotiations during the third meeting of the Conference of Parties to the UNFCCC (COP-3) in 1997 to enlist carbon services of forests under the Clean Development Mechanism (CDM) and account for emissions and removals from land use, land use change, and forestry (LULUCF) activities [16]. Reducing emissions from deforestation in developing countries was being discussed in the side events since COP-9 of the UNFCCC in 2003 under the names of "Avoided Deforestation," "Compensated reduction," and so forth [17]. When the Kyoto Protocol came into force in 2005, it focused on reductions in emissions from technological projects. This expression was used for the first time in its shortened form RED (*reducing emissions from deforestation*) during the 11th UN Conference of Parties (COP-11) in Montreal (2005) by the Coalition for Rainforest Nations led by Papua New Guinea—noting that deforestation was estimated to account for 12–15% of the overall greenhouse gas (GHG) emissions [18]. Well received at COP-11, the concept was further elaborated, expanded, and officially adopted during COP-13 in Bali, Indonesia, in 2007 in the form of REDD. The addition of "degradation" to this acronym was due to the observation that forest degradation in some developing countries was as threatening as deforestation (if not more) to the forest ecosystems. Following the debates during the 14th COP in Poznan, Poland, in 2008, it was decided that REDD should evolve to REDD+ to encompass all the initiatives that can increase the carbon absorption potential of forests.

The insertion of "+" on the acronym REDD is aimed at broadening its scope to include all operations associated with preservation, restoration, and sustainable management of forest ecosystems. The official definition of REDD+ as set by UNFCCC is "*reducing emissions from deforestation and forest degradation in developing countries, and the role of conservation, sustainable management of forests, and enhancement of forest carbon stocks in developing countries*" [19]. During COP-16 of UNFCCC in 2010 in Cancun, governments agreed to boost action to curb emissions from deforestation and forest degradation in developing countries with technological and financial support [17]. REDD+ attaches financial value to the carbon stocks stored in the forests and other incentives for developing countries by the developed countries to reduce emissions and invest in low carbon paths

to sustainable development. It goes beyond deforestation and forest degradation and includes the role of conservation, sustainable management of forests, and enhancement of carbon stocks. COP-17 in November, 2011 at Durban, South Africa, produced a landmark decision to extend the Kyoto Protocol into the second commitment period. Pursuant to Cancun decision, important decisions on REDD+ were made, namely, decisions on systems for providing information on safeguards, modalities for forests reference (emission) levels, and REDD+ financing [20]. The primary goal of REDD+ is reduction of greenhouse gas emissions, consistent with the goal of the UNFCCC to achieve "stabilization of greenhouse gas concentrations in the atmosphere at a level that would prevent dangerous anthropogenic interference with the climate system." REDD+ is expected to bring much more than emission reductions, with a properly designed mechanism contributing to multiple benefits. Depending on the location and type of REDD+ activity, these benefits potentially include poverty alleviation, indigenous people's rights, improved community livelihoods, technology transfer, sustainable use of forests resources, and biodiversity conservation [21].

REDD+ is a complex instrument of governance. It is treated by many stakeholders as a payments for ecosystems services (PES) scheme [22, 23] which seeks to translate the process of carbon sequestration through arboreal photosynthesis into real financial incentives for ecosystems managers and in so doing to promote the conservation and sustainable management of natural habitats [24, 25]. The implementation of REDD-plus requires forest-governance reforms through inclusive processes that build on existing forest-governance systems [26].

India has played an important role in REDD+ negotiations and has been instrumental in shaping the REDD+ mechanism by emphasising the role of conservation and sustainable forest management in mitigating carbon emissions [16].

3. Green Facts of India: A Megadiverse Country

The forests of India have long been an important part of its culture and a defining feature of its landscape [27]. India with a wide range of climate, geography, and culture is unique among biodiversity-rich nations and is known for its diverse forest ecosystems and megabiodiversity. It ranks as the 10th most forested nation in the world [28], with 23.4% (76.87 Mha) of its geographical area under forest and tree cover [1, 29]. Out of 34 global biodiversity "hot spots," four are located in India, that is, Eastern Himalayas, North-east, Sundarbans, and Western Ghats [30]. India is one of the 17 megadiverse countries (MoEF). Fifteen biodiversity-rich areas of country covering an area of approximately 74000 km^2 have been designated as biosphere reserve and four of them, namely Nilgiri, Nanda Devi, Sundarbans, and Gulf of Mannar, have been recognised by UNESCO underworld network of biospheres [31]. With only 2.4% of the land area, India accounts for 7 to 8 percent of the recorded species of the world [32]. This biodiversity is of immense economic, ecological, social, and cultural value. Approximately 275 million people in India (27% of the total population) are known to live in the forest fringes and earn bulk of their livelihood from forests [33–36]. FSI defines forests as "all the lands, more than one hectare in area, with a tree canopy density of more than 10%." Champion and Seth [37] classified India's forests into four major ecosystems groups, namely, tropical, subtropical, temperate, and alpine. These major groups are further divided into 16 types. Of the 16 forests types, tropical dry deciduous forests form the major percentage that is 38% of the forest cover in India.

3.1. Key Challenges to Indian Forestry Sector. India's economic growth in the last decade has raised several concerns in terms of its present and future resource demands for material and energy [38]. With 18% of global livestock population over 2.4% of world's geographic area and 17% of world's population rigorous biotic pressure being faced by Indian forests as only due to these forests, nearly 30% of fodder needs of the cattle population and 40% of domestic fuel wood needs of the people are being catered. The difference in demand and supply is also broadening for fuel wood, timber, and fodder. In Eastern and Northeastern India, forest degradation is mainly a result of shifting cultivation practices over an area about 1.2 Mha. The National Forest Commission had done a splendid task to conquer the issues forests are facing by recommending allocation of a minimum 2.5% of national budget to the forestry sector. Apprehensions over inadequate role of elected Panchayati Raj institutes vis-a-vis Joint Forest Management Committees (JFMCs) in forest management, limited participation of non-profit-making voluntary sector, control over minor forest products, and implementation of Forest Right Act and PESA are some concerns frequently voiced with little recognition for the incredible efforts made to maintain forest cover in the current conflicting scenarios [31]. As per the wood budget for the year 1996, 86 million tonnes of fuel wood is being unsustainably removed from forests [39]. India's livestock population of 467 million grazes on 11 Mha of pastures. This implies that an average of 42 animals graze on a hectare of land compared to a threshold level of 5 [40]. In absence of adequate grazing land, nearly a third of the fodder requirement is met from the forest resources in the form of grazing and cut fodder for stall feeding. An estimated 100 million cow units graze in forests annually whereas the sustainable level is only 31 million [41]. Additionally, graziers collect an estimated 175–200 million tonnes of green fodder annually [42]. Grazing has been reported in 67% of the national parks and 83% of the wildlife sanctuaries surveyed [40, 43].

3.2. Deforestation. Around 3000 B.C., nearly 80% of India was forested [44, 45]. Deforestation has occurred in the tropics throughout history [46, 47]. Growing population, widespread poverty, and limited employment opportunities in agricultural and industrial sector have resulted in heavy pressure on forests, primarily due to unsustainable extraction of fuel wood and over grazing resulting in forest degradation [48]. Forest vegetation sequesters carbon while at the same time deforestation and degradation of standing forests leads

TABLE 1: Area under different legal categories of forest in India during 1946-47 [36, 52].

Legal status	Area (million ha)	% of total area
Reserve forest	25.32	96.79
Protected forest and unclassified forest	0.84	3.21
Total	26.16	100

to release of stored carbon [49]. The total forest cover has been declining globally. It stood at 3.95 Mha or about 30% of the global land area in 2006. The gross deforestation rate is estimated to have declined between 2000 and 2005 compared to the decade of the 1990s, but it still amounted to a gross and net loss of 12.9 and 7.3 Mha/year, respectively [50]. Deforestation refers to conversion of area having forest cover to other uses, for example, croplands, pastures, or urban land. Degradation [51], on the other hand, refers to reduction in productivity and/or diversity of a forest due to unsustainable harvesting (removals exceeding replacements and changes in species composition), fire (except for fire-dependent forest systems), pests and diseases, removal of nutrients, and pollution or climate change (e.g., changes in productivity, total organic matter, and forest composition). In India, the forest cover is now relatively stable and, therefore, deforestation is not currently a significant issue in formulation of carbon sequestration policies [40].

3.3. Stabilization of Carbon Stocks in Indian Forests. India recognized the enormous importance of the forest resources and land use, land-use change, and forestry (LULUCF) activities in contributing towards GHG emissions. Strong policy skeleton in India made the conservation of forests more focused [17]. The forestry sector recognizes its increasing role to provide sustained benefits to the people and strives to attain it by integrating new frontiers of knowledge and science in planning, management, research and capacity building with forest management [31]. A major part of forest land at the time of independence came under reserved forests [21]. During 1952–1976, forests were recognized only for the commercial interest, (Table 1), that is, for eradication of valuable timber for fast growth of industrial sector, nation's development and other purposes [36, 52].

Table 2 presents the change in forest cover over the years 1987 to 2011 indicating that the forest cover has increased marginally from 64.08 million ha in 1987 to 69.2 million ha in 2011.

India is one of the few countries where deforestation rates have been reduced and regulated and forest cover has nearly stabilised, unlike most other developing countries [28]. Thus it is important to understand the likely factors contributing to the observed stabilization of forest carbon stocks in India. The factors include legislations, forest conservation and afforestation programmes, and community awareness and participation [64]. Realizing the need for conservation and regeneration, several programmes have been implemented at the government and nongovernment levels. The forest policy

TABLE 2: Profile of forests in India from 1987 to 2011 [29, 53–63].

	Recorded forest area (in Mha)					Forest cover (in Mha)				
Year	Reserved forests	Protected forests	Unclassed forests	Total forests area	% of geographical area	Dense forests	Open forests	Mangroves	Total	% of geographical area
1987	40.18	21.73	13.27	75.18	22.8	36.14	27.66	0.4	64.08	19.49
1989	40.18	21.73	13.27	75.13	22.8	37.84	25.74	0.42	63.88	19.43
1991	41.49	23.30	12.20	77.008	23.4	38.50	24.99	0.42	63.93	19.45
1993	41.49	23.30	12.20	77.008	23.4	38.55	25.02	0.42	63.93	19.45
1995	41.65	22.33	12.53	76.52	23.38	38.57	24.93	0.45	63.88	19.43
1997	41.65	22.33	12.53	76.52	23.38	36.72	26.13	0.48	63.33	19.27
1999	41.65	22.33	12.53	76.52	23.38	37.73	25.50	0.48	63.72	19.39
2001	42.33	21.72	12.78	76.84	23.38	41.68	25.87	0.45	67.55	20.55
2003	39.99	23.84	13.63	77.47	23.57	39.05	28.77	0.44	67.83	20.64
2009	43.05	20.62	13.27	76.95	23.41	40.25	28.84	0.46	69.09	21.02
2011	42.25	21.39	13.30	76.95	23.81	40.42	28.78	0.46	69.20	21.05

in India was first introduced in colonial period, and later on changes were made in postcolonial period.

The few important acts, instruments, and rules governing the protection and conservation of forests include The Indian Forest Act, 1927; National Forest Policy, 1952; The Indian Wildlife (Protection) Act, 1972, amended in 1993; Forest (Conservation) Act, 1980, amended in 1988; Forest (Conservation) Rules, 1981, amended in 1992; The National Forest Policy, 1988; Joint Forest Management (JFM), 1990; Biological Diversity Act, 2002; Forest (Conservation) Rules, 2003; Biological Diversity Rules, 2004; and National Environment Policy, 2006. Of these, major contributory acts which majorly caused the stabilization of Indian forest cover are the following.

(i) Forest (Conservation) Act, 1980, amended 1988: this act is one of the most effective pieces of legislation contributing to reduction in deforestation. This was enacted to reduce the discriminate diversion of forest land for nonforestry purposes and to help regulate and control the land-use changes in the existing forest area. With this act, the deforestation and conversion of forest lands to nonforest use were effectively checked [64]. The rate of conversion of forests to nonforestry uses has declined drastically to around 15,500 ha per annum since 1980 compared to 1,50,000 ha per annum prior to 1980 [42]. This act stipulates prior central government approval before any forest land is sought to be diverted for nonforestry purpose [40].

(ii) The National Forest Policy, 1988: India's National Forest Policy was formulated four years before the Earth Summit and embodies all elements—social, environment, and economic—of sustainable forest management. Its aim was to ensure environmental stability and maintenance of ecological balance, including atmospheric equilibrium, which is vital for sustenance of all life forms—humans, animals, and plants [16].

(iii) Joint Forest Management (JFM), 1990: The Forest Policy 1988 set the stage for participatory forest management in India. The JFM programme documented the rights of the protecting communities over forests lands. The local communities and the forest department together plan and implement forest recognition programmes and the communities are rewarded for their efforts in protection and management. The total area covered under JFM programme is over 15 Mha. This has allowed protection of forests plantation, potentially contributing to conservation of existing forests and carbon stocks [64].

3.4. Forest Degradation. Forest degradation is a complex process, and its drivers may be completely different from those for deforestation, thus presenting greater challenges in assessment or monitoring as compared to deforestation [65]. Widespread forest degradation in the developing countries remains poorly understood or quantified [66]. Change in forest composition because of selective overexploitation, loss of natural regeneration, low growing stock, and low productivity are important parameters resulting in low carbon content in the existing forests [40]. Shifting cultivation, fires, and overgrazing have resulted in the elimination of susceptible species and in making selected tolerant species more abundant [51]. The dynamics and causes of deforestation and forest degradation are multifaceted and complex and vary from place to place.

Drivers of deforestation and forest degradation in Indian perspective fall into two categories: first, who those are planned and projected in accordance with policies, legal framework, management plans, and so forth, and second, those who are spontaneous, beyond government and management control, and usually not accounted for. Planned and unplanned withdrawals from forests require proper understanding and management tools including transparent governance, effective enforcement, and appropriate mitigation actions [67]. The distinction between direct and underlying causes and between human- and naturally induced change is often not as clear as it might appear. In reality, there are long, complex chains of causation that can bring about deforestation or the degradation of forests [68].

If appropriate policy instruments and management options including effective legal framework and site specific mitigation measures are introduced then the impacts of planned or controlled drivers could be minimized. Challenge lies in addressing and managing unplanned drivers and activities which are mainly a direct outcome of local people's dependence on the adjoining forest areas to meet their livelihood needs like necessity of fuel wood, fodder, grazing, food supplements, and to a very small extent an illegal mining within forest and so forth [67]. Forest productivity is the net annual increment per unit forest area. The productivity of Indian forests is low [40]. The growing stock of forests area is estimated to be around 58.96 $m^3 ha^{-1}$ in ISFR for the year 2009 and 2011, which is far lower when compared to the global average of 130.7 $m^3 ha^{-1}$ and South and Southeast Asian average of 98.6 $m^3 ha^{-1}$ for the corresponding period [69, 70]. The National Forest Commission report 2006 indicated that around 41 per cent of total forest in the country is already degraded, 70 percent of the forests have no natural regeneration, and 55 per cent of the forests are prone to fire [71]. As per the study of Nayak et al. [70] the factors affecting forest degradation in India mainly include critical livelihood, demand and supply gap of forests products, forests fires, overgrazing, illegal felling, and diversion of forest land for nonforest uses due to competing land-use demand for development and so forth [13, 61, 63, 71–73].

3.5. Forest Degradation and Deforestation: Addressal Mechanism. Reserving forests implies foregoing the benefits that would have been generated by exploiting the resources or from adopting alternative land-use practices. Further, the onsite benefits of forests are lower than the potential benefits of alternative land uses [16]. The strategies and approaches to evolve a mechanism to tackle forest degradation and deforestation will be critical to attain the maximum benefits of incentives for the stored carbon stocks. Creation of appropriate awareness amongst stakeholders in deciding the level of their participation and commitment including that of the local community for protection of existing forests might be the leading action for checking deforestation and forest degradation. The stakeholders need to be effectively and appropriately informed, enthused, and empowered to take apt decision for conservation of forests. This should be supported by the government programmes and policies, which should advocate and provide the relevant alternative resources to the community so that the dependence of local community on forest resources could be reduced. As per Indian submission to SUBSTA, UNFCCC, these programmes could include

(i) alternative cheap cooking fuel supply,

(ii) promoting non-conventional energy sources,

(iii) low-cost permanent housing facilities,

(iv) improving agriculture and livestock productivity,

(v) free education for children,

(vi) better infrastructural facilities including health,

(vii) effective use of modern communication audiovideo tools for creating awareness among community.

Striking a balance between the need to enhance food production for growing population and to cut short deforestation requires increase in agricultural production without further deforestation. This can be addressed through better land planning and extensive investment in technology to increase yields of existing farmlands [67].

4. Safeguards for REDD+ Implementation

Since 2009, the UNFCCC negotiations have increasingly taken up the concerns regarding potential negative effects of REDD+ on the biodiversity of forest ecosystems, *safeguards* and *benefits* being key words in this matter [15]. The implementation of REDD+ actions can pose a number of risks or negative impacts including conversion of natural forests to plantations and other land uses of low biodiversity, loss of traditional territories, erosion of rights with exclusion from land, and loss of traditional livelihoods [74]. At COP-17 in Durban (Decision 12/CP.17 Para 5) it was agreed that a summary of information on how safeguards are being addressed and respected should be provided periodically in national communication channels agreed on by the COP [16].

The new term safeguards was introduced during AWG-LCA intersessions (Bonn, August 2009): "... *[safeguards to protect biological diversity in host countries, including safeguards against conversion of natural forests to forest plantations, should be established].*" The underlying principle behind this was to deal with the dilemma of not having a mandate to explicitly include biodiversity and the simultaneous need to ensure that REDD would not create incentives that could offset the biodiversity objectives of the CBD. In later AWG-LCA negotiations *cobenefits* were replaced by *safeguards* (nonpaper 18 and 39, FCCC/AWGLCA/2009/L.7/Add.6). At COP15 the text on this matter reads and "... *the following safeguards should be [promoted] [and] [supported] (e) Actions that are consistent with the conservation of natural forests and biological diversity, ensuring that actions [...] are not used for the conversion of natural forests, but are instead used to incentivize the protection and conservation of natural forests and their ecosystem services, and to enhance other social and environmental benefits*" [15].

In many REDD+ countries, discussions on safeguards are in their infancy and represent only a minor component of the overall REDD+ policy dialogue [75]. Safeguards can be most effectively addressed if explicit consideration is given to biodiversity concerns during all of the planning and design, implementation, and assessment stages of the REDD+ process [76]. The REDD+ framework that is part of Cancun Agreements includes a number of safeguard provisions that are to be addressed and respected throughout the implementation of REDD+ activities [17]. The UNFCCC, in the Cancun Agreement, articulated seven social and environmental safeguards for REDD+ [77]. These are

(1) actions that complement or are consistent with the objectives of national forest programmes and relevant international conventions and agreements;

(2) transparent and effective national forest governance structures, taking into account national legislation and sovereignty;

(3) respect for the knowledge and rights of indigenous peoples and members of local communities, by taking into account relevant international obligations, national circumstances and laws, and noting that the United Nations General Assembly has adopted the UN Declaration on the Rights of Indigenous Peoples;

(4) the full and effective participation of relevant stakeholders, in particular indigenous peoples and local communities;

(5) actions that are consistent with the conservation of natural forests and biological diversity, ensuring that actions referred to in paragraph 70 of this decision are not used for the conversion of natural forests but are instead used to incentivize the protection and conservation of natural forests and their ecosystem services and to enhance other social and environmental benefits;

(6) actions to address the risk of reversals;

(7) actions to reduce the displacement of emissions.

The Subsidiary Body for Scientific and Technical Advice (SBSTA) has been tasked with providing guidance on systems for information on how REDD safeguards are being addressed and respected throughout the implementation of REDD+ [78]. Implementation of safeguards should be country based and not imposed externally. Safeguards need to be flexible and reflect national status and not formulated as additionality [17]. Free, prior, and informed consent (FPIC) could be an important element of REDD+ safeguard programs, ensuring that activities are implemented in a way that fully respects the rights of affected communities [21, 79].

In order to effectively monitor safeguards in India, it will be necessary to evolve a clearly defined set of indicators and criteria for parameters such as forest governance structures, respect for rights of indigenous peoples, and full and effective participation of relevant stakeholders, along with a system to monitor these [20, 29]. India has well-established forest governance system, responsible for forest management as per the Indian government policy, legal institutions, and regulatory framework. The governance system impounds the dynamism of forest behaviour. The management of forests gets inputs from the various local, regional, and national level bodies, constituting with members from the local villages. In India, there are safeguards already in place to protect the customary rights and traditions of tribes, forest dwellers, and other local communities. Policy and legal instruments exist in the form of joint management programmes, the Forest Rights Act, and the Biological Diversity Act, whose provisions ensure the rights of local communities and enable them to be key players in the local level governance of the natural resources [16].

5. Developing MRV System for REDD+

Transparent, accountable, and sustainable monitoring, reporting, and verification (MRV) systems are essential for any REDD+ framework. With the prospect of a global agreement on forest preservation on the horizon, establishing functional MRV systems is one of the major goals of the so-called REDD Readiness [80]. MRV relates to both actions on the ground (i.e., that change forest carbon stocks) and REDD+ transactions (i.e., compensation and financial transactions or transfers) [81]. A robust monitoring mechanism is essential for successful implementation. As part of the Cancun Agreement, countries are supposed to develop a robust and transparent national forest-monitoring system with the capacity to consistently and accurately monitor changes in forest cover and carbon stocks over time. There is a general consensus that this forest monitoring system would be a combination of remote sensing and ground-based systems [20, 65]. Any system that intends to compensate countries for avoided deforestation and degradation requires a reference level for recent forest carbon stocks against which future performance can be measured. The choice of reference level will determine not only the effectiveness of REDD+ in terms of climate impact, but also the potential financial benefits to recipients of REDD+ funding and hence incentives for countries to participate in the scheme [82, 83].

India is among the few countries to regularly use satellite-based remote sensing technology since the 1980s in detecting and assessing forest cover changes [63]. Under the second National Communication to UNFCCC, forest biomass carbon and soil organic carbon were estimated in India, and the procedure which was followed for the programme has the potential of being developed and adopted as a REDD+ methodology for assessing changes in forest carbon stocks over a stipulated period [17]. India is of the view that the reference level (RL)/reference emission level (REL) needs to be fixed in an open and transparent manner following the procedure decided by the parties for the purpose [16]. India will have to develop a robust system for monitoring of carbon stocks in forests through a network of permanent monitoring plots to provide adequately accurate subnational estimates of carbon stock changes [20].

The Government of India formally adopted Community-based Forestry Management/Joint Forest Management resolution on July 1, 1990, in pursuance of its National Forest Policy, 1988. It laid down broad guidelines for an institutional arrangement involving the local people to jointly protect and manage the forest resources in return for benefits from it [40]. Joint forest management (JFM) has facilitated protection and regeneration of existing forests and rise of forest plantations, which will contribute to conservation of existing forests as also the carbon stocks. This approach matches well with the objectives of REDD+ programmes being implemented in many REDD+ countries [17].

6. Financing of REDD+ Mechanism

Essentially, REDD+ is an investment focusing on retaining or enhancing natural capital and provides an opportunity to enable countries to move towards realizing green development [84]. Multilateral, bilateral, and private funding mechanisms are now supporting different REDD+ activities at various levels. Multilateral mechanisms like UN-REDD programme, World Banks' Forest Caron Partnership Facility (FCPF), and Forest Investment Programme (FIP) are supporting capacity building activities in many developing countries for the effective implementation of REDD+ [85]. REDD+ investments are focused on maintaining or enhancing natural capital, either through investments in forests or through slowing, halting, or reversing drivers of deforestation and forest degradation [84]. REDD+ finance mechanism must be effective: they must contribute to tangible and independent third-party-verifiable stabilizations of atmospheric concentrations of greenhouse gases; efficient: they must result in "value for money" and allow both private-sector and public-sector institutions to participate on fair terms; and equitable: at a minimum, they must avoid exposing to greater risk the poor and most marginalized rural communities whose livelihoods depend on forests, avert the distortion of forest products markets, and allow broad participation on equitable terms at the national and international levels [26].

India supports a mix of market and global funds to finance REDD+ activities. Central funding should compensate for maintenance of forest carbon stocks whereas money for compensating change in carbon stocks (due to decrease in deforestation and degradation or increase in forest cover) could be generated by selling carbon credits in the international markets [73, 85]. Separate financial approaches need to be adopted for providing positive incentives for the two types of carbon stocks under REDD+ regime, that is, for (a) change in carbon stocks (with subcategories for incremental carbon stocks and reduced deforestation) and (b) baseline carbon stocks [17]. The market-based approaches that would be developed for incentivizing removals and emission reductions to be separate from the CDM market and conservation of forest carbon stocks could be incentivized through non-market-based mechanisms [16]. Any REDD finance mechanism—market-based or fund based—needs to properly address the concepts of carbon stock and flow, uncertainty, and discount rates [26].

6.1. Green India Mission. In the context of the overall objective of the National Action Plan on Climate Change (NAPCC), the environmental service which is of utmost importance, is the carbon sequestration potential of the forests. NAPCC correctly recognizes increase and improvement in forest and tree cover as a potential mitigation option. Indian initiative to have a "national mission for a green India" as one of the eight national missions is a very pragmatic step that fits most appropriately into the country's concern on climate change, as also into its overall developmental planning [86]. The government has put in place Green India Mission with a budget of Rs 46,000 crores (approximately USD 10 billion) over a period of 10 years. The mission will help in improving ecosystem services in 10 million ha of land and increase flow of forest-based livelihood services and income of about 3 million forest dependent households [27].

The GIM acknowledges the influence that the forestry sector has on environmental amelioration and inclusive development [87]. The Green India Mission puts the "greening" in the context of climate adaptation and mitigation, aiming to enhance ecosystem services like hydrological services and biodiversity, including carbon sequestration and storage (in forests and other ecosystems), in addition to provisioning services like fuel, fodder, timber, and NTFPs, while also addressing the livelihood issues of people living in and around forests [88]. In addition to other objectives, the mission will also seek positive incentives from REDD+ mechanism. The experience of implementing concept of JFM can successfully be replicated in protecting the rights of the tribal and other forest-dwelling communities while involving them in the proposed REDD+ mechanism. The mission is the practical demonstration of India's support for policy of conservation, sustainable management of forests and increase in forest cover as a means of reducing emissions from deforestation. However, achievement of incremental targets laid in the mission would not be possible fully unless the mission receives the supplemental financial support from the future REDD+ mechanism. Similarly, there is likelihood of financial support for maintenance of baseline stocks through REDD+ mechanism [86].

The Forest Carbon Partnership (FCPF) aims to prepare forest countries for REDD+ implementation—known as REDD+ Readiness [17]. The FCPF assists developing countries in their efforts to reduce emissions from deforestation and forest degradation and foster conservation, sustainable management of forests, and enhancement of forest carbon stocks (all activities commonly referred to as "REDD+") by providing value to standing forests. The Interim REDD+ Partnership created in Paris during the May 2010 conference on forests and climate change was made up of an initial group of six developed nations who have pledged to provide US$4.5 billion to assist developing countries jumpstart REDD+ activities [19]. The Eliasch Review [89] estimated that the one-time cost requirements for 40 countries to reform policy and build capacity for REDD totalled US$4 billion.

Whether REDD is financed through a voluntary fund, a market mechanism, a hybrid mechanism (e.g., revenues from permit auctions in a cap and trade system), or a combination thereof, it must generate finance at the appropriate scale and to the appropriate stakeholders and sustain it over time [90].

7. REDD+ Strategy of India

The REDD+ text in the Cancun Agreements has came out from the text that had been discussed for many years in the UNFCCC negotiations. The text sets the stage for a nationally driven phased approach to a REDD+ mechanism. The international framework involves a three-phase approach for an REDD+ mechanism for developing countries like India: (1) development of national strategies or action plans and capacity building; (2) implementation of national strategies or action plans that could involve REDD+ pilot projects; and (3) mobilization of funds from developed countries, with financing mechanisms [21].

The REDD+ strategy plan for India containing three phases is still under preparation. These phases are as follows

    (i) Phase I: *preparation for adoption of REDD+,*

    (ii) Phase II: *understanding implementation of REDD+,*

    (iii) Phase III: *full scale result-based implementation of REDD+.*

The strategy will be implemented in a coherent and mutually supportive way with the relevant Aichi Biodiversity Targets (5, 7, 11, 14, and 15) and the Strategic Plan for Biodiversity 2011–2020 through various elements of REDD+. The REDD+ strategy plan for India will include mechanism for addressing direct benefit for biodiversity as well as benefit sharing to indigenous and local communities [91].

The incentives so received from REDD+ would be passed to the local communities involved in protection and management of the forests. This will ensure sustained protection of our forests against deforestation [27]. In Indian context, the forest will not be managed for "carbon services" alone, but for all the ecosystem services that are flowing from the forests to the local community. The incentives for carbon services will append to the benefits that the local communities are already receiving from the forest ecosystems [91]. It is estimated that a REDD+ programme for India could provide capture of more than 1 billion tonnes of additional CO_2 over the next 30 years and provide more than USD 3 billion as carbon service incentives under REDD+ [27]. The country specific safeguards will be a part of national strategy with a view to ensure full participation of local communities and all other stakeholders. India's national REDD+ strategy aims to enhance the quantum of forest ecosystem services that flow to the local communities by enhancing and improving the forest and tree cover of the country [91].

Hitherto, India has not prepared a REDD+ strategy as required under the UNFCCC; hence a dedicated institutional structure needs to be put in place to formulate and implement a national REDD+ strategy with a clearly defined mandate, roles, and responsibilities [20]. The issues to be addressed through REDD+ are complex in nature. The implementation of REDD+ frameworks will be strongly influenced by the presence or lack of robust legal framework. The existence of a legal framework that addresses environmental, social and economic issues that reflect the international legal framework of norms and standards is crucial for India to adopt standards on REDD+ [74].

8. Conclusion

Currently, there is significant concern over global warming, green house gas emissions, and their possible impacts on society. For successful implementation of REDD+ in India, technological assistance, readiness assistance, and continued political momentum are essential. A fully fledged national REDD+ strategy needs to be developed and implemented along with action plans with an immediate start of pilot activities in selected areas in collaboration of partners at national and international levels. An enormous capacity

building at every level of forest hierarchy is needed for successful REDD+ implementation. An appropriate multilateral financial mechanism needs to be developed to operationalize capacity building programmes. The urgency of national REDD+ institutional framework stems from the fact that the implementation of REDD+ will involve tremendous coordination efforts. Pilot projects for REDD+ can be initiated in joint forest management areas. For effective implementation of REDD+ programmes there is need for forests-related legislation in place to enable REDD+ related project development which also includes definitions of ownership, rights, and obligations with respect to the sale and trade of carbon or Payment for Environmental Services (PES) accrued through REDD+ projects.

References

[1] J. Kishwan, R. Pandey, and V. K. Dadhwal, "Emission removal capability of India's forest and tree cover," *Small Scale Forestry*, vol. 11, no. 1, pp. 61–72, 2012.

[2] C. Streck, R. Sullivan, S. J. Toby, and R. Tarasofsky, Eds., *Climate Change and Forestry-Emerging Policy and Market Opportunities*, Royal Institute of International Affairs, Brookings Institution Press, Baltimore, Md, USA, 2008.

[3] B. G. Mackey, H. Keith, S. L. Berry, and D. B. Lindenmayer, "Green Carbon—the role of natural forests in carbon storage, part 1-A green carbon account of Australia's South-Eastern eucalypt forest, and policy implications," in *The Fenner School of Environment & Society*, The Australian National University E Press, Australia, 2008.

[4] S. L. Brown, P. Schroeder, and J. S. Kern, "Spatial distribution of biomass in forests of the eastern USA," *Forest Ecology and Management*, vol. 123, no. 1, pp. 81–90, 1999.

[5] J. T. Houghton, Y. Ding, and D. J. Griggs, Eds., *Climate Change 2001: The Scientific Basis. Contribution of Working Group I to the Third Assessment Report of the Intergovernmental Panel on Climate Change*, Cambridge University Press, Cambridge, UK, 2001.

[6] R. A. Houghton, K. T. Lawrence, J. L. Hackler, and S. Brown, "The spatial distribution of forest biomass in the Brazilian Amazon: a comparison of estimates," *Global Change Biology*, vol. 7, no. 7, pp. 731–746, 2001.

[7] C. L. Goodale and M. J. Apps, "Forest sinks in the Northern hemisphere," *Ecology Applied*, vol. 3, pp. 891–899, 2002.

[8] X. Zhang, M. Wang, and X. Liang, "Quantitative classification and carbon density of the forest vegetation in Lüliang Mountains of China," *Plant Ecology*, vol. 201, no. 1, pp. 1–9, 2009.

[9] Food Agriculture Organization, *Carbon Sequestration in Dryland Soils*, Food and Agriculture Organization of the United Nations, Rome, Italy, 2004.

[10] Sir Nicholas Stern, *Stern Review: The Economics of Climate Change*, Cambridge University Press, Cambridge, UK, 2006.

[11] H. J. Albers and E. J. Z. Robinson, "Reducing emissions from deforestation and forest degradation," *Encyclopaedia of Energy, Natural Resources and Environmental Economics*, vol. 2, pp. 78–85, 2013.

[12] H. Gundimeda, P. Sukhdev, R. K. Sinha, and S. Sanyal, "Natural resource accounting for Indian states—illustrating the case of forest resources," *Ecological Economics*, vol. 61, no. 4, pp. 635–649, 2007.

[13] A. Aggarwal, V. Paul, and S. Das, *Forest Resources: Degradation, Livelihoods, and Climate Change*, Looking Back to Change Track, TERI, New Delhi, India, 2009.

[14] V. K. Bahuguna, K. Mitra, D. Capistrano, and S. Saigal, *Root to Canopy: Regenerating Forests Through Community State Partnerships*, Winrock International India, Commonwealth Forestry Association, New Delhi, India, 2004.

[15] T. Pistorius, C. B. Schmitt, D. Benick, and S. Entenmann, *Greening REDD+: Challenges and Opportunities for Forest Biodiversity Conservation*, University of Freiburg, Germany, 2010.

[16] R. Sud, J. V. Sharma, and A. K. Bansal, *International REDD+ Architecture and Its Relevance for India*, The Energy and Resources Institute (TERI), New Delhi, India, 2012.

[17] V. R. S. Rawat, "REDD Plus in India: from negotiations to implementation," in *Proceedings of the Pre-Congress Workshop of 1st Indian Forests Congress*, ICFRE, Dehradun, India, 2011.

[18] G. R. Van Der Werf, D. C. Morton, R. S. Defries et al., "CO_2 emissions from forest loss," *Nature Geoscience*, vol. 2, no. 11, pp. 737–738, 2009.

[19] EDRI, *From RED To REDD+*, Africa Regional Dialogue on Forests, Governance & Climate Change, 2010.

[20] N. H. Ravindranath, N. Srivastava, I. K. Murthy, S. Malaviya, M. Munsi, and N. Sharma, "Deforestation and forest degradation in India—implications for REDD+," *Current Science*, vol. 102, pp. 1117–1125, 2012.

[21] IISD, *Safeguards and Multiple Benefits in a REDD+ Mechanism*, International Institute of Sustainable Development, Canada, 2011, www.iisd.org.

[22] T. Clements, "Reduced expectations: the political and institutional challenges of REDD+," *ORYX*, vol. 44, no. 3, pp. 309–310, 2010.

[23] M. Collins, E. A. Macdonald, L. Clayton, I. Dunggio, D. W. Macdonald, and E. J. Milner-Gulland, "Wildlife conservation and reduced emissions from deforestation in a case study of Nantu Wildlife Reserve, Sulawesi: 2. An institutional framework for REDD implementation," *Environmental Science and Policy*, vol. 14, no. 6, pp. 709–718, 2011.

[24] S. Engel, S. Pagiola, and S. Wunder, "Designing payments for environmental services in theory and practice: an overview of the issues," *Ecological Economics*, vol. 65, no. 4, pp. 663–674, 2008.

[25] C. McDermott, *REDD+ Biodiversity Safeguards: Strength in Diversity—A Study of the Capacity of the UNFCCC to Deliver Equitable, Effective, and Efficient Forest Conservation Policies for Developing Countries*, Oxford University School of Geography and the Environment, 2011, Candidate Number: 53172.

[26] The Forests Dialogue, *Investing in REDD-Plus Consensus Recommendations on Frameworks For the Financing and Implementation of REDD-Plus*, TFD Review, A TFD Publication, 2010, www.theforestsdialogue.org.

[27] Ministry of Environment and Forests, India's Forests and REDD+, Ministry of Environment and Forests, Government of India.

[28] "Global Forest Resources Assessment 2005: Progress towards Sustainable Forest management," FAO Forestry Paper 147, Food and Agriculture Organization of the United Nations, Rome, Italy, 2006.

[29] *State of Forest Report*, Forest Survey of India, Dehradun, India, 2009.

[30] B. Pisupati, *Safeguarding India's Biological Diversity: The Biological Diversity Act*, Farmer's Forum, India's Agriculture Magazine, 2011.

[31] ICFRE, Indian Forest Congress, Ministry of Environment and Forests, 2011, http://ify-india.icfre.gov.in/indiaforest.html.

[32] Y. Gokhale, *Cover Story*, Terragreen Magzine, 2010.

[33] World Bank, "New global poverty estimates in India," World Bank, 2001, http://www.worldbank.org.in/wbsite/external/countries/southasiaext/indiaextn/0,menupk:295589~pagepk:141159~pipk:141110~thesitepk:295584,00.html.

[34] M. Poffenberger, *Community and Forest Management in South Asia—A Regional Profile of the Working Group on Community Involvement in Forest Management*, Forest, people and policies, IUCN, Switzerland, 2000.

[35] World Bank, "An article of World Bank on India: Alleviating Poverty through Forest Development," 2006, World Bank, http://www.worldbank.org/ieg).

[36] B. Sinha, C. P. Kala, and A. S. Katiyar, *Enhancing Livelihoods of Forest Dependent Communities Through Synergizing FDA Activities with Other Development Programmes*, Indian Institute of Forest Management, Bhopal, India, 2010.

[37] H. G. Champion and V. K. Seth, *A Revised Survey of the Forest*, Types of India, Government of India, 1968.

[38] S. J. Singh, F. Krausmann, S. Gingrich et al., "India's biophysical economy, 1961-2008. Sustainability in a national and global context," *Ecological Economics*, vol. 76, pp. 60–69, 2012.

[39] Forest Survey of India, *Demand and Supply of Fuelwood, Timber and Fodder in India*, Forest Survey of India, Dehradun, India, 1996.

[40] P. P. Singh, "Exploring biodiversity and climate change benefits of community-based forest management," *Global Environmental Change*, vol. 18, no. 3, pp. 468–478, 2008.

[41] R. Pachauri and R. K. Batra, Eds., *Directions, Innovations, and Strategies for Harnessing Action for Sustainable Development*, Tata Energy Research Institute, New Delhi, India, 2001.

[42] National Forestry Action Plan, *Ministry of Environment and Forest*, Government of India, New Delhi, India, 1999.

[43] S. Singh, India: Assessing Management Effectiveness of Wildlife Protected Areas in India, 2001, http://www.iucu.org/themes/forests/protected area/India.pdf.

[44] T. N. Khushoo, *Environmental Priorities in India and Sustainable Development*, Indian Science Congress, Calcutta, India, 1986.

[45] F. Warner, *Indo Swedish Forestry Programme II: 1982-83 TO, 1986-87*, Background Document, Swedish Embassy Dev Cooperation office, New Delhi, India, 1982.

[46] J. F. Richards, "Global patterns of land conversion," *Environment*, vol. 26, no. 9, pp. 6–34, 1984.

[47] M. Williams, *Forests: The Earth as Transformed by Human Actions*, Cambridge University Press, Cambridge, UK,, 1990.

[48] M. Joshi and P. P. Singh, *Tropical Deforestation and Forest Degradation: A Case Study from India*, XII World Forestry Congress, Quebec City, Canada, 2003.

[49] J. Sathaye, K. Andrasko, and P. Chan, "Emissions scenarios, costs, and implementation considerations of REDD-plus programs," *Environment and Development Economics*, vol. 16, no. 4, pp. 361–380, 2011.

[50] Food Agriculture Organization, *State of the World's Forests*, Food and Agriculture Organization of the United Nations, Rome, Italy, 2007.

[51] TERI, *Looking Back to Think Ahead: GREEN India 2047*, Tata Energy Research Institute, New Delhi, India, 1998.

[52] N. H. Ravindranath, R. K. Chaturvedi, and I. K. Murthy, "Forest conservation, afforestation and reforestation in India:

Implications for forest carbon stocks," *Current Science*, vol. 95, no. 2, pp. 216–222, 2008.

[53] State of Forest Report, Forest Survey of India, Dehradun, 1987.

[54] State of Forest Report, Forest Survey of India, Dehradun, 1989.

[55] State of Forest Report, Forest Survey of India, Dehradun, 1991.

[56] State of Forest Report, Forest Survey of India, Dehradun, 1993.

[57] State of Forest Report, Forest Survey of India, Dehradun, 1995.

[58] State of Forest Report, Forest Survey of India, Dehradun, 1997.

[59] State of Forest Report, Forest Survey of India, Dehradun, 1999.

[60] State of Forest Report, Forest Survey of India, Dehradun, 2001.

[61] State of Forest Report, Forest Survey of India, Dehradun, 2003.

[62] State of Forest Report, Forest Survey of India, Dehradun, 2005.

[63] State of Forest Report, Forest Survey of India, Dehradun, 2011.

[64] D. M. Bhat, K. S. Murali, and N. H. Ravindranath, "Formation and recovery of secondary forests in India: a particular reference to Western Ghats in South India," *Journal of Tropical Forest Science*, vol. 13, no. 4, pp. 601–620, 2001.

[65] S. Bajracharya, *Community carbon forestry: remote sensing of forest carbon and forest degradation in Nepal [M.S. thesis]*, International Institute for Geo-information Science and Earth Observation, Enscheda, The Netherlands, 2008.

[66] J. O. Niles, S. Brown, J. Pretty, A. Ball, and J. Fay, *Potential Mitigation and Income in Developing Countries from Changes in Use and Management of Agricultural and Forest Lands*, Center of Environment and Society, Occasional , University of Essex, 2001.

[67] United Nations Framework Convention on Climate Change, Submission by India to SBSTA, UNFCCC on Agenda Item 4: Methodological guidance for activities relating to reducing emissions from deforestation and forest degradation and the role of conservation, sustainable management of forests and enhancement of forests carbon stocks in developing countries (UNFCCC Document FCCC/SBSTA/L. 25 dated 3 Dec 2011), 2011.

[68] United Nations Framework Convention on Climate Change, The Cancun Agreements: Outcome of the work on the Ad Hoc Working Group on Long-Term Cooperative Action under the Convention. Report of the Conference of the parties on its sixteenth session, held in Cancun from November 29-December 10, 2010. FCCC/CP/2010/7/Add. 1., 2011.

[69] Food Agriculture Organization, "Global Forest Resource Assessment 2010," FAO Forestry Paper 163, Rome, Italy, 2010.

[70] B. P. Nayak, P. Kohli, and J. V. Sharma, *Livelihood of Local Communities and Forest Degradation in India: Issues For REDD+*, The Energy and Resources Institute, New Delhi, India, 2012.

[71] Ministry of Environment and Forests, *Report of the National Forest Commission*, Ministry of Environment and Forests, Government of India, New Delhi, India, 2006.

[72] P. Davidar, S. Sahoo, P. C. Mammen et al., "Assessing the extent and causes of forest degradation in India: where do we stand?" *Biological Conservation*, vol. 143, no. 12, pp. 2937–2944, 2010.

[73] Ministry of Environment and Forests, "Asia-pacific forestry sector outlook study II: India country report," Working Paper APFSOS II/WP/2009/06, FAO, Bangkok, Thailand, 2009.

[74] N. Moss, R. Nussbaum, and J. Muchemi, *REDD+ Safeguards—Background Paper Prepared For REDD+ Partnership*, Workshop on Enhancing Coordinated Delivery of REDD+: Emerging Lessons, Best Practices and Challenges, Cancún, Mexico, 2010.

[75] D. M. Mwayafu and J. W. Kisekka, *Promoting and Implementing REDD+ Safeguards at National Level in East Africa*, REDD-net

programme, Norwegian Agency for Development Cooperation (NORAD), 2012.

[76] T. A. Gardner, N. D. Burgess, N. Aguilar-Amuchastegui et al., "A framework for integrating biodiversity concerns into national REDD+ programmes," *Biological Conservation*, vol. 154, pp. 61–71, 2012.

[77] United Nations Framework Convention on Climate Change, The Cancun Agreements 1/CP. 16., 2010.

[78] Greenpeace, REDD: A Common Approach to Safeguards, 2011.

[79] United Nations Commission on Human Rights, "Sub-Commission on the Promotion and Protection of Human Rights," in *Proceedings of the 22nd session Working Group on Indigenous Populations*, New York, NY, USA, 2004.

[80] B. Palmer Fry, "Community forest monitoring in REDD+: the 'M' in MRV?" *Environmental Science and Policy*, vol. 14, no. 2, pp. 181–187, 2011.

[81] A. Angelsen, *Realising REDD+: National Strategy and Policy Options*, Centre for International Forestry Research (CIFOR), Bogor, Indonesia, 2009.

[82] A. Angelsen, S. Brown, C. Loisel, L. Peskett, C. Streck, and D. Zarin, *Reducing Emissions from Deforestation and Forest Degradation (REDD): An Options Assessment Report*, Meridian Institute, Washington, DC, USA, 2009.

[83] E. Mattsson, U. M. Persson, M. Ostwald, and S. P. Nissanka, "REDD+ readiness implications for Sri Lanka in terms of reducing deforestation," *Journal of Environmental Management*, vol. 100, pp. 20–40, 2012.

[84] P. Sukhdev, R. Prabhu, P. Kumar et al., *REDD+ and a Green Economy: Opportunities for a mutually Supportive relAtionship*, UN-REDD Programme Policy Brief, 2012.

[85] A. Aggarwal, "Implementation of forest rights act, changing forest landscape, and politics of REDD+ in India," *Resources, Energy, and Development*, vol. 8, no. 2, pp. 131–148, 2011.

[86] Chapter on Climate Change and Forests, Ministry of Environment and Forests, Government of India.

[87] Planning Commission, *Interim Report of the Expert Group on Low Carbon Strategies for Inclusive Growth*, Government of India, 2011.

[88] National Action Plan on Climate Change, Prime Minister's Council on Climate Change, Government of India, undated, http://pmindia.nic.in/Pg01-52.pdf.

[89] J. Eliasch, *Climate Change: Financing Global Forests*, Office of Climate Change, 2008, http://www.illegal-logging.info/sites/default/files/uploads/Fullreporteliaschreview1.pdf.

[90] C. A. Harvey, B. Dickson, and C. Kormos, "Opportunities for achieving biodiversity conservation through REDD," in *Conservation Letters*, pp. 1–9, Wiley Periodicals, 2009.

[91] United Nations Framework Convention on Climate Change, Inputs of Climate change Division on Agenda Item 7 of SUBSIDIARY BODY ON SCIENTIFIC, TECHNICAL AND TECHNOLOGICAL ADVICE for Sixteenth meeting at Montreal, 30 April 5 May 2012, 2012.

6

Mixed Species Allometric Models for Estimating above-Ground Liana Biomass in Tropical Primary and Secondary Forests, Ghana

Patrick Addo-Fordjour[1,2] and Zakaria B. Rahmad[1]

[1] *School of Biological Sciences, Universiti Sains Malaysia, 11800 Pulau Penang, Penang, Malaysia*
[2] *Department of Theoretical and Applied Biology, College of Science, Kwame Nkrumah University of Science and Technology (KNUST), Kumasi, Ghana*

Correspondence should be addressed to Patrick Addo-Fordjour; paddykay77@yahoo.com

Academic Editors: W. de Vries, N. Frascaria-Lacoste, Z. Kaya, and T. L. Noland

The study developed allometric models for estimating liana stem and total above-ground (TAGB) biomass in primary and secondary forests in the Asenanyo Forest Reserve, Ghana. Liana biomass was determined for 50 individuals for each forest using destructive sampling. Various predictors involving liana diameter and length were run against liana biomass in regression analysis, and R^2, RMSE, and Furnival's index of fit (FI) were used for model comparison. The equations comprised models fitted to untransformed and log-transformed data. Forest type had a significant influence ($P < 0.05$) on liana allometric models in the current study, resulting in the development of forest-type-specific equations. There were significant and strong linear relationships between liana biomass and the predictors in both forests ($R^2 > 0.970$). Liana diameter was a better predictor of biomass than liana length. Generally, the models which were based on log-transformed data showed better fit (higher FI values) than those fitted to untransformed data. Comparison of the site specific models in the current study with previously published models indicated that the models of the current study differed from the previous ones. This indicates the need for forest specific equations to be used for accurate determination of above-ground liana biomass.

1. Introduction

Lianas are important structural component of tropical forests [1]. They perform a number of ecological functions which help to sustain tropical forest ecosystems [2]. Lianas add substantially to plant assemblages in tropical forests in terms of number of species [3] and stem density [4]. Apart from contributing directly to species diversity in tropical forests, lianas also play a number of roles which contribute to maintain diversity of other organism [5]. Due to relatively high dominance of lianas in tropical forests, they also contribute a lot to forest biomass, especially in heavy liana infested forests. Specifically, lianas can accumulate as high as 30% of total above-ground biomass in tropical forest ecosystems [2]. Comparatively, lianas devote much less biomass for stem support than tress, and therefore, they are able to allocate more biomass for their growth compared with trees [2, 5].

Consequently, lianas have higher biomass growth than trees [6]. Lianas allocate more biomass for leaf production than the amount of biomass allocated to their stems. Because lianas allocate less biomass to their stems they produce less dense wood compared to trees [7].

Estimating tropical forest biomass and determining its dynamics are important aspects of tropical forest ecology. These are more important in areas where changes in composition and structure of forests are apparent. Although changes in biomass levels in secondary forests may occur often due to persistent human disturbance, primary forest biomass may also undergo changes due to other factors which bring about changes in forest composition, structure, diversity, and productivity [8, 9]. In view of this, it is necessary for biomass studies to be conducted in both tropical primary and secondary forests. However, at the moment, most biomass assessment studies have been limited to secondary forests, neglecting tropical primary forests. Available data on the amounts of biomass within a forest is not only important in determining the amounts of carbon stored by that forest [10] but also essential in assessing the productivity, structure, and conditions of that forest [11]. In addition, biomass data on secondary forests could be useful in explaining the effects of deforestation and carbon sequestration on the global carbon balance [10]. As indicated above, lianas already contribute substantially to total above-ground biomass in heavily liana-infested tropical forests (cf. [4]). As human disturbance continues to increase in tropical forests, especially in developing countries, lianas would most likely continue to increase in abundance, which could ultimately lead to an increase in the amount of biomass they store. In spite of the ability of lianas to store high amounts of biomass in tropical forest ecosystems, they have been ignored in most biomass assessment studies in the world [12]. It appears that many experts in the area are oblivious of the significant contribution lianas can make to forest biomass. This could explain why most recent calls about the need to increase the number of biomass and carbon stock assessments of tropical forests have all been centered on trees only [13].

The easiest and most practical way to determine liana biomass in tropical forests is to employ allometric equations. The use of allometric equations in estimating plant biomass does not only avoid forest destruction but also allows for the estimation of large forest areas. This is because data on the estimators of plant biomass in allometric models can easily be obtained for large areas of forests within a relatively short period of time without harvesting them. Nonetheless, allometric models can sometimes yield biomass estimates that do not reflect the biomass content of a forest [14]. Due to the apparent lack of knowledge about the importance of lianas in storing biomass, and their neglect in forest biomass assessments, only a few liana allometric equations have been developed for them. The limited number of liana allometric equations is also partly due to the difficulty in accessing the whole length of lianas from trees [15]. In spite of this challenge, conscious efforts should be made to develop many liana allometric equations that can be used to accurately estimate the increasing liana biomass in tropical forests [2]. This would ensure that biomass estimates of tropical forests are a true reflection of the actual biomass contents of forests. The availability of plant part allometric equations enable biomass allocation of plant parts to be determined and assessed. Knowledge of biomass allocation changes in different plant parts can be used to assess changes in plant structure and biogeochemical cycles in tropical forests (cf. [16]). In spite of this, only one study [17] has developed allometric equations for liana leaves and total above-ground part. The present study therefore sought out to develop allometric equations for different liana parts.

Ghana has some of the most complex and biodiversity-rich tropical forests in which lianas feature prominently [21, 22]. Nevertheless, there is no allometric equation for lianas in Ghana and also in the whole of Africa. The current study developed allometric models for the estimation of above-ground biomass of lianas in primary and secondary forests within the Asenanyo Forest Reserve, Ghana. Allometric

TABLE 1: Number of individuals of liana species in the primary and secondary forests used for the allometric equations.

Species	Number of individuals	
	Primary forest	Secondary forest
Acacia pentagona	3	3
Adenia rumicifolia	2	—
Afrobrunnichia erecta	2	1
Agelaea paradoxa	2	2
Alafia barteri	4	2
Alafia whytei	3	3
Calycobolus africanus	2	2
Calycobolus heudelotii	2	2
Castanola paradoxa	—	2
Combretum paradoxum	2	2
Combretum sp.	3	2
Dalbergia hostilis	2	3
Dalbergiella welwitschii	2	1
Gogronema latifolium	—	2
Griffonia simplicifolia	3	2
Landolphia hirsuta	2	2
Leptoderris micrantha	2	2
Manniophyton fulvum	—	2
Millettia chrysophylla	4	3
Motandra guineensis	2	2
Neuropeltis sp.	1	2
Parquetina nigrescens	2	—
Paullinia pinnata	2	2
Salacia elegans	2	2
Salacia columna	—	2
Strophanthus barteri	1	2

equations in the present study were developed for both stem and total above-ground components of lianas. Total above-ground biomass of lianas is usually made up of stem and leaf (shoot) biomass components. Therefore, development of stem and total above-ground biomass equations in the current study makes it possible to also estimate liana leaf biomass. This would enable changes in relative contribution of various liana parts to total above-ground biomass to be assessed from time to time. This information together with other similar ones from other plant life forms, such as trees, can be useful in assessing forest ecosystem dynamics which can help in developing forest management strategies.

2. Materials and Methods

2.1. Study Area. The study was conducted in the Asenanyo Forest Reserve in the Nkwawie District, Ghana (06°26′23″N, 02°06′28″W). The forest reserve lies within the moist semideciduous forest zone in Ghana and has both primary and secondary forests. The secondary forest has undergone selective and illegal logging as well as silvicultural treatments in the past, and the relics of these human activities are still evident. The dominant species in the forest are *Celtis mildbraedii*, *Triplochiton scleroxylon*, *Albizia zygia*, and *Cedrella odorata*.

Mixed Species Allometric Models for Estimating above-Ground Liana Biomass in Tropical Primary and Secondary Forests, Ghana

69

TABLE 2: Summary of allometric properties of liana individuals in the two forest types (Asenanyo Forest Reserve) used for the study.

Parameter	Primary			Secondary		
	Min.	Max.	Mean	Min.	Max.	Mean
Diameter (cm)	1.20	13.20	6.34	1.20	13.00	5.78
Length (m)	5.07	47.55	18.33	4.96	52.00	16.12
Stem biomass (kg)	3.00	25.10	11.20	2.05	21.60	10.65
Total above-ground biomass (kg)	3.55	27.15	12.04	2.21	24.80	11.63

TABLE 3: Six previously published allometric equations (total above-ground liana biomass) used in comparing the current allometric equation. All the equations are based on the model form; total above-ground biomass = $\exp[c + \alpha \ln(\text{Diameter})]$.

Equation	c	α
Gehring et al. (2004) [17]	−1.547	2.640
Gerwing and Farias (2000) [12]	0.147	2.184
Putz (1983) [18]	0.036	1.806
Hozumi et al. (1969) [19]	−1.347	2.391
Beekman (1981) [20]	−1.459	2.566
Schnitzer et al. (2006) [4]	−1.484	2.657

The average elevation of the forest reserve is 162 m a.s.l. Daily temperatures range from 20 to 32.9°C, and the average annual rainfall is 1856 mm. Relative humidity for the area is high (91%).

2.2. Field Sampling and Biomass Measurements. A total of 22 and 24 liana species were destructively sampled from the primary and secondary forest, respectively (Table 1), for development of mixed species allometric equations (from August to December 2012). Sampling was purposely conducted in the rainy season when leaf biomass was highest [17]. Lianas were harvested from different habitats (flatlands, slopes, and undulating lands) which were comparable in the primary and secondary forests. Because in each forest type sampling occurred in flatlands, slopes, and undulating lands which were comparable with those in the other forest type, habitat biased sampling was avoided. The sampling within different habitats in each forest type was to ensure that the models produced could be a reflection of habitat variations in the forests. Liana diameter (at 1.3 m from the rooting base) was measured before harvesting the individuals, whereas their length was measured after they were harvested. A total number of 100 liana individuals (primary forest: 50 individuals; secondary forest: 50 individuals) were harvested in the study (Table 1), and their allometric characteristics are indicated in Table 2. Both single and multiple stem liana individuals were included in the study. Liana leaves and stems were separated from each other, and they were sun dried to constant weights over different periods of time (leaves: 3 weeks; stems: 3 months). The constant dry weights of the species were recorded as their biomass.

2.3. Data Analyses. Data analyses involved data exploration and model fitting using liana diameter, $[\text{diameter}]^2$, length,

$\log_{10}[\text{diameter}]$, and $\log_{10}[\text{diameter}]^2$ as estimators of liana biomass to obtain models that best fit the data. A series of models were developed using original untransformed and log-transformed ($\log_{10}$) data. In all cases, only models that complied with regression assumptions (homogeneity of variance, linearity, normality, and nonautocorrelation) and showed high goodness of fit ($R^2 > 0.97$) were retained. Homogeneity of variance and linearity of data were assessed using residual plots while autocorrelation, and normality were verified with Durbin-Watson statistics and probability plots, respectively. The Furnival's index (FI), root mean square error (RMSE), and coefficient of determination (adjusted R^2) were used for model selection and comparison. The FI was used to compare models that had different response variables while the RMSE and R^2 were used for models with the same response variables. The RMSE and R^2 could not be used to compare models with different response variables because they have the potential of producing misleading results for that purpose [23, 24]. For this reason, the Furnival's index [25] was used to compare untransformed and log-transformed models. The index was computed as follows:

$$\text{FI} = \frac{1}{[f'(Y)]} \sqrt{\text{MSE}}, \quad (1)$$

where $f'(Y)$ is the derivative of the dependant variable with respect to biomass, MSE is the mean square error of the fitted equation, and the square bracket ($[\cdot]$) is the geometric mean. Comparatively, models with lower FI and RMSE values have better goodness of fit. On the other hand, the higher the R^2 value of a model the better its goodness of fit.

Due to downward bias which usually occurs when log biomass are back transformed to arithmetic units [26], a correction factor (CF) indicated below [27] was calculated for the models, which could be used to correct them.

Consider

$$\text{CF} = \exp^{((\text{SEE}*2.303)^2/2)}, \quad (2)$$

where SEE is the standard error of the estimate.

Linear regression analyses were conducted to determine the relationships between liana biomass and the response variables (diameter, $[\text{diameter}]^2$, length, $\log_{10}[\text{diameter}]$, or $\log_{10}[\text{diameter}]^2$) in the case of both untransformed and log-transformed data. Analysis of covariance (ANCOVA) was conducted to examine possible differences in regression slopes of models between the forest types. Forest type (primary and secondary) was used as the main factor, whereas the response variables (diameter, $[\text{diameter}]^2$,

TABLE 4: Allometric equations of mixed species for estimating liana stem biomass (kg) in the primary forest.

#	Equation	c ($\pm$SE)	α ($\pm$SE)	R^2 (adjusted)	RMSE	FI
1	Stem biomass = $c + \alpha D$	1.643 ± 0.128	1.770 ± 0.020	0.994	0.471	0.47
2	Stem biomass = $c + \alpha L$	2.304 ± 0.212	0.294 ± 0.006	0.981	0.821	0.82
3	$\text{Log}_{10}(\text{Stem biomass}) = c + \alpha (\log_{10} D)$	1.004 ± 0.013	0.801 ± 0.008	0.996	0.036	0.35
4	$\text{Log}_{10}(\text{Stem biomass}) = c + \alpha (\log_{10} D^2)$	1.004 ± 0.013	0.958 ± 0.009	0.996	0.036	0.35

#: Equation number; D: Liana diameter; L: Liana length.

TABLE 5: Allometric equations of mixed species for estimating total above-ground biomass (TAGB) (kg) in the primary forest.

#	Equation	c ($\pm$SE)	α ($\pm$SE)	R^2 (adjusted)	RMSE	FI
5	TAGB = $c + \alpha D$	1.703 ± 0.131	1.915 ± 0.021	0.994	0.484	0.48
6	TAGB = $c + \alpha L$	2.425 ± 0.234	0.318 ± 0.007	0.980	0.903	0.90
7	$\text{Log}_{10}(\text{TAGB}) = c + \alpha (\log_{10} D)$	1.077 ± 0.012	0.850 ± 0.007	0.996	0.034	0.35
8	$\text{Log}_{10}(\text{TAGB}) = c + \alpha (\log_{10} D^2)$	1.077 ± 0.012	0.979 ± 0.009	0.996	0.034	0.35

#: Equation number; D: Liana diameter; L: Liana length.

TABLE 6: Allometric equations of mixed species for estimating liana stem biomass (kg) in the secondary forest.

#	Equation	c ($\pm$SE)	α ($\pm$SE)	R^2 (adjusted)	RMSE	FI
9	Stem biomass = $c + \alpha D$	-0.341 ± 0.119	1.727 ± 0.016	0.996	0.414	0.41
10	Stem biomass = $c + \alpha L$	0.765 ± 0.282	0.446 ± 0.011	0.972	1.054	1.05
11	$\text{Log}_{10}(\text{Stem biomass}) = c + \alpha (\log_{10} D)$	0.201 ± 0.009	1.115 ± 0.011	0.994	0.022	0.20
12	$\text{Log}_{10}(\text{Stem biomass}) = c + \alpha (\log_{10} D^2)$	0.201 ± 0.009	0.498 ± 0.006	0.994	0.022	0.20

#: Equation number; D: Liana diameter; L: Liana length.

TABLE 7: Allometric equations of mixed species for estimating total above-ground biomass (TAGB) (kg) in the secondary forest.

#	Equation	c ($\pm$SE)	α ($\pm$SE)	R^2 (adjusted)	RMSE	FI
13	TAGB = $c + \alpha D$	-0.360 ± 0.124	1.901 ± 0.017	0.996	0.433	0.43
14	TAGB = $c + \alpha L$	0.774 ± 0.294	0.491 ± 0.011	0.975	1.100	1.10
15	$\text{Log}_{10}(\text{TAGB}) = c + \alpha (\log_{10} D)$	0.236 ± 0.009	1.128 ± 0.012	0.994	0.023	0.22
16	$\text{Log}_{10}(\text{TAGB}) = c + \alpha (\log_{10} D^2)$	0.236 ± 0.009	0.514 ± 0.006	0.994	0.023	0.22

#: Equation number; D: Liana diameter; L: Liana length.

length, $\log_{10}[\text{diameter}]$, or $\log_{10}[\text{diameter}]^2$) were used as the covariables.

In the current study, the overall best total above-ground biomass models were determined for the forest types according to the Furnial's index of fit. These were compared with previous total above-ground liana biomass models indicated in Table 3 [4, 12, 17–20], using paired t-tests. The equations were applied to data sets collected from the forests from which the current allometric equations were developed. The data are comprised of 92 (from 32 species) and 100 (from 32 species) liana individuals in the primary and secondary forests, respectively, (diameter range; primary forest: 2–10.7 cm and secondary forest: 2 to 14 cm). Some of the data pairs were transformed (square root and log transformations) to meet t-test assumptions.

Regression analyses and t-test were performed using GenStat software (VSN International Ltd., Hemel Hempstead, UK) whereas ANCOVA was run with Minitab 15 Software (Minitab Inc.). All analyses were conducted at a significance level of 5%.

3. Results and Discussion

Same set of allometric equations can be developed for use in both primary and secondary forests in areas where forest type does not influence the equations significantly [15, 17]. However, in this study, there were significant differences in the slopes of regression models between the primary and secondary forest types (ANCOVA; $P < 0.05$). This suggested forest-specific differences in regressions between the primary and secondary forests, resulting in the development of forest-specific models for the primary and secondary forests.

In each forest, a total of four different models were developed for the estimation of liana stem and total above-ground biomass (Tables 4, 5, 6, and 7). These models were developed based on untransformed (models 1-2, 5-6, 9-10, and 13-14) and log-transformed (3-4, 7-8, 11-12, and 15-16) data. There were strong and significant linear relationships between liana biomass and the various predictors in all the models developed (Tables 4–7; Figures 1 and 2). Although liana diameter and length were good predictors of liana

Mixed Species Allometric Models for Estimating above-Ground Liana Biomass in Tropical Primary and Secondary Forests, Ghana

71

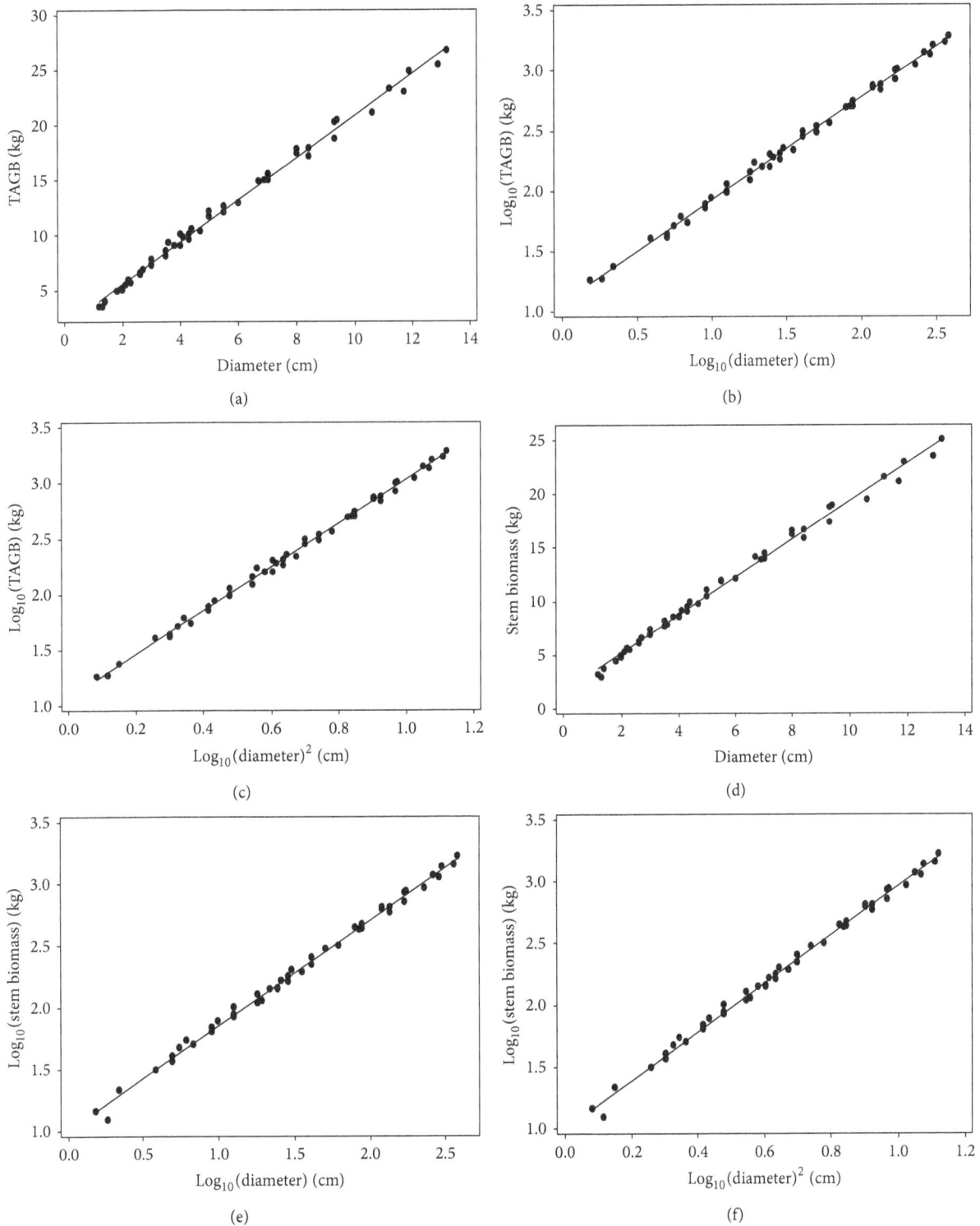

FIGURE 1: Allometric relationships between liana biomass (total above-ground biomass, TAGB: (a), (b), and (c); stem biomass: (d), (e), and (f)) and diameter in the primary forest. Relationships based on raw and log-transformed data are shown.

biomass in the models fitted to data on arithmetic scale, liana diameter (R^2 = 0.994–0.996; RMSE = 0.414–0.484) was slightly a better predictor of liana biomass than liana length (R^2 = 0.972–0.981; RMSE = 0.821–1.100) in the current study. The use of liana allometric equations that use

length as a predictor of biomass has a practical challenge. Measuring liana length on the field is impossible unless they are harvested. Therefore, the allometric models of the current study which use liana diameter as a predictor of biomass are recommended for use in liana biomass determination.

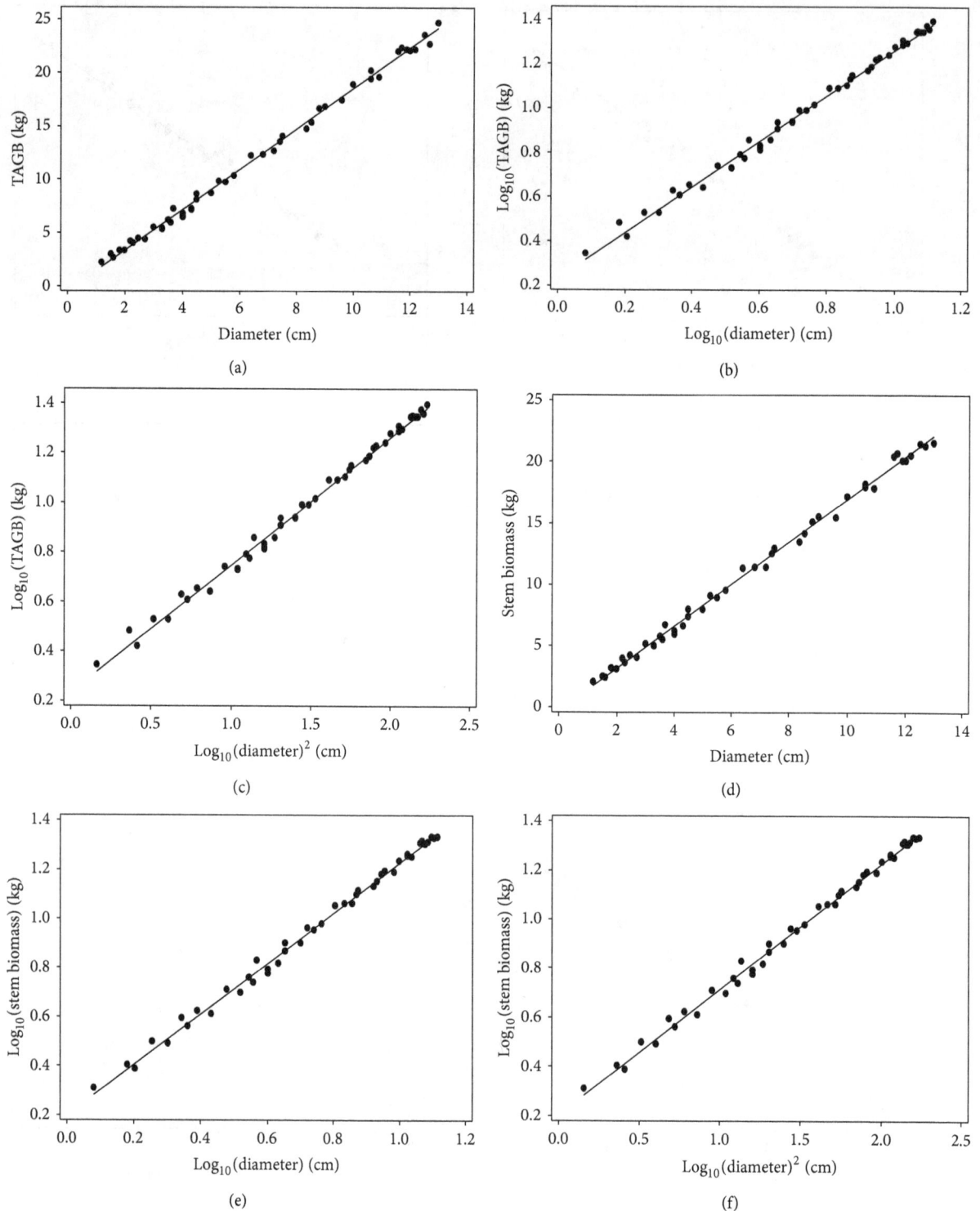

FIGURE 2: Allometric relationships between liana biomass (total above-ground biomass, TAGB: (a), (b), and (c); stem biomass: (d), (e), and (f)) and diameter in the secondary forest. Relationships based on raw and log-transformed data are shown.

Logarithmic transformation of data resulted in increased homogeneity of variance compared with data on arithmetic scale (Figures 1 and 2). The decline in homogeneity of variance of the log-transformed models is consistent with previous studies [13, 28].

On the basis of Furnival's index of fit, the log-transformed allometric equations performed better than the equations fitted to data on original arithmetic scale for both liana stem and total above-ground biomass models. The higher goodness of fit of the log-transformed models provides

TABLE 8: Correction factor (CF) of logarithmic models for stem and total above-ground biomass of lianas.

Model	CF	
	Stem	Total
Primary		
$\text{Log}_{10}(\text{Biomass}) = c + \alpha\,(\log_{10}D)$	1.0035	1.0030
$\text{Log}_{10}(\text{Biomass}) = c + \alpha\,(\log_{10}D^2)$	1.0035	1.0030
Secondary		
$\text{Log}_{10}(\text{Biomass}) = c + \alpha\,(\log_{10}D)$	1.0013	1.0014
$\text{Log}_{10}(\text{Biomass}) = c + \alpha\,(\log_{10}D^2)$	1.0013	1.0014

TABLE 9: Comparison of mean estimated total above-ground biomass (per species) between model 7 and previous models (see Table 3) with t-test in the primary forest. The value in parenthesis represents the mean biomass for the current equation, whereas those outside the parenthesis represent the means of the previous equations.

Pair	Mean	P value
This study versus Gehring et al. (2004) [17]	50.96 (126.96)	<0.001
This study versus Gerwing and Farias (2000) [12]	119.80 (126.96)	0.502
This study versus Putz (1983) [18]	63.27 (126.96)	<0.001
This study versus Hozumi et al. (1969) [19]	39.12 (126.96)	<0.001
This study versus Beekman (1981) [20]	48.18 (126.96)	<0.001
This study versus Schnitzer et al. (2006) [4]	55.97 (126.96)	<0.001

TABLE 10: Comparison of mean estimated total above-ground biomass (per species) between model 15 and previous models (see Table 3) with t-test in the secondary forest. The value in parenthesis represents the mean biomass for the current equation, whereas those outside the parenthesis represent the means of the previous equations.

Pair	Mean	P value
This study versus Gehring et al. (2004) [17]	130.10 (40.97)	<0.001
This study versus Gerwing and Farias (2000) [12]	254.20 (40.97)	<0.001
This study versus Putz (1983) [18]	115.96 (40.97)	<0.001
This study versus Hozumi et al. (1969) [19]	90.49 (40.97)	0.003
This study versus Beekman (1981) [20]	120.10 (40.97)	<0.001
This study versus Schnitzer et al. (2006) [4]	144.00 (40.97)	<0.001

evidence in support of another study which indicated that log-transformed models show better fit than models of untransformed data [29]. Based on the FI, models 3 and 4 in the primary forest and models 11 and 12 in the secondary forest were the best allometric models for liana stem biomass. In the same way, models 7 and 8 in the primary forest and models 15 and 16 in the secondary forest were the best models for total above-ground liana biomass. Since the log-transformed models that use diameter are simpler than the models that employ [diameter]^2, models 3, and 11 are recommended for liana stem biomass estimation in primary and secondary forests, respectively, in line with the Ockham's Razer principle [30]. The principle suggests the selection of simplest models if there are many models that have equal goodness of fit. Likewise, models 7 and 15 are recommended for estimating total above-ground biomass of lianas in primary and secondary forests, respectively. The correction factor values determined for the various log-transformed allometric equations (Table 8) were very low (primary forest: 0.3–0.35%; secondary forest: 0.13-0.14%), indicating that downward bias of the models was negligible. A similar finding was reported by Addo-Fordjour and Rahmad [15].

One of the best total above-ground models in the primary forest (model 7) was compared with some previously

published models by applying them on a data set taken from the same primary forest. The results indicated that on the whole, most of the previous equations underestimated total above-ground biomass of lianas in this study (Table 9). The mean total above-ground biomass per liana species estimated by five of the previous equations were significantly lower (Table 9; $P < 0.05$) than that predicted by the current study. Nonetheless, one of the previous allometric equations estimated a mean total above-ground biomass that was similar ($P = 0.502$) to that of the current allometric equation. Another allometric equation selected as one of the best total above-ground model in the secondary forest of this study (model 15) was compared with the previously mentioned equations in the same manner, using a data set collected from the secondary forest. All the previous equations highly over-predicted total above-ground biomass of lianas in the secondary forest (Table 10). The mean total above-ground biomass per species estimated by all the previous equations differed significantly from that estimated by the current model (Table 10; $P < 0.05$). The equation of Schnitzer et al. [4] which is considered as somewhat general (because it was developed from large data sets from four different countries) also estimated total above-ground biomass that differed considerably from those of the current models in both the primary and secondary forests. Similarly, in a study in Malaysia, a universal equation of Brown [31] highly overestimated total above-ground biomass of trees by as much as 100% compared with the equation developed in that study [32]. The wide disparities between the estimation powers of the site specific and general allometric equations of plants observed in this study as well as that of Kenzo et al. [32] indicate that general allometric equations may not provide accurate biomass estimation in specific forests. The differences in estimation of total above-ground biomass between the previous and the current models may be due to differences in liana wood density in the various forests from which the equations were developed [32, 33]. The comparison of the previous equations to the current ones has revealed the need for site specific models to be encouraged for accurate determination of liana biomass in tropical forests. However, where site specific allometric equations are not available, care must be taken when choosing allometric equations for forests. As much as possible, equations from the same region or continent should be given preference to equations from different continents [15].

4. Conclusion

The current study developed allometric relationships between liana biomass, diameter, [diameter]^2, and length for the estimation of stem and total above-ground biomass. Forest type had a significant influence on liana allometric models in the current study, resulting in the development of forest-type-specific equations. Models were developed on data fitted to log-transformed and untransformed data. In both forest types, log-transformed data fitted better compared to untransformed data. Comparison of the site specific models in the current study with previously published models indicated that the models of the current study differed from the previous ones.

Conflict of Interests

The authors wish to state that they do not have any direct financial relation whatsoever with the content of this paper, that might lead to a conflict of interest for any of them. Consequently, they declare no conflict of interests.

Acknowledgments

This study was supported by TWAS-USM Postgraduate Fellowship and Research University Grant (RU) (1001/PBI-OLOGI/815086). The authors are grateful to Mr. Abu Husin from the Forest Research Institute Malaysia and Mr. Ntim Gyakari of the Forestry Commission of Ghana for their assistance in plant identification. The authors finally thank Mr. S. M. Edzham from the School of Biological Sciences, USM, Malaysia, for his immense assistance on the field.

References

[1] L. Kammesheidt, A. Berhaman, J. Tay, G. Abdullah, and M. Azwal, "Liana abundance, diversity and tree infestation in the Imbak Canyon conservation area, Sabah, Malaysia," *Journal of Tropical Forest Science*, vol. 21, no. 3, pp. 265–271, 2009.

[2] S. A. Schnitzer and F. Bongers, "The ecology of lianas and their role in forests," *Trends in Ecology and Evolution*, vol. 17, no. 5, pp. 223–230, 2002.

[3] P. Addo-Fordjour, A. K. Anning, E. A. Atakora, and P. S. Agyei, "Diversity and distribution of climbing plants in a semi-deciduous rain forest, KNUST Botanic Garden, Ghana," *International Journal of Botany*, vol. 4, no. 2, pp. 186–195, 2008.

[4] S. A. Schnitzer, S. J. DeWalt, and J. Chave, "Censusing and measuring lianas: a quantitative comparison of the common methods," *Biotropica*, vol. 38, no. 5, pp. 581–591, 2006.

[5] Y. Tang, R. L. Kitching, and M. Cao, "Lianas as structural parasites: a re-evaluation," *Chinese Science Bulletin*, vol. 57, no. 4, pp. 307–312, 2012.

[6] Z. Q. Cai, L. Poorter, K. F. Cao, and F. Bongers, "Seedling growth strategies in bauhinia species: comparing lianas and trees," *Annals of Botany*, vol. 100, no. 4, pp. 831–838, 2007.

[7] B. Kusumoto, T. Enoki, and Y. Kubota, "Determinant factors influencing the spatial distributions of subtropical lianas are correlated with components of functional trait spectra," *Ecological Research*, vol. 28, no. 1, pp. 9–19, 2013.

[8] J. Chave, D. Coomes, S. Jansen, S. L. Lewis, N. G. Swenson, and A. E. Zanne, "Towards a worldwide wood economics spectrum," *Ecology Letters*, vol. 12, no. 4, pp. 351–366, 2009.

[9] D. Tilman, P. B. Reich, J. Knops, D. Wedin, T. Mielke, and C. Lehman, "Diversity and productivity in a long-term grassland experiment," *Science*, vol. 294, no. 5543, pp. 843–845, 2001.

[10] Q. M. Ketterings, R. Coe, M. van Noordwijk, Y. Ambagau', and C. A. Palm, "Reducing uncertainty in the use of allometric biomass equations for predicting above-ground tree biomass in mixed secondary forests," *Forest Ecology and Management*, vol. 146, no. 1–3, pp. 199–209, 2001.

[11] D. B. MacKay, P. M. Wehi, and B. D. Clarkson, "Evaluating restoration success in urban forest plantings in Hamilton, New Zealand," *Urban Habitats*, vol. 6, no. 1, 2011.

[12] J. J. Gerwing and D. L. Farias, "Integrating liana abundance and forest stature into an estimate of total aboveground biomass for an eastern Amazonian forest," *Journal of Tropical Ecology*, vol. 16, no. 3, pp. 327–335, 2000.

[13] J. R. Moore, "Allometric equations to predict the total above-ground biomass of radiata pine trees," *Annals of Forest Science*, vol. 67, no. 8, article 806, 2010.

[14] M. Segura and M. Kanninen, "Allometric models for tree volume and total aboveground biomass in a tropical humid forest in Costa Rica," *Biotropica*, vol. 37, no. 1, pp. 2–8, 2005.

[15] P. Addo-Fordjour and Z. B. Rahmad, "Development of allometric equations for estimating above-ground liana biomass in tropical primary and secondary forests, Malaysia," *International Journal of Ecology*, vol. 2013, Article ID 658140, 8 pages, 2013.

[16] M. A. Cairns, S. Brown, E. H. Helmer, and G. A. Baumgardner, "Root biomass allocation in the world's upland forests," *Oecologia*, vol. 111, no. 1, pp. 1–11, 1997.

[17] C. Gehring, S. Park, and M. Denich, "Liana allometric biomass equations for Amazonian primary and secondary forest," *Forest Ecology and Management*, vol. 195, no. 1-2, pp. 69–83, 2004.

[18] F. E. Putz, "Liana biomass and leaf area of a "tierra firme" forest in the Rio Negro Basin, Venezuela," *Biotropica*, vol. 15, no. 3, pp. 185–189, 1983.

[19] K. Hozumi, K. Yoda, S. Kokawa, and T. Kira, "Production ecology of tropical rain forests in South-Western Cambodia. I. Plant biomass," *Oecologia*, vol. 145, pp. 87–99, 1969.

[20] F. Beekman, *Structural and Dynamic Aspects of the Occurrence and Development of Lianes in the Tropical Rain Forest*, Department of Forestry, Agricultural University, Wageningen, The Netherlands, 1981.

[21] P. Addo-Fordjour, Z. B. Rahmad, J. Amui, C. Pinto, and M. Dwomoh, "Patterns of liana community diversity and structure in a tropical rainforest reserve, Ghana: effects of human disturbance," *African Journal of Ecology*, vol. 51, no. 2, pp. 217–227, 2013.

[22] P. Addo-Fordjour, P. El Duah, and D. K. K. Agbesi, "Factors influencing liana species richness and structure following

Mixed Species Allometric Models for Estimating above-Ground Liana Biomass in Tropical Primary and Secondary Forests, Ghana

75

anthropogenic disturbance in a tropical forest, Ghana," *ISRN Forestry*, vol. 2013, Article ID 920370, 11 pages, 2013.

[23] B. R. Parresol, "Assessing tree and stand biomass: a review with examples and critical comparisons," *Forest Science*, vol. 45, no. 4, pp. 573–593, 1999.

[24] R. J. Hyndman and A. B. Koehler, "Another look at measures of forecast accuracy," *International Journal of Forecasting*, vol. 22, no. 4, pp. 679–688, 2006.

[25] G. M. Furnival, "An index for comparing equations used inconstructing volume tables," *Forest Science*, vol. 7, pp. 337–341, 1961.

[26] G. L. Baskerville, "Use of logarithmic regression in the estimation of plant biomass," *Canadian Journal of Forest Research*, vol. 2, no. 1, pp. 49–53, 1972.

[27] D. G. Sprugel, "Correcting for bias in log-transformed allometric equations," *Ecology*, vol. 64, no. 1, pp. 209–210, 1983.

[28] H. P. Piepho, "Data transformation in statistical analysis of field trials with changing treatment variance," *Agronomy Journal*, vol. 101, no. 4, pp. 865–869, 2009.

[29] C. Wang, "Biomass allometric equations for 10 co-occurring tree species in Chinese temperate forests," *Forest Ecology and Management*, vol. 222, no. 1–3, pp. 9–16, 2006.

[30] D. J. C. Mackay, *Information Theory, Inference and Learning Algorithms*, Cambridge University Press, Cambridge, UK, 2003.

[31] S. Brown, "Estimating biomass and biomass change in tropical forests. a primer," Forestry Paper 134, Food and Agriculture Organization of the United Nations, Rome, Italy, 1997.

[32] T. Kenzo, T. Ichie, D. Hattori et al., "Development of allometric relationships for accurate estimation of above- and below-ground biomass in tropical secondary forests in Sarawak, Malaysia," *Journal of Tropical Ecology*, vol. 25, no. 4, pp. 371–386, 2009.

[33] T. Kenzo, R. Furutani, D. Hattori et al., "Allometric equations for accurate estimation of above-ground biomass in logged-over tropical rainforests in Sarawak, Malaysia," *Journal of Forest Research*, vol. 14, no. 6, pp. 365–372, 2009.

7

Vulnerability of Trees to Climate Events in Temperate Forests of West Germany

Stefanie Fischer[1] and Burkhard Neuwirth[2]

[1] *Department of Geography, University of Bonn, Meckenheimer Allee 166, 53115 Bonn, Germany*
[2] *DeLaWi Tree-Ring Analysis, Preschlinallee 2, 51570 Windeck, Germany*

Correspondence should be addressed to Stefanie Fischer; fischer@giub.uni-bonn.de

Academic Editors: M. S. Di Bitetti, M. Kanashiro, F. Le Tacon, and P. Newton

An improved understanding of the spatiotemporal climate/growth relationship of our forests is of particular importance for assessing the consequences of climate warming. A total of 67 stands of beech (*Fagus sylvatica* L.), pedunculate oak (*Quercus robur* L.), sessile oak (*Quercus petraea* (Matt.) Liebl.), Scots pine (*Pinus sylvestris* L.), and spruce (*Picea abies* Karst.) from sites located in the transition zone from the lowlands to the low mountain ranges of West Germany have been analysed. A combination of pointer year and cluster analysis was used to find groups with similar growth anomaly patterns over the 1941–2000 period. Shifted reaction patterns especially characterise differences in the growth behaviour of the clusters. These are controlled by different reactions to the climate conditions in winter and spring and are determined by a complex system of forcing factors. Results of this study reflect the enormous importance of the length of the growing season. Increasing the duration of the vegetation period climate warming can change the climate/growth relationship of trees, thereby confounding climate reconstructions which use tree rings. Since forcing factors have been detected that are more important than the tree species, we recommend the application of growth-specific approaches for the analysis of tree species' vulnerability to climate.

1. Introduction

Whereas tree growth at the timberline is mostly limited by only one specific dominant factor [1], growth of temperate forest regions is influenced by a multitude of biotic and abiotic factors [2–4]. This is caused by predominant temperate climate conditions and the fact that mostly native tree species are growing in the range of their natural distribution areas [5]. Nevertheless, climate control is still a crucial forcing factor for annual tree-ring growth in lower altitudes [3]. The 20th century warming trends are extraordinary [6] and steady since the late 1970s [7] and have lengthened the duration of the growing season [8]. An increase of severe climate extremes such as heat waves is inherent with these [9], whereas changes in precipitation and dryness extremes are less clearly linked [10]. However, a better understanding of the spatiotemporal climate/growth relationships, including the identification of the environmental drivers, is of particular importance [11–13] to understand climate-induced changes in forest

productivity with regard to different tree species and site characteristics. Previous studies have shown that analyses of tree-ring width at lower elevation sites are suitable for climatological interpretations [4, 12, 14–17]. Comprehensive network analyses have been made by Neuwirth et al. [12] investigating the interannual climate/growth relationship of temperate and humid forest stands within a wide region of mid-latitude Europe. Covering an altitudinal range of 10–2300 m a.s.l., the investigation area is affected by strong environmental gradients, leading to a comparison of temperate with limiting conditions [12]. To further enhance comprehension of the complex climate/growth relationship in temperate forests, Schweingruber and Nogler [18] recommended the analysis of tree growth at a more regional scale. Friedrichs et al. [4] investigated slight variations in oak growth within relatively homogeneous growth patterns on a regional scale. Homogeneous subsets caused by smaller environmental gradients could be identified by grouping splined tree-ring series. Babst et al. [19] detected 15 groups with

FIGURE 1: Topographical situation of the research area and location of the investigated sites. Black circles denote beech stands, grey circles pedunculate oak stands, white circles sessile oak stands, black triangles Scots pine stands, and white triangles spruce stands.

similar high-frequency growth variability due to temperature and precipitation anomalies for a supraregional network covering the European continent. Lacking from previous studies is an analysis of multispecies growth anomalies within single years as a response to abrupt changes in climatic conditions on a regional scale.

Pointer years are an accepted tool to analyse high-frequency changes in radial tree growth, which are primarily climatologically induced. They can be interpreted as expressions of rapid environmental changes [20]. Since trees are optimally adapted to their site-specific environmental conditions [5, 18], positive pointer years reflect optimal growth conditions [12], with almost all growth factors close to optimum [21]. Extreme environmental conditions lead to reductions of tree-growth [20]. Negative pointer years are a suitable method to detect single environmental factors, which deviate strongly from optimal conditions [21]. As such, our analyses are focused on evaluating negative pointer years, which can be seen as reactions to external disturbances. The pointer year analysis was conducted on the basis of measured tree-ring width values, in contrast to visual approaches described by Kaennel and Schweingruber [22].

The aim of this study is to detect forcing factors for the climate/growth relationships of different tree species in temperate forests. The analysis is based on a multispecies tree ring network of beech (*Fagus sylvatica* L.; FASY), pedunculate oak (*Quercus robur* L.; QURO), sessile oak (*Quercus petraea* Matt *Liebl.*; QUPE), Scots pine (*Pinus sylvestris* L.; PISY), and

spruce (*Picea abies* Karst.; PCAB), covering the transition of the lowlands to the low mountain ranges of West Germany. In line with this investigation, a particular approach was applied, combining pointer year analysis with cluster analysis.

2. Material and Methods

2.1. Research Area and Tree-Ring Data. The research area is situated 6.2–9.1°E and 50.1–51.9°N and covers the transition from the lowlands to the low mountain ranges of West Germany (Figure 1). Climatic conditions of the region are characterised by northwesterly atmospheric flows [23]. They are predominantly homogeneous with some local differentiations, such as a tendency to more oceanic conditions in the western area and more continental conditions in the eastern parts of the research area. The mean annual temperature over the 1961–1990 period was 9.0°C with a minimum of 0.83°C in January and a maximum of 17.1°C in July. The average annual precipitation was 816 mm, ranging from 53 mm in February to 79 mm in July. The length of the growing season was shortened with increasing altitude and latitude [8, 24].

The dataset contains 67 sites of a continuously extended dendroecological network [4, 16, 17, 25–27] including stands of beech, pedunculate oak, sessile oak, Scots pine, and spruce. Sites range in altitude from 40 to 710 m a.s.l. and represent a variety of gradients (0–45°) and aspects. Soil analyses detect cambisols and luvisols as predominant soil types which are also typical for temperate forest regions in West

Germany [28]. The available water capacity ranges from low (60–110 mm) to very high (>240 mm) (Table 1).

Tree-ring series were prepared using standard procedures [29, 30]. Tree-ring widths were measured with a programme for tree-ring analyses (Time Series Analysis and Presentation; TSAP) in a resolution of 1/100 mm followed by synchronising the series carried out with TSAP [31] and a programme which is used to control the quality of the cross-dating and the tree-ring chronologies (COFECHA) [32]. Tree-ring series were averaged to site chronologies and afterwards to tree-species as well as cluster chronologies. The chronology of all sites including cluster 1 corresponds to the master chronology across all sites.

The four parameters Gleichläufigkeit, NET, Rbar, and EPS were calculated on the basis of the undetrended chronologies as measures of homogeneity for each site, tree-species and cluster chronology. The Gleichläufigkeit is a sign test of the synchronous year-to-year changes of a single series [3] whereas NET is a combination of the Gleichläufigkeit and the coefficient of variation and additionally supplies information about the growth level of the series. Small NET values indicate high signal strength of the mean tree-ring chronology with a maximum of 0 [33]. A threshold of 0.7 was defined for Gleichläufigkeit and a threshold of 0.8 was used for NET [33]. The interseries correlation Rbar and the expressed population signal (EPS) are based on the correlation coefficient between the single series and were calculated over 30 years, lagged by 15 years. EPS is a function of Rbar and the series replication. EPS values increase with sample size and interseries correlation [34] and should remain above the commonly applied threshold of 0.85 [35]. Site chronologies were considered to be homogeneous when at least two parameter values satisfied the defined thresholds.

2.2. Climate Data. To investigate climate/growth relationship, gridded ($0.5° \times 0.5°$) datasets of monthly temperature and precipitation series [36] were used. The climate data from all grid points covering the research area were averaged and prepared as anomalies with respect to the 1961–1990 reference period. Climate data were classified using thresholds of $\pm 1z$, $\pm 1.5z$, and $\pm 2z$ to divide them into "hot," "very hot," and "extremely hot" as well as "cold," "very cold," and "extreme cold," respectively.

2.3. Methods. The detection of stands with similar growth behaviour in single years within the multi-tree-species network was carried out over the 1941–2000 period and combined two different techniques, a pointer year and a cluster analysis.

To emphasize annual growth reactions in answer to extreme environmental conditions, comparable values for extreme growth years were calculated after Cropper [37]. Raw tree-ring series were standardized by using a 13-year moving average and calculating ratios. Afterwards a z-transformation was applied to normalise these values resulting in the Cropper series with a mean of 0 (μ) and a standard deviation of 1 (σ).

Cluster analysis was performed to detect sites showing similar increment patterns due to climatic influences,

modified by the site's ecologic conditions [4, 17, 38, 39]. Site-specific Cropper series were grouped, stepwise, using hierarchical cluster analysis. To induce a clear separation of the resulting clusters, Ward's method was applied, using the squared Euclidian distance as a measure of similarity [17, 40]. The intergroup variance was then maximised while the intragroup variance was minimised [39]. An indicator for determining the number of clusters was a jump in the squared Euclidian distance between two steps of the cluster analysis, bigger than all those preceding [41].

For each cluster, the so-called cluster plots were calculated by averaging all corresponding site-specific Cropper series. Years with values above $+1\sigma$ and below -1σ were defined as pointer years of which these years, reflected by at least 80% of the sites of a cluster, were defined as "characteristic" for the cluster. Characteristic pointer years were classified with respect to standard deviation (σ) units. Anomalies larger than $\pm 1\sigma$ were defined as "weak," $\pm 1.28\sigma$ as "strong," and $\pm 1.645\sigma$ as "extreme" pointer values. Attributes such as tree species or site ecological aspects appearing only by one of the clusters which were combined in the following grouping step were defined as differentiating criteria. For each grouping step, characteristic pointer years for corresponding clusters as well as differentiating criteria have been detected. For each cluster and for each tree species, a specific variance value of growth anomalies was calculated, by averaging the Cropper value variances per year over the 1941–2000 period (VaC).

Analysis of the climate/growth relationship of each group was carried out by interpreting cluster-characteristic growth anomalies with the corresponding climatic conditions. To maximise the quality of the interpretation of the cluster-specific climate/growth relationship, only these climatic influences could be used with high probability, which could be detected in several cluster-specific pointer years [21]. Growth reactions due to the same climatic impact occurring in one cluster in the same year as the climatic event and in another cluster one year later were defined as shifted reactions.

3. Results

3.1. Homogeneity. The parameters Gleichläufigkeit, NET, Rbar, and EPS confirm the homogeneity of all analysed site chronologies (Table 1). Chronology statistics for cluster 1 refer to the overall homogeneous growth behaviour within the research area, reflected by values of NET, Gleichläufigkeit, Rbar, and EPS, which suffice the recommended thresholds (Table 2; for detailed information about the clusters, see Figure 2 and subsection Structure of the clusters). Mean parameter values for the cluster-specific chronologies are superior to these for the species-specific chronologies throughout. Only the mean EPS value is slightly lower for the cluster-specific chronologies, since EPS values are calculated including series replication. Regarding the chronology parameters as well as the VaC values for a single tree species and clusters, a better differentiation is effected at the cluster level.

Whereas coniferous trees show the highest NET and nearly the highest VaC values, the eastern coniferous cluster

TABLE 1: Geographical position and attributes of the investigated stands.

Site	Longitude	Latitude	Species	Elev.	Asp.	Grad.	AWC	n	AGR	s	GLK	NET	Rbar	EPS	Clu
1	7.0472	50.6706	QURO	185	15	2	4	12	181	70	0.78	0.62	0.53	0.92	8
2	7.0472	50.6700	QURO	185	15	2	4	12	198	60	0.79	0.51	0.52	0.92	8
3	7.0372	50.6836	QURO	165	20	2	4	14	190	63	0.78	0.55	0.57	0.95	8
4	6.4194	50.9272	QURO	105	0	0	4	16	222	73	0.77	0.58	0.50	0.91	8
5	6.4239	50.9169	QURO	103	0	0	3	16	171	61	0.79	0.58	0.57	0.95	8
6	6.8031	51.0447	QUPE	45	180	2	3	14	235	97	0.77	0.66	0.44	0.91	8
7	6.7975	51.3147	QURO	40	0	0	3	15	236	83	0.78	0.59	0.26	0.84	8
8	7.8864	51.9069	QURO	60	0	0	4	17	162	55	0.78	0.57	0.57	0.96	8
9	8.9681	51.1708	QUPE	290	135	30	1	16	149	50	0.79	0.55	0.45	0.93	9
10	7.2492	50.6694	QUPE	370	320	40	2	18	147	69	0.73	0.74	0.36	0.91	9
11	7.2475	50.6703	QUPE	375	180	45	n/a	16	152	59	0.74	0.64	0.47	0.93	9
12	7.2217	50.6836	QUPE	280	150	20	n/a	7	178	77	0.81	0.59	0.47	0.81	9
13	6.3250	50.6653	QUPE	320	340	15	3	10	143	47	0.70	0.63	0.44	0.87	9
14	6.5656	50.4400	QUPE	560	90	5	3	13	149	55	0.75	0.63	0.31	0.84	9
15	6.2794	50.6806	QUPE	460	0	0	1	13	149	59	0.76	0.65	0.54	0.93	9
16	6.3992	50.6244	QUPE	400	300	40	3	14	151	45	0.81	0.48	0.50	0.93	9
17	6.3603	50.5708	QUPE	500	150	15	2	13	128	36	0.78	0.49	0.49	0.92	9
18	8.1175	50.7272	QUPE	450	19	40	2	16	116	49	0.78	0.65	0.50	0.90	9
19	8.2269	50.8661	QUPE	440	180	25	1	14	118	39	0.81	0.53	0.65	0.96	9
20	6.8439	50.7917	QURO	155	240	5	2	12	300	95	0.80	0.52	0.61	0.95	9
21	7.7169	50.9825	QUPE	385	270	15	3	12	190	76	0.78	0.63	0.48	0.92	9
22	7.1111	51.2297	QUPE	260	230	30	1	14	167	73	0.80	0.66	0.61	0.94	9
23	6.0631	51.7503	QUPE	50	180	3	3	11	108	31	0.78	0.50	0.51	0.93	9
24	8.0228	51.1028	QUPE	470	130	35	1	16	133	49	0.79	0.58	0.56	0.93	9
25	8.6736	51.5831	QURO	295	350	2	3	11	188	67	0.76	0.61	0.21	0.67	9
26	7.5303	50.7933	QURO	175	290	15	n/a	8	207	66	0.75	0.58	0.57	0.90	9
27	8.9619	51.1694	FASY	340	360	45	n/a	13	173	92	0.80	0.75	0.55	0.90	10
28	8.9669	51.1703	FASY	310	150	30	n/a	15	178	70	0.78	0.64	0.51	0.94	10
29	9.0117	51.1936	FASY	280	180	35	n/a	13	140	68	0.80	0.70	0.58	0.92	10
30	8.9583	51.1672	FASY	420	345	3	n/a	16	152	47	0.84	0.48	0.66	0.97	10
31	7.2361	50.6628	FASY	230	270	10	n/a	16	297	120	0.73	0.69	0.47	0.91	11
32	7.2489	50.6689	FASY	360	320	40	n/a	14	232	82	0.75	0.63	0.51	0.92	11
33	7.2469	50.6700	FASY	340	160	45	n/a	17	160	65	0.80	0.63	0.55	0.94	11
34	6.3250	50.6653	FASY	320	340	15	3	16	207	85	0.76	0.66	0.44	0.90	11
35	6.4569	50.6086	FASY	530	315	15	1	16	99	34	0.77	0.58	0.50	0.94	11
36	6.4900	50.6061	FASY	470	60	10	1	15	209	60	0.83	0.47	0.63	0.96	11
37	6.2794	50.6806	FASY	440	45	1	1	13	230	84	0.78	0.60	0.49	0.91	11
38	6.3597	50.5711	FASY	490	150	15	2	14	149	47	0.82	0.51	0.57	0.94	11
39	6.3611	50.5717	FASY	480	150	10	2	15	149	51	0.79	0.56	0.57	0.95	11
40	8.1186	50.7289	FASY	500	140	30	n/a	16	182	58	0.80	0.54	0.55	0.95	11
41	8.2253	50.8656	FASY	440	180	25	1	18	188	61	0.79	0.56	0.53	0.94	11
42	6.8439	50.7917	FASY	155	240	5	2	10	420	141	0.82	0.54	0.52	0.91	11
43	7.1111	51.2300	FASY	260	330	30	1	12	195	65	0.79	0.56	0.48	0.90	11
44	8.0222	51.1022	FASY	455	130	40	1	15	185	61	0.77	0.58	0.54	0.94	11
45	8.1422	51.4425	FASY	350	310	8	2	19	157	56	0.77	0.59	0.49	0.94	11
46	8.1297	51.4469	FASY	360	280	3	2	19	156	56	0.74	0.63	0.42	0.92	11
47	8.1297	51.4469	FASY	350	140	3	n/a	10	151	48	0.79	0.54	0.49	0.89	11
48	6.8489	51.5633	FASY	65	30	1	1	15	172	70	0.80	0.62	0.53	0.94	11
49	9.0842	51.1564	QUPE	350	180	30	1	13	138	45	0.82	0.50	0.57	0.93	12
50	9.0842	51.1572	QUPE	365	180	20	1	6	88	40	0.81	0.63	0.60	0.90	12
51	9.0842	51.1581	QUPE	380	180	25	1	7	139	38	0.82	0.43	0.58	0.90	12

TABLE 1: Continued.

Site	Longitude	Latitude	Species	Elev.	Asp.	Grad.	AWC	n	AGR	s	GLK	NET	Rbar	EPS	Clu
52	9.0761	51.1561	QUPE	400	215	10	n/a	13	140	40	0.84	0.43	0.58	0.92	12
53	9.0769	51.1556	QUPE	390	150	20	n/a	10	165	46	0.81	0.46	0.59	0.93	12
54	6.8731	50.1236	PCAB	470	255	8	n/a	22	492	19	0.74	0.70	0.58	0.96	13
55	7.2128	50.7122	PCAB	170	70	5	n/a	17	251	104	0.74	0.71	0.41	0.91	13
56	6.4917	50.6056	PCAB	470	60	25	n/a	15	144	79	0.75	0.81	0.43	0.91	13
57	8.1133	50.7311	PCAB	520	45	5	n/a	16	210	71	0.79	0.57	0.45	0.92	13
58	8.2261	50.8661	PCAB	440	180	25	1	7	205	84	0.79	0.63	0.49	0.86	13
59	8.4408	51.1997	PCAB	710	340	10	n/a	14	242	109	0.76	0.71	0.42	0.88	13
60	9.0844	51.1569	PISY	360	180	20	n/a	13	113	41	0.80	0.56	0.50	0.92	14
61	8.9872	51.1553	PCAB	340	315	30	n/a	12	168	63	0.80	0.59	0.46	0.90	14
62	8.9897	51.1572	PISY	350	270	40	n/a	12	104	34	0.84	0.50	0.58	0.94	14
63	6.4919	50.6058	PISY	490	210	20	n/a	15	164	70	0.74	0.73	0.30	0.85	15
64	6.5783	50.5083	PISY	490	270	35	n/a	11	107	40	0.73	0.69	0.45	0.90	15
65	6.4769	50.6067	PISY	415	280	30	n/a	13	126	45	0.77	0.60	0.43	0.90	15
66	7.5697	50.7867	PISY	275	260	30	n/a	14	195	106	0.71	0.91	0.45	0.91	15
67	8.5839	51.8875	PISY	110	330	0–5	n/a	9	136	49	0.73	0.67	0.40	0.84	15

Elev.: elevation, Asp.: aspect [°], Grad.: gradient [°], AWC: available water capacity: 1 = low (60–110 mm); 2 = mean (110–170 mm); 3 = high (170–240 mm); 4 = very high (>240 mm); n/a: not applicable, n: number of trees per site, AGR: average growth rate, s: standard deviation, GLK: Gleichläufigkeit, NET: parameter of signal strength, Rbar: interseries correlation (calculated over 30 years lagged by 15 years), EPS: expressed population signal (calculated over 30 years lagged by 15 years), Clu: cluster.

TABLE 2: Chronology and Cropper's series statistics of cluster 1 (master chronology; including all sites), species-specific and cluster-specific chronologies.

	n	AGR	s	v	GLK	NET	Rbar	EPS	VaC
Cluster 1	63	1.70	0.80	0.47	0.73	0.74	0.38	0.97	0.57
QURO	10	1.71	0.44	0.25	0.84	0.41	0.55	0.90	0.40
QUPE	20	1.38	0.36	0.27	0.76	0.51	0.47	0.95	0.46
FASY	22	2.05	0.74	0.37	0.81	0.55	0.61	0.97	0.31
PCAB	7	2.19	1.29	0.57	0.81	0.76	0.55	0.90	0.49
PISY	7	0.93	0.35	0.37	0.81	0.57	0.36	0.79	0.46
Clu 8	8	1.71	0.33	0.19	0.86	0.34	0.70	0.93	0.35
Clu 9	18	1.45	0.42	0.29	0.77	0.53	0.51	0.94	0.45
Clu 10	4	2.02	0.45	0.22	0.89	0.33	0.64	0.87	0.15
Clu 11	18	2.05	0.78	0.38	0.82	0.57	0.65	0.97	0.28
Clu 12	5	1.27	0.27	0.21	0.94	0.28	0.85	0.97	0.10
Clu 13	6	2.29	1.37	0.58	0.81	0.77	0.56	0.87	0.46
Clu 14	3	1.14	0.40	0.35	0.89	0.46	0.64	0.84	0.28
Clu 15	5	0.93	0.41	0.43	0.82	0.61	0.38	0.75	0.42
x_Species	13	1.65	0.64	0.36	0.81	0.56	0.51	0.90	0.42
x_Clu	8	1.61	0.55	0.33	0.85	0.49	0.62	0.89	0.31

X_Species and x_Clu represent mean values for species-specific and cluster-specific chronologies. Values of n, AGR, s, v, GLK, NET, Rbar, and EPS are calculated on the basis of the undetrended chronologies. n: number of trees per site, AGR: average growth rate, s: standard deviation, v: coefficient of variation, GLK: Gleichläufigkeit, NET: parameter of signal strength, Rbar: inter-series correlation (calculated over 30 years lagged by 15 years), EPS: expressed population signal (calculated over 30 years lagged by 15 years). VaC: mean of the annual variance of the Cropper values per year over the 1941–2000 period calculated for the Cropper series.

14 shows much better values compared to spruce cluster 13 and Scots pine cluster 15. VaC is especially low for pedunculate oak and beech and the highest for spruce. At the cluster level, eastern oak cluster 12 shows the lowest VaC value, followed by eastern beech cluster 10, low mountain range beech cluster 11, and lowland oak cluster 8. All evaluated parameters are most favourable for cluster 12, thereby revealing its comparatively high homogeneity. However, low

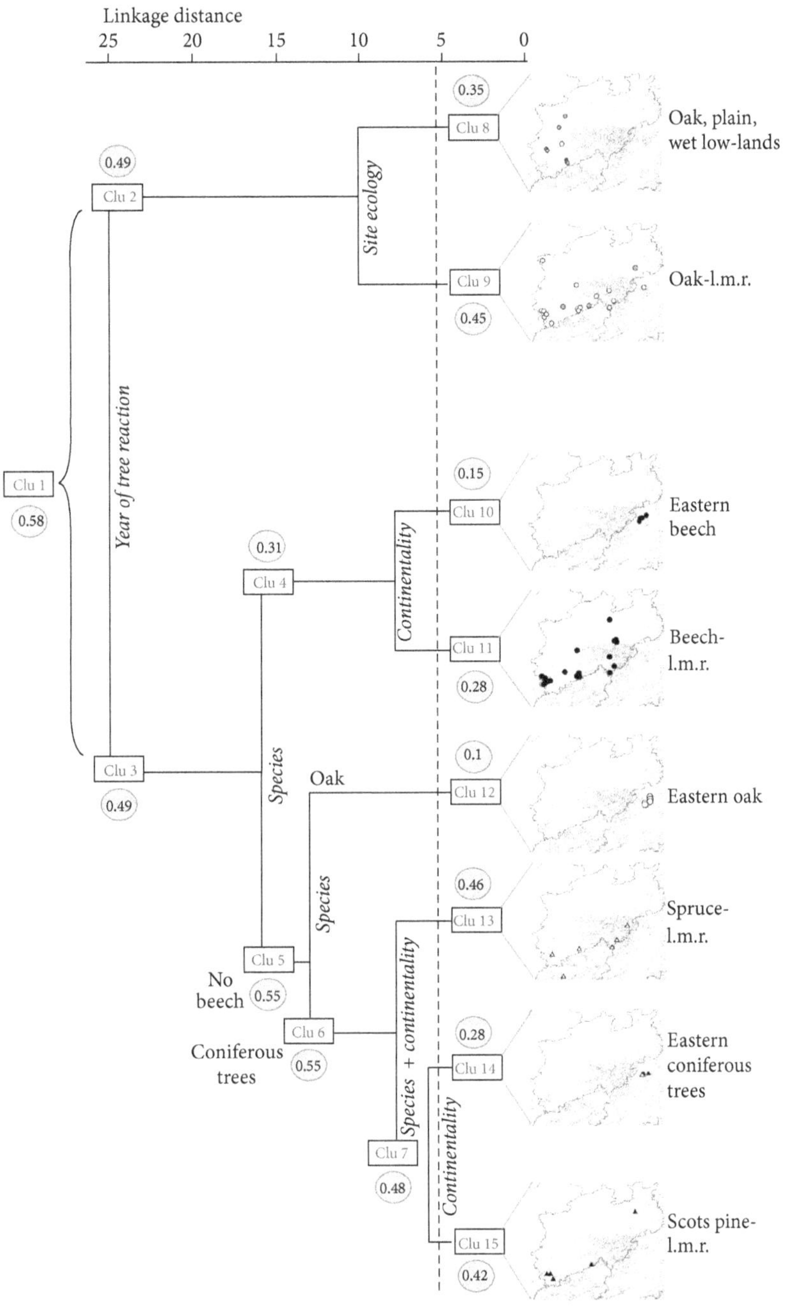

FIGURE 2: Dendrogram of the cluster analysis showing the relationship between the different clusters (Clu 1–15), ecological attributes, like tree species or topographical position, as well as differentiating criteria between the clusters (in italics); l.m.r.: low mountain ranges. Numbers in circles denote the cluster-specific mean variance over the annual variances of the Cropper values (VaC). Dashed line marks the linkage distance of five which was chosen for the selection of the final clusters (Clu 8–15). Maps show the topographical positions of the corresponding sites for each final cluster.

mountain range oak cluster 9 together with spruce cluster 13 shows the highest VaC values, Scots pine cluster 15 the lowest Rbar and EPS values.

3.2. Structure of the Clusters. Each step of the cluster analysis results in the formation of clusters, combining sites with similar pointer year patterns. In general, VaC increases inversely with the number of clusters. In the following, 15

clusters will be described and interpreted. While the bigger clusters 1–7 are defined as succeeding clusters, cluster 8–15 are selected as final clusters. At a linkage distance of five, eight final clusters are differentiated, namely, three oak clusters (8, 9, 12), two beech clusters (10, 11), one spruce (13), one Scots pine (15), and one coniferous tree (14) cluster (Figure 2). Cluster 10, 12, and 14 are situated in the eastern part of the research area. Cluster 1 includes all investigated sites and

VaC is comparably high (0.58). One grouping step before, cluster 2 and 3 can be differentiated by their reaction pattern concerning the year of tree reaction. Both of them have a VaC value of 0.49. As parts of cluster 2, oak clusters 8 and 9 distinguish themselves by site ecological aspects. The oaks from plain wet lowlands (cluster 8) have a lower VaC value than those from low mountain range regions (cluster 9). One plain wet lowland site (site 23, Table 1) is classified into cluster 9. As parts of cluster 3, clusters 4 and 5 differ in tree species. Beech cluster 4 has a smaller VaC value than the mixed species cluster 5. As parts of cluster 4, eastern beech cluster 10 and low mountain range cluster 11 can be distinguished by continentality. As parts of cluster 5, clusters 12 and 6 are differentiated by tree species. Cluster 12 consists of eastern oaks and has the lowest VaC value of all groups (0.1). One eastern oak stand (site 9, Table 1) is classified into cluster 9. The variance value of coniferous tree cluster 6 is still high (0.55). One grouping step before, low mountain range spruce cluster 13 differs from cluster 7 in both tree species and continentality. Cluster 13 shows the highest VaC value of the final clusters. As parts of cluster 7, the eastern coniferous tree cluster 14 and the low mountain range Scots pine cluster 15 also distinguish themselves by tree species and continentality. Clusters 10, 12, and 14 are located east of $8°57'$ E, with VaC values comparably lower in relation to the more westerly-situated clusters.

3.3. Climatological Interpretation of the Characteristic Pointer Years. Characteristic pointer years are classified according to their main climatic conditions in six pointer year types (PYTs) characterised by (1) a hot previous year and a cold summer of the current year, (2) coldness + dryness, (3) coldness + wetness/lack of solar radiation, (4) a hot previous year and a hot and dry summer, (5) dryness in combination with drought in summer of the current year, and (6) dryness in spring, especially during a hot May. In addition to these typical attributes, further climatic criteria can be detected as having influenced the tree's reactions in these years (Figure 3).

Years of PYT 1 (previous year hot, summer of the current year cold) are 1947/1948, 1959/1960, 1984, 1996, and 2000. In 1947/1948, the whole previous year growing season is hot/extremely hot in combination with dryness during the previous summer. Winter 1947/1948 is extraordinarily warm and wet, with temperature anomalies exceeding values of $+1z$ and precipitation anomalies of $+2.09z$ (January). Current year's April is hot and below-average summer temperatures occur in June. The previous year growing season of 1959/1960 is dry/extremely dry, but temperature anomalies are clearly lower than those in 1947/1948. Winter temperatures are slightly above average. Summer coldness is detected in July, and August and October are extremely wet. 1984 is also characterised by very/extremely high precipitation values in previous year's May, both May and September in combination with coldness. Only previous year's June and August show extreme temperature values; summer coldness begins in May. Temperatures during winter months are slightly above average. Climate conditions in 1996 are also

dry/extremely dry from previous year's autumn until January, plus large parts of the growing season coupled with cold in winter, spring, May, and September. Previous year's temperature anomalies do not appear over the entire previous years' growing season. In 2000, the entire previous year's growing season is hot/extremely hot; below-average temperatures first occur in July.

1956, 1962/1963, 1973, and 1991 belong to PYT 2 (cold + dry).

1942, 1968, and 1981 are classified as PYT 3 (cold + wet/lack of solar radiation). Since high precipitation values were normally accompanied by lower temperatures and reduced solar radiation, positive precipitation anomalies in August 1941 and July 1942, August and September 1968, and June and July 1980, as well as in March, June, August, and October 1981, reflect a lack of solar radiation and lower temperatures in these months. In 1942, additional dryness in February and April is observed.

1976 and 1983 belong to PYT 4 (previous year hot + hot and dry summer). In 1976, the whole growing season is dry/very dry while drought is most pronounced in June (temperature: $+2.38z$; precipitation: $-1.9z$). In 1983, the entire previous year's growing season is hot/very hot in combination with a dry previous year's July; April and May are extremely wet.

1946/1947, 1958/1959, and 1963/1964 are defined as PYT 5 (dryness in combination with drought in summer of the current year). In 1947 and 1959, the whole growing season is hot and dry. All of these years are additionally characterised by dry/extremely dry, in 1946/1947 and 1963/1964 also extremely cold winters. However, a further differentiation of the years belonging to PYT 5 is possible. Whereas in 1963/1964 dryness is particularly pronounced already in December, it begins in 1947 in August and 1959 in February. While in 1946/1947 and 1958/1959 nearly the whole growing season is hot/extremely hot, in 1963/1964 only May and June show comparably lower temperature anomalies. October in 1964 is very cold.

1988, 1989, and 1990 belong to PYT 6 (dryness in a hot May). 1988 additionally shows an extremely wet March and a dry April and June, 1989 a dry January, and 1990 a very dry and hot previous year's May.

3.4. Ecological and Climatological Interpretation of the Clusters. Each of the clusters is determined by its characteristic pointer years as well as specific ecological criteria (Table 3). 1976 and 1996 are characteristic for each cluster, with the exception of 1976 in cluster 2 and cluster 9 and 1996 in cluster 10. Final clusters show a more homogeneous increment pattern than all sites including cluster 1, resulting in smaller VaC values and a predominantly higher number of characteristic pointer years, respectively (Figure 2, Table 3).

Several controlling factors for climate/growth relationships can be detected, most importantly shifted reactions of the clusters, followed by tree species and site ecological or topographical aspects. Stands located east of $8°57'$ E form individual groups showing in general stronger reactions in response to dryness than those of the western part of the

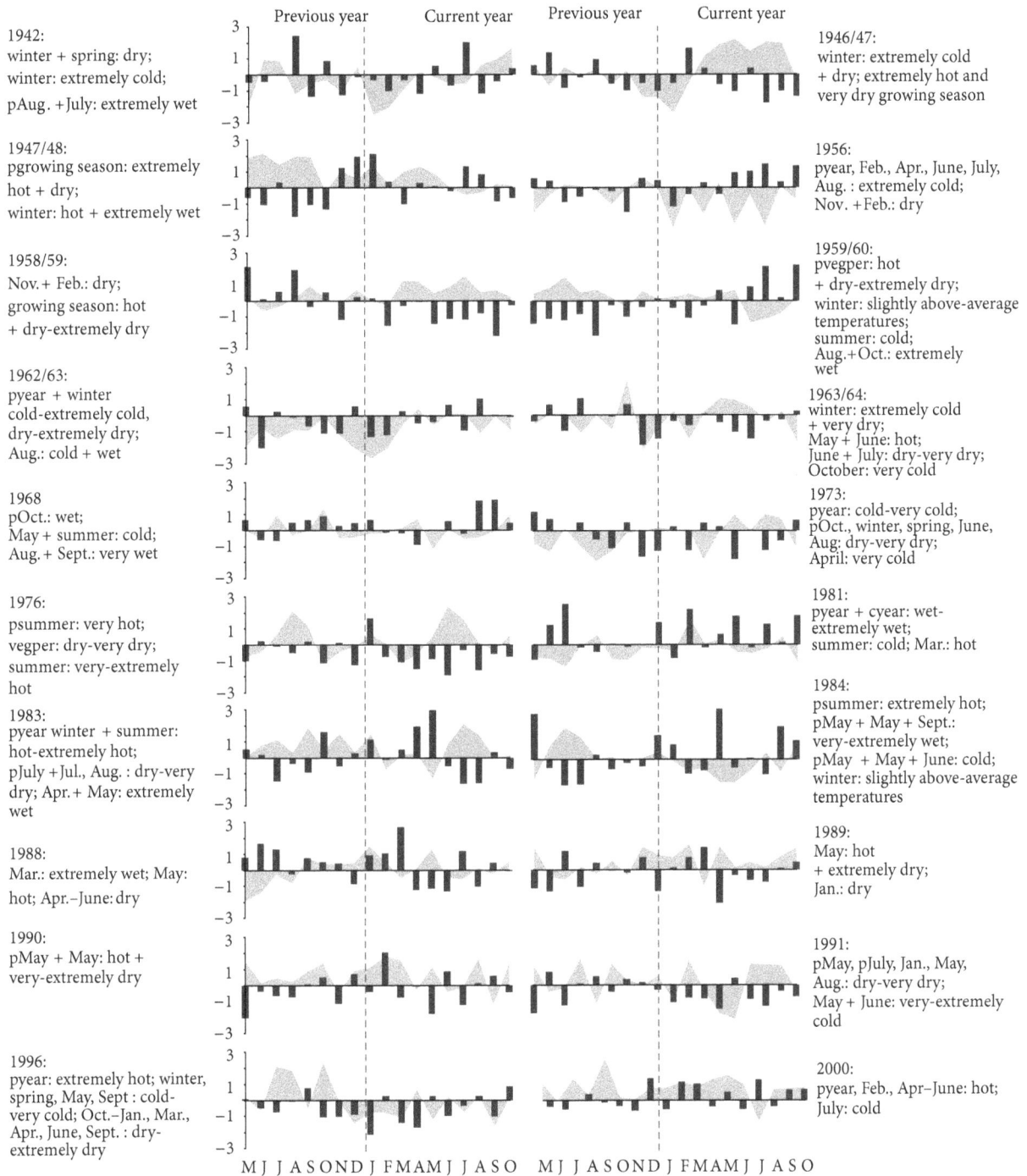

FIGURE 3: Climate anomalies of detected characteristic pointer years from previous year's May to October of the current year. Black bars mark precipitation, grey areas temperature values. Dashed lines denote the transition of previous year to the year of growth, p = previous year, c = current year.

research area. In the following, the interpretation of the clusters begins with the largest (cluster 1) and ends with the defined clusters. For cluster 1–7, a description of typical ecological attributes is not meaningful because of the high number of corresponding sites showing diverse topographical and ecological features (Table 3). Cluster 1 contains all investigated sites representing five tree species. There are only

two negative pointer years, 1976 and 1996, in which more than 80% of the sites show Cropper's values smaller than −1σ. Both of these are strong negative pointer years and appear in each cluster with a different intensity. In general, the investigated temperate forests show strong negative growth anomalies in response to the climatic conditions of PYT 1 (1996) and PYT 4 (1976).

TABLE 3: Cluster-specific attributes in terms of tree species, site ecological aspects, and pointer years.

Clu	Species	Site ecological description[1]	Char. pointer years
Clu 1	FASY, QUPE, QURO, PISY, PCAB		*1976, 1996*
Clu 2	QUPE, QURO		*1942, 1959, 1996* 1947, 1981
Clu 3	FASY, QUPE, QURO, PISY, PCAB		**1976** *1948, 1996* 1960
Clu 4	FASY		**1948, 2000** *1976, 1996*
Clu 5	QUPE, QURO, PISY, PCAB		**1976** *1996*
Clu 6	PISY, PCAB		**1976** *1996*
Clu 7	PISY, PCAB		**1996** *1976* 1973
Clu 8	QUPE, QURO	Plain lowlands, wet	**1947** *1959, 1963, 1968, 1976, 1981, 1996*
Clu 9	QUPE, QURO	Low mountain ranges, 155–560 m a.s.l., 2 lowland sites with bigger gradient, 1 plain wet lowland site	**1996** *1942*
Clu 10	FASY	East of 8°57′, 280–420 m a.s.l., dry	**1948, 1983** *1964, 1976, 2000*
Clu 11	FASY	Low mountain ranges, 155–530 m a.s.l.	**1948, 2000** *1960, 1976, 1990, 1996*
Clu 12	QUPE	East of 8°57′, slope sites	**1976** *1948, 1960, 1973, 1996* 1964, 1988
Clu 13	PCAB	Low mountain ranges, 170–710 m a.s.l.	**1976** *1948, 1984, 1996*
Clu 14	PISY, PCAB	Coniferous trees, east of 8°57′, 340–360 m a.s.l., slope sites	**1964, 1976** *1989, 1996* 1973
Clu 15	PISY	Low mountain ranges, 110–490 m a.s.l., western expositions	**1996** *1956, 1976, 1991* 1960, 1973

[1] Descriptions of the characteristic site ecological aspects are only reasonable for final clusters. Extreme pointer years are designated by bold font, strong pointer years by italic font, and weak pointer years by regular font numbers.

The differentiation of clusters 2 and 3, the parts of cluster 1, can only be explained by their shifted reactions concerning 1947/1948 and 1959/1960, neither by tree species, nor by site ecological or topographical aspects. Beside 1996 in cluster 2 and 3 and 1976 in cluster 3, there are no negative pointer years that appear in more than 80% of sites.

Because of obvious differences in interannual increment pattern in terms of shifted growth reactions, further pointer years are interpreted as characteristic. Cluster 2 consists of 26 oak stands from the lowlands and the low mountain ranges and one oak site from the eastern part of the research area. There are sessile oak as well as pedunculate oak sites. Five years are classified as characteristic for this cluster: 1942 (70% of the sites react), 1947 (65% of the sites react), 1959 (78% of the sites react), 1981 (65% of the sites react), and 1996. Negative growth reactions occur in cluster 2 when climate conditions in the winter are extremely cold and dry followed by a dry spring, such as that in 1942 and 1996. Lower temperatures and high precipitation values, along with a lack of solar radiation, in August 1941, July 1942, July 1980, and June and October 1981 also reduce the tree growth. Extremely cold and very dry winter months provoke immediate growth reductions in response to a hot/extremely hot and dry/extremely dry growing season (PYT 5; 1947, 1959).

As well as five sessile oak sites, cluster 3 contains all the investigated stands of beech, Scots pine, and spruce. This group is characterised by 1948 (78% of the sites react), 1960 (68% of the sites react), 1976, and 1996. Compared to cluster 2 there is a shifted reaction in 1948 and 1960 to the hot and dry summer/growing season in 1947 and 1959. Thus, climatic conditions of PYT 1 (1947/1948, 1959/1960, 1996) and PYT 4 (1976) are growth reducing.

Whereas cluster 2 is affected by cold and dryness during the winter months followed by a hot and dry summer, as well as lower temperatures and a lack of solar radiation in previous year autumn and current year summer, growth of cluster 3 is less influenced by the climate conditions in winter. However, it is more susceptible to summer heat and dryness in the year prior to ring formation.

The mixed oak clusters 8 and 9, parts of cluster 2, are differentiated by site ecological and topographical aspects. Cluster 8 contains plain oak stands in the lowlands characterised by soil wetness. Cluster 9 includes oak stands of the low mountain ranges as well as two pedunculate oak stands from the lowlands on more inclined sites, one plain wet lowland oak stand (site 23), and one eastern oak stand (site 9). 1996 is characteristic for both of the clusters but for cluster 9 more important (extreme negative pointer year). Cluster 9 is characterised only by two negative pointer years, 1942 and 1996, along with a comparatively high VaC value (cluster 9: 0.45; cluster 8: 0.35). This comparably lower homogeneity reflects the varied ecological spectrum of lowland and low mountain range sites belonging to cluster 9. Beside winter dryness and cold during 1996 and 1942, lower temperatures and high precipitation values, along with a lack of solar radiation, in August 1941 and July 1942, as well as the dry April are crucial for cluster 9.

For cluster 8 seven pointer years are classified as characteristic (Table 3). Climatic conditions of PYT 1 (1996), PYT 4 (1976), and PYT 5 (1947, 1959) lead to strong, even extreme (1947) growth reductions. While cold, particularly in the previous year (1963, 1981) and during winter (1963), that influences tree growth in a negative manner, high precipitation values, along with a lack of solar radiation, in previous year's summer and March (1981), and summer and autumn months (1968, 1981) cause stress.

Both of the groups are strongly negatively influenced by coldness as well as by a lack of solar radiation in previous and current year's summer. Whereas climatic conditions in March are important for cluster 8, April moisture controls the growth of cluster 9. Hot and dry conditions during the growing season after a hot previous year's summer restrict the growth of cluster 9 less.

The differentiation of clusters 4 and 5, as parts of cluster 3, is apparently conditioned by tree species. While cluster 4 includes all beech sites, cluster 5 contains five sessile oak sites situated in the east of the research area as well as all conifer sites.

Cluster 4 and 5 are both characterised by 1976 and 1996; additionally, cluster 4 is extremely negatively affected by 1948 and 2000. More than 80% of the beech sites react with a negative growth anomaly smaller than -1σ when previous climate conditions correspond to PYT 1 (1947/1948, 1996, 2000) and 4 (1976). In 1960 only 76% of the beech sites react with a negative pointer year. While summer in 1960 is comparably cold, similar to 2000, the previous year's summer is less hot than those during the years 1947/1948, 1976, 1996, and 2000. Hence, the most important factor for growth reduction in beech is a hot previous year's summer. The influence of previous year's heat and dryness seems to be crucial for beech growth.

Cluster 5 clearly shows less homogenous growth behaviour than beech cluster 4, which is reflected by a comparatively high VaC value (0.55 versus 0.31 for cluster 4). Additionally, there are no further characteristic pointer years other than 1976 and 1996 that are characteristic for nearly all clusters.

Beech clusters 10 and 11, as parts of cluster 4, are differentiated by the degree of continentality. Cluster 10 contains all beech stands located east of $8°57'$E. These beeches show growth reductions in years of PYT 4 (1976, 1983). In 1983, the entire previous year's growing season is hot/very hot in combination with a dry previous July followed by a hot/extremely hot and dry/very dry summer. In spite of high precipitation values in April and May, cluster 10 reacts with an extreme growth reduction (-1.66σ). Water supply at the beginning of the growing season is not high enough to compensate for the following summer dryness.

Furthermore, climatic conditions of PYT 1 (1947/1948, 2000) lead to growth reductions of cluster 10. In 1947/1948, the very dry previous year's summer strengthens growth reductions. No significant negative reactions can be detected in 1959/1960, 1984, and 1996 belonging also to PYT 1. Whereas in 1947/1948 and 2000 the entire previous year's growing season is hot/extremely hot, in 1984 and 1996 only single months are extremely hot. Furthermore, high precipitation values in the previous May and May 1984 appear to lessen the influence of dryness. In 1959/1960, temperature anomalies during the previous year's growing season are clearly of lower magnitude than those in 1947/1948. Additionally, high precipitation values in August and October may have positively influenced the growth of cluster 10, which is predominantly limited by dryness. Negative growth anomalies in 1963/1964 also reveal the sensitivity of cluster

10 to dryness. Additionally, the very cold October in 1964 could shorten the growing season. Cluster 11 contains all other beech stands. These respond strongly to PYT 4. High precipitation values in March and April 1983 lead to the absence of negative growth reactions of cluster 11, revealing the lower sensitivity of cluster 11 to dryness compared to cluster 10. The strong growth reduction in 1990 refers to a vulnerability particularly to climatic conditions of PYT 6. The absence of negative reactions in 1988 and 1989, belonging also to PYT 6, reflects the importance of climate conditions in the previous May for cluster 11.

Climate conditions of PYT 1 are, however, crucial for growth reductions of cluster 11 (1948, 1960, 1996, 2000). Hence, summer coldness is of high importance for growth reductions of cluster 11, whereas growth of cluster 10 is more heavily influenced by permanent dryness, especially in previous and current years' summers and by coldness at the end of the growing season. Previous year's heat is crucial for the growth performance of both beech clusters; the growth of cluster 10 is also controlled by the duration and intensity of the hot period.

The differentiation of the parts of cluster 5, cluster 6 (coniferous trees), and cluster 12 (five sessile oak sites) is also determined by tree species. Cluster 12 contains all oak stands situated east of $8°57'$E, with the exception of site 9. Climate conditions are more continental than those in other parts of the research area. Whereas cluster 6, according to cluster 5, only shows the characteristic pointer years 1976 and 1996 in combination with an even high VaC value (0.55), oak cluster 12 is characterised further by 1948, 1960, 1964, 1973, and 1988 (VaC 0.1). Negative growth reactions of cluster 12 are caused by climatic conditions of PYT 1 (1947/1948, 1959/1960, 1996) in combination with a very dry beginning of the growing season of the current year (1996) and PYT 4 (1976). Probably, high precipitation values in August and October 1960, along with a lack of solar radiation, also lower the growth performance of eastern oak. Dryness during the growing season leads to negative growth anomalies of cluster 12 (1964, 1973, 1988). In 1973, dryness already begins in previous year's October with a maximum in July ($-1.81z$) affecting a stronger growth reduction than detected for 1964 and 1988. The extremely wet March in 1988 ($+2.65z$) primarily lowers the dryness in the following growing season, but a simultaneous negative impact due to too little solar radiation cannot be excluded here. In 1976 the growth reduction is the strongest (-1.72σ). Thus, dryness especially in the summer months of the current year and in the late summer/autumn of the year prior to ring formation is the most important growth limiting factor for oak sites in the region eastwards of $8°57'$E. Coldness, a lack of solar radiation, or dryness at the beginning of the vegetation period can enhance growth reductions.

Clusters 13 and 7, composed of clusters 14 and 15, are parts of cluster 6. They can be differentiated by tree species in combination with the degree of continentality. Cluster 13 consists of all spruce stands located west of $8°57'$E and shows 1976 (PYT 4), 1948, 1984, and 1996 (PYT 1) as characteristic pointer years. In 1983 (PYT 4), high precipitation values in

April and May lead to an absence of negative growth reactions of cluster 13. Similar to cluster 11, climatic conditions of PYT 1 reduce growth of cluster 13 (1947/1948, 1984, 1996). The extremely high precipitation values in previous year's May and the current May as well as in September 1984, along with a lack of solar radiation and below-average temperatures, have growth-reducing effects. In 1960 and 2000, also characterised by summer coldness, cluster 13 does not react with negative growth anomalies. Whereas coldness begins in 1984 and 1996 in May, and in 1948 in June, below-average temperatures first occur in 1960 and 2000 during July.

Cluster 7 mainly includes stands of Scots pine. VaC is comparably high (0.48), reflecting low homogeneous growth reactions of the corresponding sites also shown in the low number of characteristic pointer years, 1973 (PYT 2), 1976 (PYT 4), and 1996 (PYT 1). All of these years show below-average precipitation values over large parts of the growing season, leading to the assumption that dryness is the most important growth-limiting factor for cluster 7.

Cluster 14 and 15 are differentiated by topography in terms of the degree of continentality. Cluster 14 consists of all spruce and Scots pine stands located east of $8°57'$E. Cluster 15 contains all other Scots pine stands. For both clusters, VaC is smaller than that for cluster 7 (cluster 14: 0.28; cluster 15: 0.42; cluster 7: 0.48).

Cluster 14 shows negative growth anomalies in response to the climatic conditions of PYT 1 (1996) and PYT 4 (1976). Additionally, in 1996, winter and April are very/extremely dry. Because of wet conditions in April and May 1983, no negative growth anomaly occurs in this year. Beside 1996, no negative growth anomalies occur in any other year of PYT 1. All of these years show above-average winter temperatures, especially 1947/1948. Dry conditions in winter and summer reduce growth in the years 1964 and 1973; in 1989 this is further influenced by an extremely dry May. In 1964, the growing season is probably shortened by very low temperatures in October. Additionally, extremely cold conditions in winter 1963 lead to an extreme growth reduction in eastern coniferous growth (-1.77σ). Only clusters in the eastern part of the research area (cluster 10, 12, 14) are affected by the climatic conditions of 1964. Among these clusters, coniferous tree cluster 14 shows the strongest reaction, followed by clusters 10 (-1.55σ) and 12 (-1.18σ). Growth of eastern coniferous tree cluster 14 is mainly limited by dryness in summer and May. Winter dryness causes a lack of moistness at the beginning of the vegetation period, whereas above-average winter temperatures enhance growth performance.

The climatic conditions of PYT 1 (1959/1960, 1996) and PYT 4 (1976) limit the growth of Scots pine cluster 15. Reasons for the absence of negative growth anomalies in 1983 (PYT 4) and the other years belonging to PYT 1 are similar to those described previously for cluster 14. The growth anomaly in 1959/1960 detected in cluster 15 is possibly caused by a lack of solar radiation in combination with below-average temperatures in August and October. These high precipitation values in turn have a positive influence on the growth of cluster 14, limited mainly by dryness. Similarly, eastern beech cluster 10 does not show a negative growth anomaly in 1960. Climate conditions of PYT 4 lower growth of cluster 15. An extremely cold February and a cold/extremely cold growing season (1956) also reduce growth, as do dryness in winter (1973) and the growing season (1973, 1991) in combination with a very/extremely cold spring or summer.

Tree growth of cluster 15 is less strongly limited by dryness compared to cluster 14 and more reduced by coldness during winter and growing season. Winter temperatures are crucial for the growth performance of clusters 14 and 15.

4. Discussion

Grouping of the Cropper series leads to the detection of eight subsets of trees with specific growth anomaly patterns within an already fairly homogeneous growth behaviour on a regional scale. Nearly all of these subsets can be explained by the detected driving factors, which control the climate/growth relationships of the different groups. This is possible for a tree-ring network consisting of five tree species in the transition zone from the lowlands to the low mountain ranges in temperate forest regions of Germany. The group assignment of only one site cannot be explained by ecological and climatological features.

Results show predominantly more homogeneous growth patterns for the clusters than for the species-specific groups, confirming the progress of this particular growth-specific clustering process in contrast to widely used species-specific approaches. While in numerous studies the species-specific vulnerability to changing climatic conditions has been analysed [16, 42], results of this investigation lead to the assumption that it is not primarily the tree species but a shifted tree reaction that is the most important forcing factor for the climate/growth relationship. These findings can only be detected using this particular growth-specific clustering approach.

Whereas Neuwirth et al. [43] can detect immediate reactions of oak stands and lagged reactions of beech stands from the northwest German low mountain ranges, this study shows that shifted growth reactions are mainly not controlled by tree species as evidenced by the immediate as well as lagged reactions of oak groups. Z'Graggen [21] suggests an enhancing effect of previous year's autumn and winter dryness on spring dryness. We hypothesise that clusters which are strongly influenced by winter/spring dryness are already weakened when dryness and heat begin in summer, thereby leading to an immediate growth reduction in the same year. However, these groups, which are predominantly susceptible to summer heat and dryness, react in the year after the climatological event. Lebourgeois and Ulrich [44] detect an earlier onset of the growing season for oak stands between 10 and 15 days in the warmer, more oceanic and southerly regions of France, when compared to the beginning of the vegetation period of the more continentally located oak stands. This corroborates our assumption that the more continentally situated oak stands are less influenced by winter/spring dryness. Furthermore, the annual timing of the growing season is primarily controlled by temperature [45], and an increase of altitude delays budburst [44]. A later

beginning of the growing season due to altitude would explain the impact of climatic conditions in March 1981 for plain wet lowland cluster 8 and the climatic influence in April 1942 on higher altitude cluster 9. For cluster 8, this impact occurs in March, whereas for cluster 9 it occurs later, during April. This may also explain the fact that 1947 and 1959 are characteristic for lowland cluster 8 only, but not for cluster 9 which is, on average, located at higher elevations. In cluster 9 stands react during 1947 or 1948 and predominantly, but less than 80%, in 1959. Cluster 9 stands that react during 1947 and 1959 are located predominantly at lower altitudes than those that react during 1948 or indeed do not react during 1959. However, there are stands that show growth reduction in 1947 as well as in 1948 but are situated at lower altitudes than those that react in only 1947 or 1948. Furthermore, a species-specific differentiation such as that detected by Friedrichs et al. [4], with pedunculate oak growth controlled by winter and spring climatic conditions and sessile oak growth influenced mainly by June precipitation, can not be confirmed here. This suggests perhaps a more complex system of controlling factors of cluster 9, as indicated by the high VaC values and the low number of characteristic pointer years. For beech, Lebourgeois and Ulrich [44] detail a later beginning of the growing season. However, due to the higher elevation (~ 400/1000 m a.s.l.) of the French tree-ring network beech stands than their oak stands (200 m a.s.l.), we cannot transfer their findings to our network; in this study beech and oak stands both show a similar altitude distribution. Due to a relatively homogeneous topographical distribution of all our investigated stands, we assume that a complex interaction of forcing factors control the stronger influence of early spring climatic conditions on tree growth of cluster 2, when compared to cluster 3.

Climatic conditions in the eastern sector of the research area can be described as more continental than these in the western sector [26]. Comparably lower VaC values confirm a higher homogeneity of the easterly located clusters and reflect a stronger climate control of the corresponding stands. Noticeably, chronology statistics are most favourable for the eastern oak cluster, which reveal its strong reaction to climate and complement the findings of Bonn [15], who identifies beech as the most sensitive tree species. Beside a delayed growing season, as discussed above, the amount of water supply is smaller for easterly situated clusters. Trees growing on dryer sites are generally more sensitive to drought [2, 46], thus explaining the comparably stronger reactions of the eastern stands to dryness. The occurrence of negative growth anomalies in 1964 is typical for stands exposed to more continental conditions. While dryness is, with the exception of the winter months, not strong or extreme, its extended persistence leads to growth reductions. Neuwirth et al. [43] detect the contribution of low autumn temperatures, such as those seen in 1964, to negative growth reactions of beech in the same year. The duration of the vegetation period is generally limited by daily mean temperatures above 5°C [8]. Below-average autumn temperatures at the end of the growing season can, especially in more continental regions, lead to an earlier end of the physiological productivity and smaller ring widths.

The degree of continentality is an important driving factor for the climate/growth relationship at a regional scale and is empirically defined in this study with a threshold delineated here as east of $8°57'$. This result is similar to findings of Babst et al. [47], regarding the existence of a boundary that separates temperature- and moisture-sensitive areas on a continental scale. The eastern oak stand which does not belong to cluster 12 could be influenced by more oceanic climatic conditions because of its lower situation close to the Edersee compared with stands belonging to the eastern oak cluster (Table 1).

Growth of the oak clusters and of the spruce cluster is influenced by extremely high precipitation values, predominantly at the beginning or at the end of the vegetation period. Whereas sufficient precipitation normally affects growth positively, extremely high precipitation values have a negative influence on tree growth and can be interpreted as a resonance effect, going along with lower temperatures [38, 48] and a lack of solar radiation. A strong positive relationship between Norway spruce growth and intercepted radiation is demonstrated by several authors including Bergh et al. [49]. A clear designation of the impact of high precipitation anomalies is, however, not always possible. Clusters that are very sensitive to dryness at the beginning of the growing season may also be influenced positively by extremely high precipitation values, thereby lessening the influence of any subsequent dry months. This may result in missed negative reactions, such as that during 1984 where cluster 10 does not show a growth reduction in spite of an extremely hot and very dry previous year's summer due to extremely high precipitation during previous year's May, this compensating for the effects of the previous year's dryness.

In general, beech growth is negatively affected by previous year's heat and dryness [19]. Above-average temperatures, in combination with a deficit of precipitation in June and July of the year prior to ring formation, stimulate the formation of flower buds and potentially lead to a beech mast [50]. Negative correlations are detected between fruit production and radial growth [51]. Oswald [52] suggests that tree-ring width can be reduced by 50% in good mast years. However, other studies discuss that reserve substances that are needed for a radial tree growth are not necessarily mainly used for fructification but that climate conditions enhancing reproductive growth have a coincident limiting influence on the radial tree growth [53, 54]. In succession to Neuwirth et al. [43] and Fischer and Neuwirth [26], it was possible to differentiate beech sensitivity to climate by detecting two beech groups that both react with growth reductions to previous year's heat, but, conversely, both show cluster-specific climatic sensitivities. Whereas growth of the low mountain range beech stands is more limited by coldness during summer, growth reductions of the easterly located beech stands are more controlled by permanent dryness, particularly dryness in the previous and current year's summer and coldness at the end of the growing season.

Winter temperatures can have a positive influence on the growth performance of coniferous trees, which may be explained by supplementary rates of photosynthesis, because needles are photosynthetically active when temperatures are above freezing even in the winter months [55]. The impact of growth-reducing climatic factors can be weakened or compensated due to above-average temperatures during winter.

For spruce cluster 13, the growth-enhancing impacts of the above-average winter temperatures, such as those seen in other coniferous tree clusters, cannot be identified here. According to Neuwirth [38], below-average precipitation in March 1948 likely results in a lack of any positive impact from warm and wet winter conditions during 1947/1948. His results suggest that, after mild winters, the precipitation volumes around the onset of the growing season are important for the growth performance of spruces below 750 m a.s.l. While solar radiation has a strong positive influence on spruce growth, coldness in May and summer can reduce growth, as can drought, particularly in June [56].

Western Scots pine cluster 15 is limited by dryness, albeit less strongly than the more easterly situated coniferous tree cluster 14; however, it is also limited by cold during the growing season. Winter coldness precludes additional photosynthetic activity in both clusters 14 and 15. Clusters characterised by a more continental location could be found for all of the broadleaved tree species. Only coniferous tree species form one easterly situated cluster. This is likely caused by the comparably low number of coniferous stands. Further investigation, including more Scots pine and spruce stands, is now necessary to gather more results supported by statistical evidence.

5. Conclusion

We investigate an extended tree-ring width network of five tree species in temperate forest regions of West Germany. We adopt a particular approach whereby we group sites with similar growth anomaly patterns. Here we show for the first time that growth clusters with homogeneous interannual increment patterns regarding growth anomalies in single years can be detected on a regional scale. Climate/growth relationships of these groups are controlled by several ecological factors, such as tree species, soil moisture, or continentality. Overall, the tree species is not the most important controlling factor. Our analyses show that, first of all, shifted reactions due to differing influences of coldness and dryness in winter and spring distinguish growth behaviour of the clusters. The tree reaction to climatic conditions in early spring is determined by a complex system of controlling factors and cannot clearly be described by just one forcing factor, such as altitude, continentality, or tree species. The length of the growing season clearly plays a major role. Increasing the duration of the vegetation period due to climate warming is likely to change the climate/growth behaviour of the trees, considerably complicating the climate reconstructions. General statements concerning tree species' reactions to climate are not necessarily valid for stands in temperate forest regions with various topographical and ecological features. Growth controlling factors that are more important than the tree species can only be detected using a particular growth-specific clustering approach. Analyses of climate sensitivity in temperate forest regions should therefore be based on growth-specific approaches rather than species-specific grouping approaches.

Acknowledgments

The authors thank Dr. M. Röös (Nationalpark Eifel), A. Frede (Nationalpark Kellerwald-Edersee), and U. Schulte (Landesbetrieb Wald und Holz NRW) for their helpful cooperation concerning the field work and the necessary research permits.

References

[1] C. Körner, "A re-assessment of high elevation treeline positions and their explanation," *Oecologia*, vol. 115, no. 4, pp. 445–459, 1998.

[2] H. C. Fritts, *Tree Rings and Climate*, Academic Press, London, UK, 1976.

[3] F. H. Schweingruber, *Tree Rings and Environment. Dendroecology*, Paul Haupt, Stuttgart, Germany, 1996.

[4] D. A. Friedrichs, U. Büntgen, J. Esper, D. C. Frank, B. Neuwirth, and J. Löffler, "Complex climate controls on 20th century oak growth in central-west Germany," *Tree Physiology*, vol. 29, pp. 39–51, 2009.

[5] H. Ellenberg, *Die Vegetation Mitteleuropas mit den Alpen in ökologischer, dynamischer und historischer Sicht*, Ulmer, Stuttgart, Germany, 5th edition, 1996.

[6] D. C. Frank, J. Esper, C. C. Raible et al., "Ensemble reconstruction constraints on the global carbon cycle sensitivity to climate," *Nature*, vol. 463, no. 7280, pp. 527–530, 2010.

[7] G. Foster and S. Rahmstorf, "Global temperature evolution 1979–2010," *Environmental Research Letters*, vol. 6, pp. 1–8, 2011.

[8] T. E. Skaugen and O. E. Tveito, "Growing-season and degree-day scenario in Norway for 2021–2050," *Climate Research*, vol. 26, pp. 221–232, 2004.

[9] M. Stefanon, F. D'Andrea, and P. Drobinski, "Heatwave classification over Europe and the Mediterranean region," *Environmental Research Letters*, vol. 7, pp. 1–9, 2012.

[10] B. Orlowsky and S. I. Seneviratne, "Global changes in extreme events: regional and seasonal dimension," *Climatic Change*, vol. 110, pp. 669–696, 2012.

[11] K. R. Briffa, T. J. Osborn, F. H. Schweingruber, P. D. Jones, S. G. Shiyatov, and E. A. Vaganov, "Tree-ring width and density data around the Northern Hemisphere: part 2, spatio-temporal variability and associated climate patterns," *The Holocene*, vol. 12, no. 6, pp. 759–789, 2002.

[12] B. Neuwirth, F. H. Schweingruber, and M. Winiger, "Spatial patterns of central European pointer years from 1901 to 1971," *Dendrochronologia*, vol. 24, no. 2-3, pp. 79–89, 2007.

[13] L. DeSoto, J. J. Camarero, J. M. Olano, and V. Rozas, "Geographically structured and temporally unstable growth responses of Juniperus thurifera to recent climate variabilityin the Iberian Peninsula," *European Journal of Forest Research*, vol. 131, pp. 905–917, 2012.

[14] C. Dittmar and W. Elling, "Jahrringbreite von Fichte und Buche in Abhängigkeit von Witterung und Höhenlagen," *Forstwissenschaftliches Centralblatt*, vol. 118, pp. 251–270, 1999.

[15] S. Bonn, "Dendroökologische Untersuchung der Konkurrenzdynamik in Buchen-Eichen-Mischbeständen und zu erwartende Modifikationen durch Klimaänderungen," *Forstwissenschaftliche Beiträge*, vol. 3, Dresden, Germany, 1998.

[16] D. A. Friedrichs, V. Trouet, U. Büntgen et al., "Species-specific climate sensitivity of tree growth in Central-West Germany," *Trees*, vol. 23, pp. 729–739, 2009.

[17] D. A. Friedrichs, B. Neuwirth, M. Winiger, and J. Löffler, "Methodologically induced differences in oak site classifications in a homogeneous tree-ring network," *Dendrochronologia*, vol. 27, no. 1, pp. 21–30, 2009.

[18] F. H. Schweingruber and P. Nogler, "Synopsis and climatological interpretation of Central European tree-ring sequences," *Botanica Helvetica*, vol. 113, no. 2, pp. 125–143, 2003.

[19] F. Babst, M. Carrer, C. Urbinati, B. Neuwirth, and D. Frank, "500 years of regional forest growth variability and links to climatic extreme events in Europe," *Environmental Research Letters*, vol. 7, no. 4, Article ID 045705, 2012.

[20] F. H. Schweingruber, D. Eckstein, F. Serre-Bachet, and O. U. Bräker, "Identification, presentation and interpretation of event years and pointer years in dendrochronology," *Dendrochronologia*, vol. 8, pp. 9–38, 1990.

[21] S. Z'Graggen, *Dendrohistometrisch-klimatologische Untersuchung an Buchen (Fagus sylvatica L.) [Ph.D. thesis]*, Universität Basel, 1992.

[22] M. Kaennel and F. H. Schweingruber, Eds., *Multilingual Glossary of Dendrochronology*, Birmensdorf, Swiss Federal Institute for Forest, Snow and Landscape Research. Paul Haupt, Stuttgart, 1995.

[23] J. W. Hurrel, Y. Kushnir, G. Ottersen, and M. Visbeck, "An overview of the North Atlantic Oscillation. Climatic significance and environmental impact," *Geophysical Monograph*, vol. 134, pp. 1–35, 2003.

[24] R. Schulz and N. Asche, "Klima, Standort, Wald. Regionales Wasserhaushaltsmodell auf Bundesebene übertragbar ?" *AFZ-DerWald*, vol. 1, pp. 20–24, 2008.

[25] B. Neuwirth, "Jahrringe als Indikator für Klima- und Umweltveränderungen in Mitteleuropa," *Passauer Kontaktstudium Geographie*, vol. 11, pp. 67–78, 2011.

[26] S. Fischer and B. Neuwirth, "Klimasensitivität der Douglasie in Eifel und Kellerwald," *Allgemeine Forst- und Jagdzeitung*, vol. 183, pp. 23–33, 2012.

[27] J. Schultz and B. Neuwirth, "A new atmospheric circulation tree-ring index (ACTI) derived from climate proxies: procedure, results and applications," *Agricultural and Forest Meteorology*, vol. 164, pp. 149–160, 2012.

[28] U. Schulte and A. Scheible, *Atlas der Naturwaldzellen in Nordrhein-Westfalen,*, B.o.s.s. Druck und Medien, Kleve, Germany, 2005.

[29] M. A. Stokes and T. L. Smiley, *An Introduction To Tree-Ring Dating*, University of Arizona Press, Tucson, Ariz, USA, 1968, reprinted 1996.

[30] J. R. Pilcher, "Sample preparation, cross-dating and measurement," in *Methods of Dendrochronology. Applications in the Environmental Sciences*, E. R. Cook and L. A. Kairiukstis, Eds., pp. 40–51, Kluwer Academic Publishers, Dordrecht, The Netherlands, 1990.

[31] F. Rinn, "TSAP-Win. Time series analysis and presentation for dendrochronology and related applications," Version 0.53 for Microsoft Windows. User Reference, Heidelberg, Germany, 2003.

[32] R. L. Holmes, "Computer-assisted quality control in tree-ring dating and measurement," *Tree-Ring Bulletin*, vol. 43, pp. 69–78, 1983.

[33] J. Esper, B. Neuwirth, and K. Treydte, "A new parameter to evaluate temporal signal strength of tree-ring chronologies," *Dendrochronologia*, vol. 19, pp. 93–102, 2001.

[34] K. R. Briffa and P. D. Jones, "Basic chronology statistics and assessment," in *Methods of Dendrochronology: Applications in the Environmental Sciences*, E. R. Cook and L. A. Kairiukstis, Eds., pp. 137–152, Kluwer Academic Publishers, Dordrecht, The Netherlands, 1990.

[35] T. M. L. Wigley, K. R. Briffa, and P. D. Jones, "On the average value of correlated time series with applications in dendroclimatology and hydrometeorology.," *Journal of Climate & Applied Meteorology*, vol. 23, no. 2, pp. 201–213, 1984.

[36] T. D. Mitchell and P. D. Jones, "An improved method of constructing a database of monthly climate observations and associated high-resolution grids," *International Journal of Climatology*, vol. 25, no. 6, pp. 693–712, 2005.

[37] J. P. Cropper, "Tree-ring skeleton plotting by computer," *Tree-Ring Bulletin*, vol. 39, pp. 47–59, 1979.

[38] B. Neuwirth, *Klima/Wachstums-Beziehungen zentraleuropäischer Bäume von 1901 bis 1971—Eine dendroklimatologische Netzwerkanalyse [Ph.D. thesis]*, Universität Bonn, 2005.

[39] M. Koprowski and A. Zielski, "Dendrochronology of Norway spruce (*Picea abies* (L.) Karst.) from two range centres in lowland Poland," *Trees*, vol. 20, no. 3, pp. 383–390, 2006.

[40] I. Leyer and K. Wesche, *Multivariate Statistik in der Ökologie*, Springer, Heidelberg, Germany, 2007.

[41] G. Bahrenberg, E. Giese, and J. Nipper, *Statistische Methoden in der Geographie 2: Multivariate Statistik*, Borntraeger, Berlin, Germany, 2nd edition, 2003.

[42] H. G. Hidalgo, J. A. Dracup, G. M. MacDonald, and J. A. King, "Comparison of tree species sensitivity to high and low extreme hydroclimatic events," *Physical Geography*, vol. 22, no. 2, pp. 115–134, 2001.

[43] B. Neuwirth, D. Friedrichs, and M. Hennen, "2003—where is the negative pointer year? A case study for the NW-German low mountain ranges," *Tree Rings in Archaeology, Climatology and Ecology*, vol. 5, pp. 113–120, 2007.

[44] F. Lebourgeois and E. Ulrich, "Forest trees phenology in the French Permanent Plot Network (Renecofor, ICP forest network)," in *Forests, Carbon Cycle and Climate Change*, D. Lousteau, Ed., pp. 158–171, Collection UpDateSciences and Technologies, 2010.

[45] A. Menzel, G. Jakobi, R. Ahas, H. Scheifinger, and N. Estrella, "Variations of the climatological growing season (1951–2000) in Germany compared with other countries," *International Journal of Climatology*, vol. 23, no. 7, pp. 793–812, 2003.

[46] E. R. Cook, M. A. Kablack, and G. C. Jacoby, "The 1986 drought in the southeastern United States: how rare an event was it?" *Journal of Geophysical Research*, vol. 93, no. 11, pp. 14–260, 1988.

[47] F. Babst, B. Poulter, V. Trouet et al., "Site and species-specific responses of forest growth to climate across the European continent," *Global Ecology and Biogeography*. In press.

[48] K. R. Briffa, F. H. Schweingruber, P. D. Jones, T. J. Osborn, S. G. Shiyatov, and E. A. Vaganov, "Reduced sensitivity of recent tree-growth to temperature at high northern latitudes," *Nature*, vol. 391, no. 6668, pp. 678–682, 1998.

[49] J. Bergh, S. Linder, and J. Bergström, "Potential production of Norway spruce in Sweden," *Forest Ecology and Management*, vol. 204, pp. 1–10, 2005.

[50] H. Wachter, "Über die Beziehung zwischen Witterung und Buchenmastjahren," *Forstarchiv*, vol. 35, pp. 69–78, 1964.

[51] C. Dittmar, W. Zech, and W. Elling, "Growth variations of Common beech (*Fagus sylvatica* L.) under different climatic and environmental conditions in Europe—A dendroecological study," *Forest Ecology and Management*, vol. 173, no. 1–3, pp. 63–78, 2003.

[52] H. Oswald, "Importance et périodicité des faînées. Influence des facteurs climatiques et sylvicoles," in *Le Hêtre*, E. Teissier du Cros, F. Le Tacon, G. Nepveu, J. Pardé, R. Perrin, and J. Timbal, Eds., pp. 207–241, Paris, France, 1981.

[53] V. Selås, G. Piovesan, J. M. Adams, and M. Bernabei, "Climatic factors controlling reproduction and growth of Norway spruce in southern Norway," *Canadian Journal of Forest Research*, vol. 32, no. 2, pp. 217–225, 2002.

[54] M. Mund, W. L. Kutsch, C. Wirth et al., "The influence of climate and fructification on the inter-annual variability of stem growth and net primary productivity in an old-growth, mixed beech forest," *Tree Physiology*, vol. 30, no. 6, pp. 689–704, 2010.

[55] H. P. Mäkinen, N. Mielikäinen, and K. Mielikäinen, "Climatic signal in annual growth variation of Norway spruce (*Picea abies*) in southern Finland," *Trees*, vol. 15, pp. 177–185, 2001.

[56] J. H. Bassman, G. E. Edwards, and R. Robberecht, "Long-term exposure to enhanced UV-B radiation is not detrimental to growth and photosynthesis in Douglas-fir," *New Phytologist*, vol. 154, no. 1, pp. 107–120, 2002.

8

Individual Growth Model for *Eucalyptus* Stands in Brazil Using Artificial Neural Network

Renato Vinícius Oliveira Castro,[1] Carlos Pedro Boechat Soares,[2] Helio Garcia Leite,[2] Agostinho Lopes de Souza,[2] Gilciano Saraiva Nogueira,[3] and Fabrina Bolzan Martins[4]

[1] Department of Forestry, Faculty of Technology, University of Brasília, Campus Darcy Ribeiro, 70904-970 Brasília, DF, Brazil
[2] Department of Forestry, Federal University of Viçosa, Campus UFV, 36570-000 Viçosa, MG, Brazil
[3] Department of Forestry, Federal University of the Valleys of Jequitinhonha and Mucuri, Campus Diamantina, 39100-000 Diamantina, MG, Brazil
[4] Natural Resources Institute, Federal University of Itajubá, Campus Itajubá, 37500-903 Itajubá, MG, Brazil

Correspondence should be addressed to Renato Vinícius Oliveira Castro; castrorvo@ymail.com

Academic Editors: J. Kaitera, P. Newton, T. L. Noland, and P. Robakowski

This work aimed to model the growth and yield of *Eucalyptus* stands located in northern Brazil, at the individual tree level, by using artificial neural networks (ANNs). Data from permanent plots were used for training the neural networks to predict tree height and diameter as well as mortality probability. Once trained, the networks were evaluated using an independent data set. The first group was composed of 33 plots (11 in each productive capacity class) and was used for artificial neural network training. In five measurements, this group totaled 8,735 cases (measurements of individual trees), as each plot had 53 trees on average throughout this evaluation. The second group was composed of 30 plots (10 in each productive capacity class) and was used for model validation. This group totaled 7,756 cases. Were tested different network architectures Multilayer Perceptron (MLP). Results revealed an underestimation bias for number of surviving trees. However, estimates of diameter, height, and volume per hectare were found to be accurate. This indicates that artificial neural networks are a viable alternative to the traditional growth and yield modeling approach in the forestry sector.

1. Introduction

Individual tree models are constituted by a set of submodels that estimate diameter and height growth as well as mortality probability through tree- and stand-related variables and through competition data [1–3]. According to Munro [4], these models may be categorized according to how they consider the competition among trees, as represented by distance-dependent and distance-independent models.

Since Newnham [5], many studies have been conducted worldwide in an attempt to improve growth models at the individual tree level and the relevant submodels. Methods to estimate parameters as well as different explanatory variables have been evaluated in an attempt to produce accurate, unbiased estimates of diameter and height growth and also tree

mortality [6, 7]. In Brazil, few studies have been conducted that used these types of growth model for *Eucalyptus* [8, 9], a genus whose planted area intended to supply the processing industry amounts to 4.75 million hectares [10]. Some models have been developed and fitted for some species, including canjerana (*Cabralea canjerana*), black cinnamon (*Nectandra megapotamica*), cedar (*Cedrela fissilis*), and araucaria (*Araucaria angustifolia*), in natural forest conditions, yet without computing all the submodels that compose an individual tree model [11–14].

Typically, estimates of equation parameters for the models are derived from linear and nonlinear regressions [15, 16]. However, other estimation techniques that include artificial neural networks have been successfully used in forest mensuration [17–22], providing just as accurate estimates as those derived through regression, models [23–25].

Artificial neural networks (ANNs) are a type of artificial intelligence system similar to human brain, having a computational capability which is acquired through learning [26]. They can be used in data classification, time series analysis or regression, and pattern recognition [27–29].

Artificial neurons are processing units composed of an activation function which is also known as transfer function. This function is applied to a linear combination among the input variables and weights that reach a given neuron and then returns an output value [19, 28].

The architecture of an artificial neural network refers to how neurons are organized. They comprise three layers: an input layer where variables are introduced to the network, a hidden or intermediate layer where most of the signal processing occurs, and an output layer where the end result is completed and presented.

MLP (Multilayer Perceptron) is the most widely used artificial neural network model for predicting continuous variables. In MLP training, network functioning is as follows: each neuron is connected to every neuron in the subsequent layer; there are no connections between neurons in the same layer; it has at least one processing hidden layer and a high connectivity level between neurons, which is defined by synaptic weights.

The input layer distributes the inputs to subsequent layers. Input nodes have liner activation functions and no thresholds. Each hidden unit node and each output node have thresholds associated with them in addition to the weights. The hidden unit nodes have nonlinear activation functions and the outputs have linear activation functions. Hence, each signal feeding into anode in a subsequent layer has the original input multiplied by a weight with a threshold added and then is passed through an activation function that may be linear or nonlinear (hidden units) [28]. Only three layer MLPs will be considered in this work since these networks have been shown to approximate any continuous function.

According to this architecture, an approximation was obtained of an unknown function $f(x)$ that describes the mapping of input (x)-output(y) pairs of a set of n training patterns [28].

The ANN method was successfully used for modeling regular mortality of individual trees of *Quercus coccínea* [30] and survival of *Pinus resinosa* [31]; for predicting some functional characteristics of ecosystems [23]; for classifying wood using CT scans [32]; for modeling growth rings using climate variables in *Pseudotsuga menziesii* [33]; for comparing different types of ANN in mortality estimates of *Picea abies* [19]; for classifying forest inventory methods [34]; for predicting forest attributes using treetop, soil, and tree size variables [35]; for estimating volume outside bark of *Pinus brutia* [20]; for estimating trunk volume of *Pinus brutia*, *Abies cilicica*, *Cedrus libani*, and *Pinus sylvestris* [36] among others. All studies being cited demonstrate the great potential of ANNs for the use in several forest engineering fields, in particular forest mensuration and management.

In Brazil, despite there being more than 4 million hectares planted with *Eucalyptus* species [10], the use of neural networks as an alternative methodology to regression models is still incipient [20, 36]; the latter is the traditional method for

TABLE 1: Characteristics of *Eucalyptus grandis* × *Eucalyptus urophylla* stands located in Monte Dourado, Pará state, Brazil.

Age (months)	24–72
Diameter at breast height: $\overline{\text{d.b.h.}}$ (cm)	4.0–29.4
Average diameter: q (cm)	7.3–18.4
Total height: Ht (m)	8.5–34.1
Dominant height: Hd (m)	13.1–34.8
Basal area (m² ha⁻¹)	4.7–27.2
Volume (m³ ha⁻¹)	23.8–353.9
Density (trees ha⁻¹)	760–1180

$\overline{\text{d.b.h.}} = \sum_{i=1}^{n} \text{d.b.h.}_i/n$, $q = \sqrt{\sum_{i=1}^{n} \text{d.b.h.}_i^2/n}$, Ht is the distance between the ground and the top of the tree, and Hd is the five trees heights of largest d.b.h. on each plot.

growth and yield modeling in the forestry sector. With that in mind, this work aimed to model the growth of *Eucalyptus* stands at the individual tree level by using artificial neural networks.

2. Material and Methods

2.1. Data. Data were obtained from 63 permanent plots, approximately 500 m² in area, containing clonal stands of *Eucalyptus grandis* × *Eucalyptus urophylla*, with initial spacing of 3 × 3 m between plants (Table 1). This study was conducted in the municipality of Monte Dourado, Pará state, on the banks of river Jari, northern Brazil (Figure 1). The equations to estimate volume with bark are presented in Table 2. Local soil types include yellow latosols, cambisols, and podzols. The local climate is characterized as equatorial, hot, and humid, with a rainfall regime (average annual rainfall of 2,115 mm) marked by two well-defined seasons, with a rainy period from January to July which accounts for about 80% of the annual precipitation and a dry period from August to December. Regards the wind regime, the average wind speed is 2 to 4 m s⁻¹. Wind blasts are nonetheless common, potentially exceeding 100 km h⁻¹ in some cases. The predominant natural vegetation is the submontane and montane dense moist forest type. The average annual temperature is 26.4°C [9, 37].

In every permanent plot, measurements were taken of the diameter at breast height (d.b.h.) using a measuring tape, and total height (Ht) of each tree, using a digital hypsometer, for five annual measurements (24, 36, 48, 60, and 72 months). The last measurement is equivalent to the age of cutting forest. The volume with bark (V) was derived using equations provided by a local forest company. The height of dominant trees (m) was used for productive capacity classification through site indices (S) [15]. The site index curves were constructed employing the guide curve method assuming the index age of 60 months.

The set of permanent plots was randomly divided into two groups. The first group was composed of 33 plots (11 in each productive capacity class) and was used for artificial

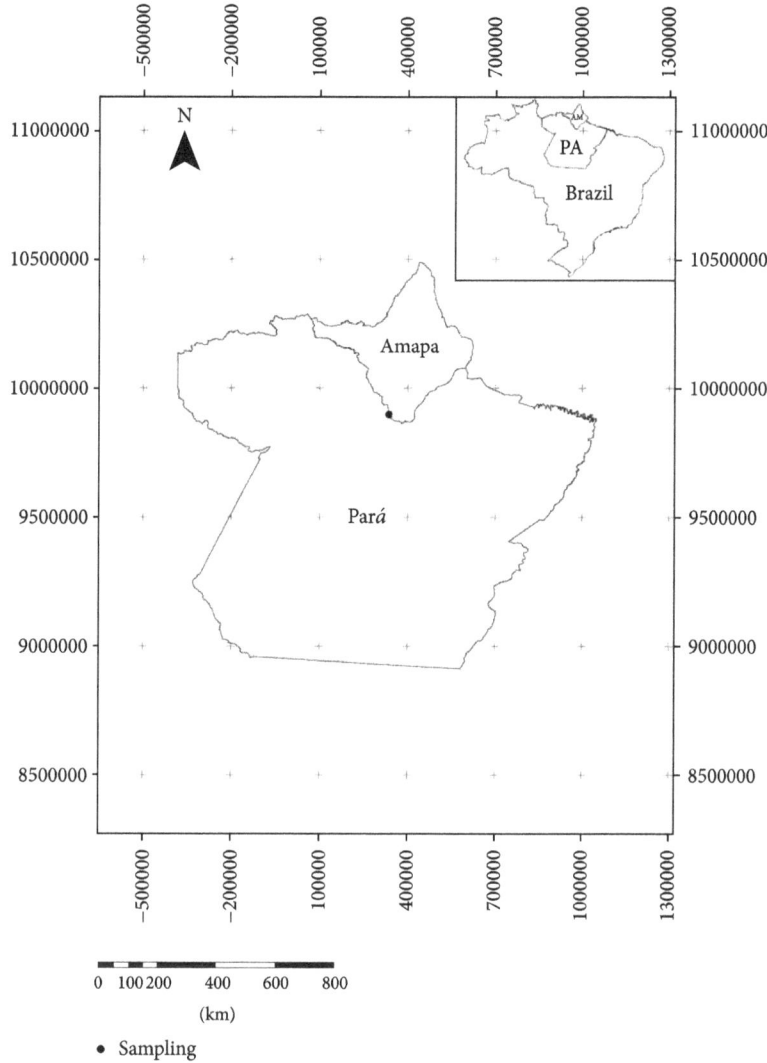

FIGURE 1: Study site. Source: adapted from Silva [38].

TABLE 2: Equations to estimate individual volume with bark.

Area code	Equation
Stratum 001	$\ln V = -11{,}145922 + 1{,}886699 \cdot \ln \text{d.b.h.} + 1{,}373333 \cdot \ln \text{Ht}$
Stratum 016	$\ln V = -9{,}850475 + 1{,}7856259 \cdot \ln \text{d.b.h.} + 1{,}0923825 \cdot \ln \text{Ht}$
Stratum 026	$\ln V = -10{,}162655 + 1{,}8427101 \cdot \ln \text{d.b.h.} + 1{,}1041629 \cdot \ln \text{Ht}$
Stratum 041	$\ln V = -10{,}636733 + 1{,}9185875 \cdot \ln \text{d.b.h.} + 1{,}2004476 \cdot \ln \text{Ht}$
Stratum 042	$\ln V = -10{,}508245 + 1{,}883954 \cdot \ln \text{d.b.h.} + 1{,}1755465 \cdot \ln \text{Ht}$
Stratum 077	$\ln V = -10{,}382205 + 1{,}8935363 \cdot \ln \text{d.b.h.} + 1{,}1283936 \cdot \ln \text{Ht}$

V: volume with bark (m^3); d.b.h.: diameters, in centimeter and Ht: hight, in meters.

neural network training. In five measurements, this group totaled 8,735 cases (measurements of individual trees), as each plot had 53 trees on average throughout this evaluation. The second group was composed of the remaining 30 plots (10 in each productive capacity class) and was used for model validation. This group totaled 7,756 cases.

The annual mortality probability (P_m) for each mensuration period was obtained by calculating the proportion of dead trees per diameter class, in-between mensuration, through the following equation [39–45]:

$$P_m = \frac{n_{j1} - n_{j2}}{\sum_{i=1}^{j} n_{j1}} \cdot 100, \tag{1}$$

where n_{j1} is the number of living trees in the jth diameter class, at the start of the period, n_{j2} is the number of living trees in the jth diameter class at the end of the period. This stand was not observed ingrowth of trees.

Distance-independent competition indices (IID$_i$) being evaluated included [2, 40, 46]

$$IID_1 = \frac{d.b.h._i^2}{\overline{D}^2},$$

$$IID_2 = \frac{Ht_i}{\overline{Ht}},$$

$$IID_3 = \frac{\left(d.b.h._i^2 Ht_i\right)}{\overline{D}^2 \cdot \overline{Ht}}, \quad (2)$$

$$IID_4 = \frac{d.b.h._i^2}{q^2},$$

$$IID_5 = BAL_i,$$

where d.b.h.$_i$ is diameter of the subject tree, $\overline{D}$ is arithmetic mean diameter of trees in the plot, Ht$_i$ is total height of the subject tree, $\overline{Ht}$ is average total height of trees in the plot, q is quadratic diameter, and BAL$_i$ is aggregate basal area of trees larger than the subject tree.

2.2. Neural Network Training. Five hundred artificial neural networks were trained for the following output variables: annual mortality probability (P_m), height at a future age (Ht$_2$), and diameter at a future age (d.b.h.$_2$), using for each output variable a set with different input variables (Table 3).

For the training of artificial neural networks, software Statistica 8.0 [47] was used, testing different architectures of Multilayer Perceptron (MLP) networks.

MLP networks are feedforward multilayer networks having one or more layers of neurons between the input and output layers, known as hidden layer [48]. These hidden layers are able to extract nonlinear data patterns [49]. According to [28], with one hidden layer an MLP network can implement any continuous function, while two hidden layers enable approximating any function. In this network model, each neuron is connected to every neuron in the subsequent layer, but there are no connections between neurons within the same layer nor there is any feedback.

The feedforward type of training was used, by the supervised method. In this procedure, the data flow algorithm moves in only one, noncyclic direction, to initially define the synapse weights, excluding the input variables with low-weight synapses, while the supervised method indicates the input and output variables [28].

The training stages, such as preprocessing, actual training, with selection of architectures and stopping methods, and postprocessing, were performed by the optimization tool Intelligent Problem Solver (IPS), from software Statistica. 500 networks were initially trained in order to estimate each variable. Without precise and accurate networks, this number would be increased.

This software normalizes data in the range 0-1 and tests various architectures and network models. In the supervised method, input and output variables are set by the user. In

TABLE 3: Output and input variables for neural network training.

Output variable	Input variables	Number of trained networks
P_m	I_1, S, CLA, IID$_1$	100
	I_1, S, CLA, IID$_2$	100
	I_1, S, CLA, IID$_3$	100
	I_1, S, CLA, IID$_4$	100
	I_1, S, CLA, IID$_5$	100
Total		500
Ht$_2$	I_1, I_2, S, Ht$_1$, IID$_1$	100
	I_1, I_2, S, Ht$_1$, IID$_2$	100
	I_1, I_2, S, Ht$_1$, IID$_3$	100
	I_1, I_2, S, Ht$_1$, IID$_4$	100
	I_1, I_2, S, Ht$_1$, IID$_5$	100
Total		500
d.b.h.$_2$	I_1, I_2, S, d.b.h.$_1$, IID$_1$	100
	I_1, I_2, S, d.b.h.$_1$, IID$_2$	100
	I_1, I_2, S, d.b.h.$_1$, IID$_3$	100
	I_1, I_2, S, d.b.h.$_1$, IID$_4$	100
	I_1, I_2, S, d.b.h.$_1$, IID$_5$	100
Total		500

P_m: annual mortality probability; Ht$_2$: height at a future age (m); d.b.h.$_2$: diameter at a future age (cm); I_1 and I_2: current and future ages (months); S: site index (m); CLA: diameter class in which the tree was (cm); d.b.h.$_1$ and Ht$_1$: current diameters (cm) and heights (m); and IID$_1$ to IID$_5$: competition indices.

the feedforward procedure, the data flow algorithm moves in only one, noncyclic direction, to initially determine the synapse weights. The back-propagation algorithm corrects the initial synapse weights so as to minimize prediction error. Therefore, in this process, initial input variables can be excluded during training for not helping (low synapse weight) minimize prediction error [28].

The definition of network architecture, that is, number of neurons per layer, number of layers, and parameterization was optimized by the tool Intelligent Problem Solver, from software Statistica.

The selection of best network for each output variable was based on the following criteria [50–53]: (a) coefficient of correlation between observed and estimated values ($r_{y\hat{y}}$), (b) coefficient of variation (CV %), (c) root mean square error (RMSE), (d) bias, (e) bias %, (f) absolute mean differences (AMD), and (g) graphic analysis of observed versus estimated values.

2.3. Validation. The validation of selected networks was done by annually projecting the tree mortality, the height and diameter of living trees, and the volume per hectare of plots until age of 72 months, according to the flowchart of basic steps and decisions for an individual tree model [2].

To verify network behavior under different growing conditions, the plots were divided into three productivity classes (high, medium, and low) based on site indices (S) ($S = 32; 26; 20$ m).

The tree mortality rule was the one that is used by Pretzsch et al., [54], whereby after estimating mortality

TABLE 4: Architecture of artificial neural networks trained to obtain the mortality probability, height and diameter at future ages, and respective statistics.

ANN	Input variables	Number of neurons (Layers)			$r_{y\hat{y}}$	CV %	RMSE	Bias	Bias %	AMD
		Input	Hidden	Output						
			Mortality probability P_m							
M_1	I_1, S, CLA, IID_1	4	9	1	0.744	43.1	0.034	0.0004	−16.91	0.0255
M_2	IID_2	1	3	1	0.579	52.3	0.041	0.0051	−15.84	0.0305
M_3	I_1, S, CLA, IID_3	4	8	1	0.799	38.6	0.030	0.0009	−12.73	0.0235
M_4	I_1, S, CLA, IID_4	4	5	1	0.711	45.2	0.037	0.0112	4.01	0.0286
M_5	I_1, CLA, IID_5	3	4	1	0.790	39.3	0.031	0.0000	−16.34	0.0244
			Height (Ht_2)							
H_1	S, Ht_1, IID_1	3	6	1	0.993	3.0	0.626	−0.0011	−0.12	0.4370
H_2	I_1, I_2, S, Ht_1, IID_2	5	6	1	0.993	2.9	0.592	−0.0089	−0.17	0.4069
H_3	I_1, I_2, S, Ht_1, IID_3	5	7	1	0.994	2.6	0.532	0.0001	−0.07	0.4048
H_4	I_1, I_2, S, Ht_1, IID_4	5	7	1	0.995	2.5	0.526	−0.0028	−0.10	0.3979
H_5	I_1, I_2, S, Ht_1, IID_5	5	9	1	0.993	2.9	0.608	0.0028	−0.14	0.4222
			Diameter (d.b.h.$_2$)							
D_1	I_1, I_2, d.b.h.$_1$, IID_1	4	5	1	0.990	4.6	0.618	−0.0048	−0.31	0.4089
D_2	I_1, I_2, S, d.b.h.$_1$, IID_2	5	7	1	0.990	4.4	0.601	−0.0118	−0.42	0.3986
D_3	I_2, d.b.h.$_1$, IID_3	3	4	1	0.993	3.8	0.523	0.0008	−0.31	0.3913
D_4	I_1, I_2, S, d.b.h.$_1$, IID_4	5	5	1	0.993	3.7	0.508	−0.0021	−0.25	0.3821
D_5	I_1, I_2, S, d.b.h.$_1$, IID_5	5	6	1	0.988	4.9	0.667	0.0013	−0.28	0.5049

P_m: annual mortality probability; Ht_2: height at a future age (m); d.b.h.$_2$: diameter at a future age (cm); I_1 and I_2: current and future ages (months); S: site index (m); CLA: diameter class in which the tree was (cm); d.b.h.$_1$ and Ht_1: current diameters (cm) and heights (m); and IID_1 to IID_5: competition indices.

probability for each tree, a random number between zero and one was generated (P_a) and compared to the estimated mortality probability (P_m). If $P_m > P_a$ the tree will die, otherwise the tree will remain alive and its dimensions should be projected to the next age through neural networks. The possibility of adding new trees was not considered.

In order to evaluate the accuracy of mortality estimates, a graphical analysis was performed of the estimated number of surviving trees in relation to observations. For the variables height and diameter, the F-test was used [55] ($\alpha = 0.05$), scatter plots were generated, and the statistics $r_{y\hat{y}}$, CV %, RMSE, bias, bias %, AMD were calculated. Estimates of volume per hectare, as projected until age of 72 months in each productivity class, were compared to observed values using Student's t-test ($\alpha = 0.05$).

3. Results

3.1. Training of Artificial Neural Networks (ANNs). The network with the best performance to estimate annual mortality probability P_m was M_5, with three neurons in the input layer (variables), four neurons in the hidden layer, and one neuron in the output layer. Although not with the best statistics (Table 4), this network provided a better trend regarding the distribution of observed values in relation to estimated values (Figure 2). With only the competition index as input variable, M_2 was the network with the poorest performance.

The best networks for the variables height and diameter were H_4 and D_4, respectively, with special mention of the estimate accuracy (Table 4 and Figure 2). The networks for

these variables comprised five input layers (variables), seven and five hidden layers, respectively, and one output layer.

3.2. Validation. The projection of the number of surviving trees per hectare indicates a slight underestimation bias in the validation plots (Figure 3). The estimates can be considered reasonable, since mortality is a random event which is difficult to measure and estimate, consequently obstructing the good performance of probability models [56].

The neural networks selected for the variables diameter and height provided accurate estimates, regardless of the productivity class (Table 5 and Figure 4), with an error increase tendency as age advanced, due to error propagation. However, no difference was found between estimated and observed values, according to the F-test ($P < 0.05$).

Estimates of volume outside bark per hectare, as projected until age of 72 months (Figure 5), did not differ statistically from observed values ($P < 0.05$), in all productive capacity classes and ages, according to the t-test. Percentage differences between estimated and observed volumetric stocks at age of 72 months were 7%, 0%, and −6% for the high, medium, and low productivity class, respectively. Similarly to height and diameter estimates (Table 5), the older the age, the higher the error in volumetric stock estimates, due to error propagation.

4. Discussion

In recent decades, a major concern in the field of forest mensuration has been to develop growth and yield models using individual trees [40, 57–60]. Modeling growth and yield

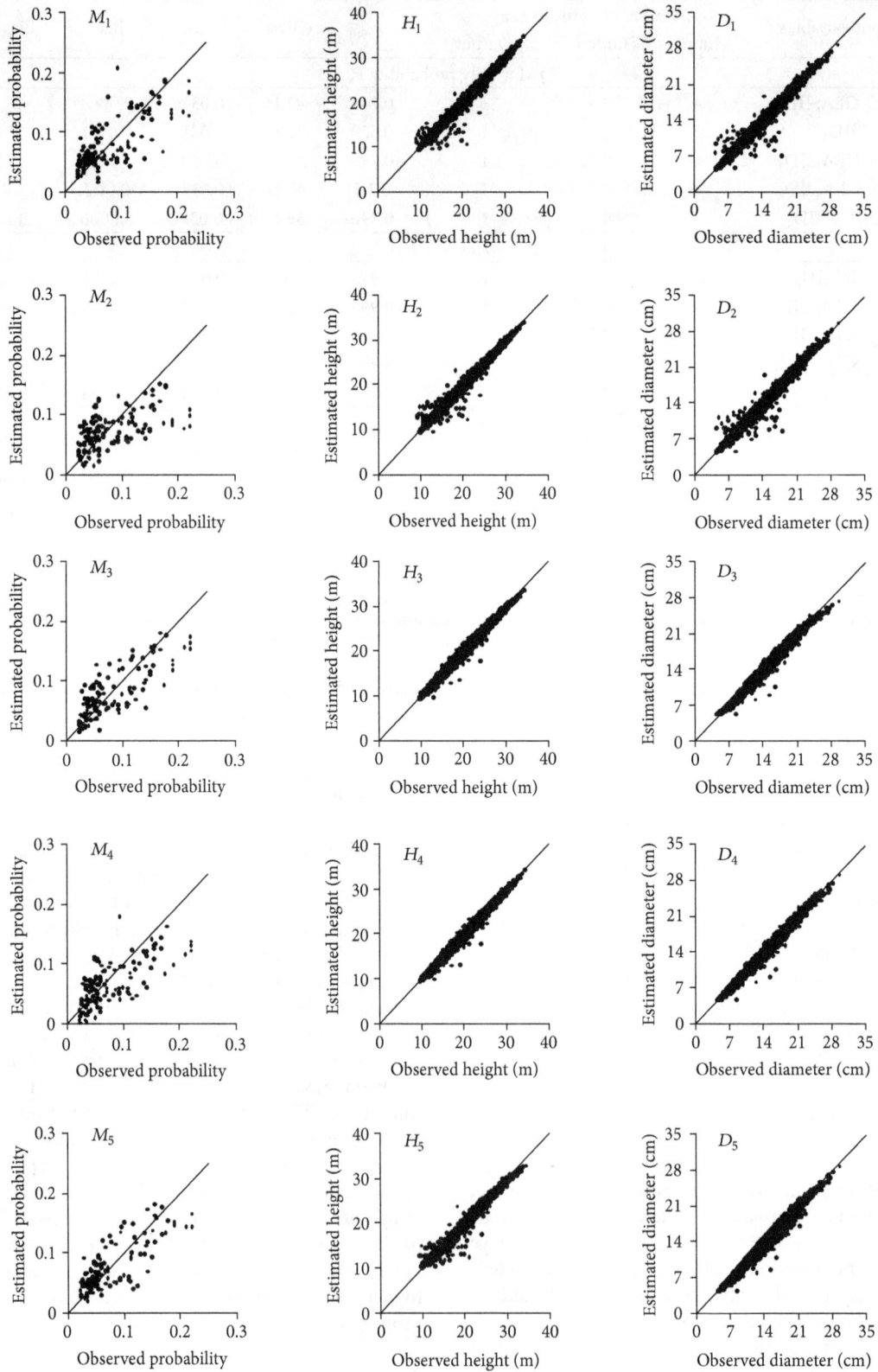

FIGURE 2: Mortality probability, height and diameter, as observed and estimated by artificial neural network training.

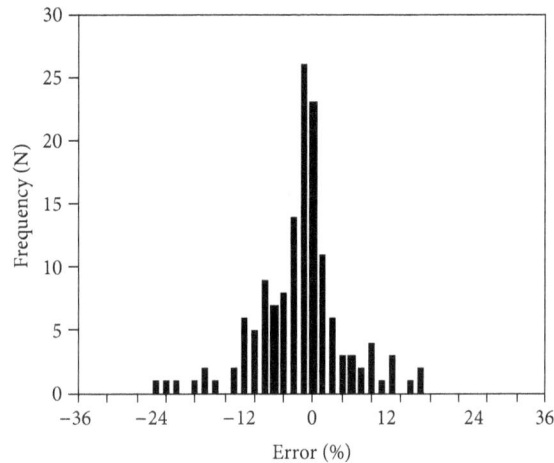

FIGURE 3: Histogram of residuals by the artificial neural network model for independent data (generalization).

TABLE 5: Statistics of height (Ht) and diameter (d.b.h.), as estimated and observed per age (months) and productive capacity class.

							Variable						
Productive capacity class	Age (months)			Ht (m)						d.b.h. (cm)			
		$r_{y\hat{y}}$	CV % (±)	RMSE	Bias	Bias %	AMD	$r_{y\hat{y}}$	CV % (±)	RMSE	Bias	Bias %	AMD
High (S = 32)	36	0.979	3.5	0.707	−0.080	−0.489	0.456	0.976	5.5	0.767	−0.081	−0.856	0.516
	48	0.965	4.6	1.040	0.027	0.003	0.741	0.961	7.1	1.065	−0.016	−0.443	0.752
	60	0.953	5.1	1.272	−0.223	−1.067	0.930	0.945	8.5	1.379	−0.184	−1.642	0.961
	72	0.934	5.9	1.540	−0.153	−0.784	1.159	0.925	10.3	1.748	−0.211	−1.890	1.248
Medium (S = 26)	36	0.981	3.4	0.621	0.090	0.494	0.453	0.981	4.8	0.624	0.147	1.246	0.461
	48	0.957	5.1	1.091	0.219	1.017	0.848	0.962	7.0	0.971	0.135	1.174	0.754
	60	0.938	6.3	1.437	0.170	0.635	1.154	0.940	9.0	1.345	0.137	0.964	1.067
	72	0.921	7.1	1.747	0.149	0.369	1.425	0.922	10.7	1.685	0.162	0.965	1.351
Low (S = 20)	36	0.966	4.1	0.635	0.025	0.101	0.486	0.975	5.4	0.565	0.061	0.559	0.416
	48	0.947	5.7	1.031	0.216	1.015	0.811	0.953	7.7	0.924	0.265	2.163	0.735
	60	0.921	7.0	1.440	0.442	2.017	1.138	0.929	9.6	1.266	0.414	3.138	1.012
	72	0.903	7.8	1.684	0.420	1.642	1.327	0.909	11.0	1.507	0.427	2.806	1.201

can be difficult and complex if relying on individual trees as the basic modeling unit [61] due to the high resolution level required [62]. This type of modeling consists of establishing different equations to predict diameter and height growth as well as mortality, including tree- and stand-related data as explanatory variables which represent the competitive status of a tree [63–65], among others.

A widely used resource in this type of modeling is regression analysis through linear and nonlinear functions [6]. The different fit functions and the functional form with which these variables are included in the model are the greatest stumbling block to modeling growth and yield at the individual tree level. Numerous studies have been conducted to compare and provide the best fit function (linear and nonlinear) for predicting growth using different explanatory variables [9, 59, 60, 65, 66]. However, fitting these models requires knowledge of the functional relationship of variables, knowledge of modeling tools, in addition to accurate estimates. Even where such requirements are met, there is neither proven advantage of one function over another nor an indication of which are the most suitable

explanatory variables to be used in the model, leading to uncertainty in growth and yield modeling for an individual tree. Additionally, the high resolution level of this model type is accompanied by problems caused by cumulative errors [62]. Therefore, using ANNs may work as an alternative to the traditional procedure for modeling an individual tree [31].

ANNs are increasingly becoming a popular tool in forest mensuration [16, 20, 36] and in estimation of individual tree growth and yield [30, 31], requiring no presuppositions about the functional form of variables or fit functions. Instead, ANNs are trained to find such relationship. An ANN is capable of achieving optimum growth estimates without mentioning the functional form of the relationship between input and output layers [23]. Also, an ANN may help identify the most critical input variables to predict diameter and height growth as well as mortality probability and provide a better understanding of the dynamic of models at the individual tree level, becoming a valuable tool for *Eucalyptus* forest management.

A large number of authors have discussed the structure, technique, and operation of ANNs [18, 26, 27, 29]. Yet only

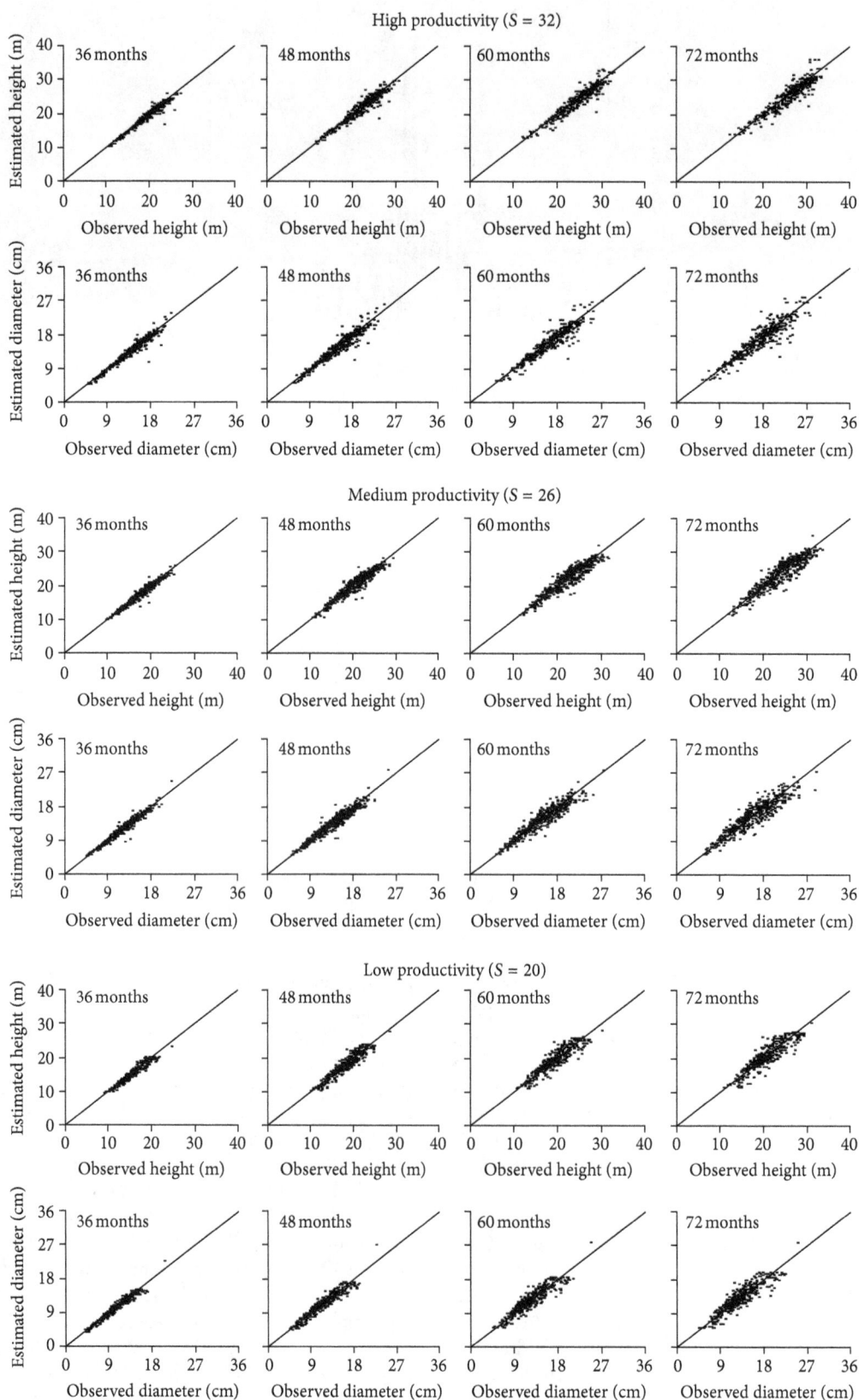

FIGURE 4: Heights and diameters, as observed and estimated per age and productivity class (high, medium, and low) by the artificial neural network model for independent data (generalization).

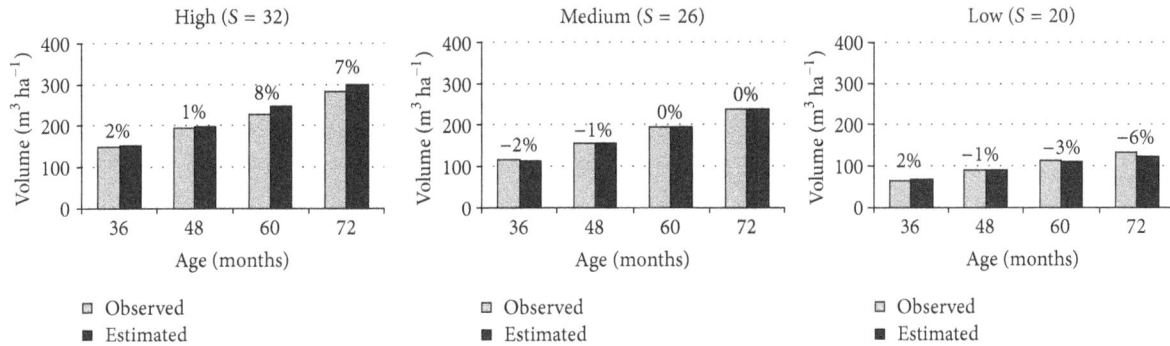

FIGURE 5: Estimates of volume outside bark ($m^3 ha^{-1}$) per age (36–72) and productivity class (high, medium, and low). Values on the bars indicate the difference.

in the past decade interest in using ANNs has been stirred up, with an increasing number of practical applications in environmental and forest modeling [67]. Applications for use in forest mensuration and management include forest classification and mapping, growth and forest dynamics modeling, spatial data analysis, analysis of forest inventory types, and dendroclimatology modeling [25, 36, 68].

As far as results found in this study are concerned, the competition index of the network selected as best for mortality probability (M_5) is BAL_i, that is, the sum of basal areas of trees larger than the subject tree. According to this index, the smaller the tree is, the less competitive it is and the greater its mortality probability is [69–71]. Hasenauer et al. [19] found better prediction results for tree mortality using neural networks, in which IID_5 (BAL) was one of the variables used in network training.

Studies involving individual tree mortality through regression models usually estimate mortality probability only as a function of the competition index as independent variable [9, 40, 69, 72]. However, besides the competition index, Guan and Gertner [30], Monserud and Sterba [71], and Palahí and Pukkala [73] recommend using tree- and stand-related variables to improve mortality estimates. In this study, the networks with the best performance included these input variables (Table 4).

Network input variables for estimating height and diameter include the IID_4, also known in literature as Basal Area Index (BAI). According to Daniels [58], this competition index was found to be the most correlated one with diameter and height growth in loblolly pine trees. Martins [9] observed that the IID_4 provided the highest partial contribution, if compared to other competition indices, in diameter and height growth models for *Eucalyptus* trees. Explanations for best performance include the ease of calculation and the biological realism of the IID_4 index.

Other than the competition index, tree- and stand-related variables such as age and site index are also widely used to express height and diameter growth at the individual tree level, using linear and nonlinear regression [53, 60, 74]. The selection of these variables in the best networks (H_4, D_4) demonstrates realism in network estimates, adequately capturing height and diameter growth (Figure 3 and Table 4).

Martins [9] used the same data as this study for modeling individual trees yet using regression models. The fitted equations to estimate mortality probability were inferior if compared to the obtained statistics using artificial neural network training. As to the fitted equations to estimate diameter and height in future ages, the equations obtained using regression models were as accurate as the trained neural networks. However, in model validation, using independent data from regression fitting and network training, the projections of tree growth and mortality using the neural network model discussed in this work were proved to be superior. The bias % values at the individual tree level for height and diameter obtained by the previously mentioned author were in the order of 1.5%, while in this work they were in the order of 0.5%.

In the validation at the stand level, the percentage errors using artificial neural networks were in the order of 0.5%, while those using regression models were in the order of 6% for total volume per hectare.

Since the same data were used, with the same methodology, it can be said that the tool artificial neural networks is effective in individual tree modeling and provides superior results if compared to regression models, particularly in the model generalization or validation stage.

Considering the results found, artificial neural networks should be studied for describing the structure and dynamics of natural tropical forests, with all of their complexity resulting from size and species diversity [3], which are usually difficult to model through traditional growth and yield methods.

5. Conclusions

Results in this study confirm the use potential of MLP artificial neural networks, through the supervised learning method, for individual tree modeling of commercial *Eucalyptus* plantations, given the precision of estimates found. Yet further investigation is required to improve this methodology, seeking to evaluate other types of artificial neural networks, as, for instance, RBF (*Radial Bases Function*) networks, and to study the effects of including qualitative input variables in the networks.

Acknowledgments

The authors wish to thank Conselho Nacional de Desenvolvimento Científico e Tecnológico (CNPQ), Brazil, for granting the scholarship and Monica Castellani for English revision.

References

[1] J. L. Clutter, J. C. Fortson, L. V. Pienaar, G. H. Brister, and R. L. Bailey, *Timber Management: A Quantitative Approach*, John Wiley & Sons, New York, NY, USA, 1983.

[2] L. S. Davis and K. N. Johnson, *Forest Management*, McGraw-Hill, New York, NY, USA, 1987.

[3] J. K. Vanclay, *Modeling Forest Growth and Yield: Aplications to Mixed Tropical Forest*, CAB International, Wallingford, UK, 1994.

[4] D. D. Munro, "Forest growth models—a prognosis," in *Growth Models for Tree and Stand Simulation*, J. Fries, Ed., pp. 1–21, Royal College of Forestry, Stockholm, Sweden, 1974.

[5] R. M. Newnham, *The development of a stand model for Douglas-fir [Ph.D. thesis]*, University of British Columbia, Canada, 1964.

[6] P. Soares and M. Tomé, "A distance dependent diameter growth model for first rotation eucalyptus plantation in Portugal," in *Empirical and Process—Bases Models for Forest Tree and Stand Growth Simulation*, A. Amaro and M. Tomé, Eds., pp. 267–270, Salamandra, 1997.

[7] F. Crescente-Campo, P. Soares, M. Tomé, and U. Diéguez-Aranda, "Modeling noncatastrophic individual tree mortality for *Pinus radiate* plantations in northwestern Spain," *Forest Ecology and Management*, vol. 257, no. 6, pp. 1542–1550, 2010.

[8] B. R. Mendes, N. Calegario, C. E. S. Volpato, and A. A. Melo, "Desenvolvimento de modelos de crescimento de árvores individuais fundamentado em equações diferenciais," *Cerne*, vol. 12, pp. 254–263, 2006.

[9] F. B. Martins, *Modelagem de crescimento em nível de árvore individual para plantios comerciais de eucaliptos [Ph.D. thesis]*, Universidade Federal de Viçosa, Brazil, 2011.

[10] ABRAF-Associação Brasileira de Florestas Plantadas, "Anuário estatístico da ABRAF: ano base 2010. Brasília," 2011, http://www.abraflor.org.br/estatisticas/ABRAF11/ABRAF11-BR.pdf.

[11] M. A. Durlo, "Relações morfométricas para *Cabralea canjerana* (Well.) Mart," *Ciencia Florestal*, vol. 11, no. 1, pp. 141–149, 2001.

[12] J. B. Della Flora, M. A. Durlo, and P. Soathelf, "Modelo de crescimento para árvores singulares—*Nectandra megapotamica* (Spreng.) Mez," *Ciencia Florestal*, vol. 14, no. 1, pp. 165–177, 2004.

[13] M. A. Durlo, F. J. Sutili, and L. Denardi, "Modelagem da copa de *Cedrela fissilis* Vellozo," *Ciencia Florestal*, vol. 14, no. 2, pp. 79–89, 2004.

[14] T. Chassot, *Modelos de crescimento em diâmetro de árvores individuais de Araucaria angustifólia (Bertol.) Kuntze na floresta ombrófila mista [Ph.D. thesis]*, Universidade Federal de Santa Maria, Brazil, 2009.

[15] J. C. C. Campos and H. G. Leite, *Mensuração Florestal: Perguntas e Respostas*, Universidade Federal de Viçosa, Viçosa, Brazil, 2009.

[16] P. Miehle, M. Battaglia, P. J. Sands et al., "A comparison of four process-based models and a statistical regression model to predict growth of *Eucalyptus globulus* plantations," *Ecological Modelling*, vol. 220, no. 5, pp. 734–746, 2009.

[17] D. Merkl and H. Hasenauer, "Using neural networks to predict individual tree mortality," in *Proceedings of the Int'l Conference on Engineering Applications of Neural Networks*, pp. 10–12, Gibraltar, UK, 1998.

[18] M. Weingartner, D. Merkl, and H. Hasenauer, "Improving tree mortality predictions of Norway Spruce stands with neural networks," in *Proceedings of the Symposiun on Integration in Environmental Information Systems*, Austria, 2000.

[19] H. Hasenauer, D. Merkl, and M. Weingartner, "Estimating tree mortality of Norway spruce stands with neural networks," *Advances in Environmental Research*, vol. 5, no. 4, pp. 405–414, 2001.

[20] M. J. Diamantopoulou, "Artificial neural networks as an alternative tool in pine bark volume estimation," *Computers and Electronics in Agriculture*, vol. 48, no. 3, pp. 235–244, 2005.

[21] E. Görgens, *Estimação do volume de árvores utilizando redes neurais artificiais [Ph.D. thesis]*, Universidade Federal de Viçosa, Brazil, 2006.

[22] M. L. M. Silva, D. H. B. Binoti, J. M. Gleriani, and H. G. Leite, "Ajuste do modelo de Schumacher e Hall e aplicações de redes neurais artificiais para estimar volumes de árvores de eucalipto," *Árvore*, vol. 33, no. 6, pp. 1133–1139, 2009.

[23] J. M. Paruelo and F. Tomasel, "Prediction of functional characteristics of ecosystems: a comparison of artificial neural networks and regression models," *Ecological Modelling*, vol. 98, no. 2-3, pp. 173–186, 1997.

[24] M. Gevrey, I. Dimopoulos, and S. Lek, "Review and comparison of methods to study the contribution of variables in artificial neural network models," *Ecological Modelling*, vol. 160, no. 3, pp. 249–264, 2003.

[25] H. G. Leite, M. L. M. da Silva, D. H. B. Binoti, L. Fardin, and F. H. Takizawa, "Estimation of inside-bark diameter and heartwood diameter for *Tectona grandis* Linn. trees using artificial neural networks," *European Journal of Forest Research*, vol. 130, no. 2, pp. 263–269, 2011.

[26] A. P. Braga, Carvalho, A. C. P. L. F, and T. B. Ludemir, "Redes neurais artificiais," in *Sistemas Inteligentes*, S. O. Rezende, Ed., pp. 141–168, Manole, Barueri, Brazil, 2003.

[27] A. K. Jain, J. Mao, and K. M. Mohiuddin, "Artificial neural networks: a tutorial," *Computer*, vol. 29, no. 3, pp. 31–44, 1996.

[28] S. Haykin, *Redes Neurais: Princípios e Prática*, Bookman, Porto Alegre, Brazil, 2001.

[29] J. M. Barreto, *Introdução às Redes Neurais Artificiais*, Universidade Federal de Santa Catarina, Florianópolis, Brazil, 2002.

[30] B. T. Guan and G. Gertner, "Using a parallel distributed processing system to model individual tree mortality," *Forensic Science*, vol. 37, pp. 871–885, 1991.

[31] B. T. Guan and G. Gertner, "Modeling red pine tree survival with an artificial neural network," *Forensic Science*, vol. 37, pp. 1429–1440, 1991.

[32] D. L. Schomoudt, P. Li, and A. L. Abbot, "Machine vision using artificial neural networks with local 3d neighborhoods," *Computers and Electronics in Agriculture*, vol. 97, pp. 101–119, 1997.

[33] Q. B. Zhang, R. I. Hebda, Q. J. Zhang, and R. I. Alfaro, "Modeling tree-ring growth responses to climatic variables using artificial neural networks," *Forest Science*, vol. 46, no. 2, pp. 229–239, 2000.

[34] C. Liu, L. Zhang, C. J. Davis, D. S. Solomon, T. B. Brann, and L. E. Caldwell, "Comparison of neural networks and statistical methods in classification of ecological habitats using FIA data," *Forest Science*, vol. 49, no. 4, pp. 619–631, 2003.

[35] S. A. Corne, S. J. Carver, W. E. Kunin, J. J. Lennon, and A. W. S. van Hees, "Predicting forest attributes in Southeast Alaska using artificial neural networks," *Forest Science*, vol. 50, no. 2, pp. 259–276, 2004.

[36] R. Özçelik, M. J. Diamantopoulou, J. R. Brooks, and H. V. Wiant Jr., "Estimating tree bole volume using artificial neural network models for four species in Turkey," *Journal of Environmental Management*, vol. 91, no. 3, pp. 742–753, 2010.

[37] R. A. Demolinari, *Crescimento de povoamentos de eucalipto não-desbastados [Ph.D. thesis]*, Universidade Federal de Viçosa, Viçosa, Brazil, 2006.

[38] M. G. Silva, *Produtividade, idade e qualidade da madeira de Eucalyptus destinada à produção de polpa celulósica branqueada [Ph.D. thesis]*, Universidade de São Paulo, São Paulo, Brazil, 2011.

[39] T. D. Keister and G. R. Tidwell, "Competition ratio dynamics for improved mortality estimates in simulated growth of forests stands," *Forest Science*, vol. 21, pp. 46–51, 1975.

[40] G. R. Glover and J. N. Hool, "A basal area ratio predictor of loblolly pine plantation mortality," *Forest Science*, vol. 25, pp. 275–282, 1979.

[41] R. C. de Miranda, J. C. C. Campos, F. de Paula Neto, and L. M. de Oliveira, "Predição da mortalidade regular para eucalipto," *Árvore*, vol. 13, pp. 152–173, 1989.

[42] S. A. Machado, A. E. Tonon, A. F. Filho, and E. B. Oliveira, "Comportamento da mortalidade natural em bracatingais nativos em diferentes densidades iniciais e classes de sítio," *Ciencia Florestal*, vol. 12, pp. 41–50, 2002.

[43] R. Maestri, C. R. Sanquetta, and J. C. Arce, "Modelagem do crescimento de povoamentos de *Eucalyptus grandis*através de processos de difusão," *Floresta*, vol. 33, pp. 169–182, 2003.

[44] L. M. B. Rossi, H. S. Koehler, C. R. Sanquetta, and J. E. Arce, "Modelagem da mortalidade em florestas naturais," *Floresta*, vol. 37, pp. 275–291, 2007.

[45] D. Zhao, B. Borders, M. Wang, and M. Kane, "Modeling mortality of second-rotation loblolly pine plantations in the Piedmont/Upper coastal plain and lower coastal plain of the southern United States," *Forest Ecology and Management*, vol. 252, no. 1–3, pp. 132–143, 2007.

[46] A. R. Stage, "Prognosis model for stand development," USDA Forest Service Research Papers INT-137, USDA, Washington, DC, USA, 1973.

[47] StatSoft Inc, STATISTICA (data analysis software system), version 8. 0, 2007.

[48] R. P. Lippmann, "An introduction to computing with neural nets," *IEEE ASSP Magazine*, vol. 4, no. 2, pp. 4–22, 1987.

[49] A. P. Braga, Carvalho, A. P. L. F, and T. B. Ludemir, *Redes Neurais Artificiais: Teoria e Aplicações*, Rio de Janeiro, Rio de Janeiro, Brazil, 2000.

[50] R. G. D. Steel and J. H. Torrie, *Principles and Procedures of Statistics*, McGraw-Hill, New York, NY, USA, 1960.

[51] P. A. Murphy and H. S. Sternitzke, *Growth and Yield Estimation for Loblolly Pine in the West Gulf*, Southern Forest Experiment Station, New Orleans, La, USA, 1979.

[52] J. Siipilehto, "A comparison of two parameter prediction methods for stand structure in Finland," *Silva Fennica*, vol. 34, no. 4, pp. 331–349, 2000.

[53] A. Monty, P. Lejeune, and J. Rondeux, "Individual distance-independent girth increment model for Douglas-fir in southern Belgium," *Ecological Modelling*, vol. 212, no. 3-4, pp. 472–479, 2008.

[54] H. Pretzsch, P. Biber, and J. Ďurský, "The single tree-based stand simulator SILVA: construction, application and evaluation," *Forest Ecology and Management*, vol. 162, no. 1, pp. 3–21, 2002.

[55] F. A. Graybill, *Theory and Application of Linear Model*, Duxbury, North Scituate, Mass, USA, 1976.

[56] D. A. Hamilton Jr., "A logistic model of mortality in thinned and unthinned mixed conifer stands of northern Idaho," *Forest Science*, vol. 32, no. 4, pp. 989–1000, 1986.

[57] I. E. Bella, "A new competition model for individual tree," *Forest Science*, vol. 17, pp. 364–372, 1971.

[58] R. F. Daniels, "Simple competition indices and their correlation with annual loblolly pine tree growth," *Forest Science*, vol. 22, pp. 454–456, 1976.

[59] H. Hasenauer and R. A. Monserud, "A crown ratio model for Austrian Forests," *Forest Ecology and Management*, vol. 84, no. 1–3, pp. 49–60, 1996.

[60] S. Vospernik, R. A. Monserud, and H. Sterba, "Do individual-tree growth models correctly represent height: diameter ratios of Norway spruce and Scots pine?" *Forest Ecology and Management*, vol. 260, no. 10, pp. 1735–1753, 2010.

[61] H. Hasenauer, "Princípios para a modelagem de ecossistemas florestais," *Revista Ciência & Ambiente*, vol. 20, pp. 53–69, 2000.

[62] Q. V. Cao, "Predictions of individual-tree and whole-stand attributes for loblolly pine plantations," *Forest Ecology and Management*, vol. 236, no. 2-3, pp. 342–347, 2006.

[63] M. Tome and H. E. Burkhart, "Distance-dependent competition measures for predicting growth of individual trees," *Forest Science*, vol. 35, no. 3, pp. 816–831, 1989.

[64] H. Hasenauer, R. A. Monserud, and T. G. Gregoire, "Using simultaneous regression techniques with individual-tree growth models," *Forest Science*, vol. 44, no. 1, pp. 87–95, 1998.

[65] M. S. González, RÍO, M. del, I. Cañellas, and G. Montero, "Distance independent tree diameter growth model for cork oak stands," *Forest Ecology and Management*, vol. 225, no. 1–3, pp. 262–270, 2006.

[66] K. Andreassen and S. M. Tomter, "Basal area growth models for individual trees of Norway spruce, Scots pine, birch and other broadleaves in Norway," *Forest Ecology and Management*, vol. 180, no. 1–3, pp. 11–24, 2003.

[67] C. H. Peng and X. Wen, *Recent Applications of Artificial Neural Networks in Forest Resource Management: An Overview*, American Association for Artificial Intelligence Technical, Menlo Park, Calif, USA, 1999.

[68] E. Dogan, B. Sengorur, and R. Koklu, "Modeling biological oxygen demand of the Melen River in Turkey using an artificial neural network technique," *Journal of Environmental Management*, vol. 90, no. 2, pp. 1229–1235, 2009.

[69] P. W. West, "Simulation of diameter growth and mortality in regrowth eucalypt forest of southern Tasmania," *Forensic Science*, vol. 27, pp. 603–616, 1981.

[70] D. W. Hann and C. H. Wang, *Mortality Equations for Individual Trees in the Mixed-Conifer Zone of Southwest Oregon*, vol. 67 of *Research Bulletin*, Forest Research Laboratory, Oregon, Wash, USA, 1990.

[71] R. A. Monserud and H. Sterba, "Modeling individual tree mortality for Austrian forest species," *Forest Ecology and Management*, vol. 113, no. 2-3, pp. 109–123, 1999.

[72] F. Crecente-Campo, P. Soares, M. Tomé, and U. Diéguez-Aranda, "Modelling annual individual-tree growth and mortality of Scots pine with data obtained at irregular measurement intervals and containing missing observations," *Forest Ecology and Management*, vol. 260, no. 11, pp. 1965–1974, 2010.

[73] M. Palahí and T. Pukkala, "Optimising the management of Scots pine (*Pinus sylvestris* L.) stands in Spain based on individual-tree models," *Annals of Forest Science*, vol. 60, no. 2, pp. 105–114, 2003.

[74] F. C. C. Uzoh and W. W. Oliver, "Individual tree height increment model for managed even-aged stands of ponderosa pine throughout the western United States using linear mixed effects models," *Forest Ecology and Management*, vol. 221, no. 1–3, pp. 147–154, 2006.

9

Estimation of Genetic and Phenotypic Parameters for Growth Traits in a Clonal Seed Orchard of *Pinus kesiya* in Malawi

Edward Missanjo, Gift Kamanga-Thole, and Vidah Manda

Malawi College of Forestry and Wildlife, Private Bag 6, Dedza, Malawi

Correspondence should be addressed to Edward Missanjo; edward.em2@gmail.com

Academic Editors: N. Frascaria-Lacoste, J. Kaitera, and Z. Kaya

Genetic and phenotypic parameters for height, diameter at breast height (dbh), and volume were estimated for *Pinus kesiya* Royle ex Gordon clonal seed orchard in Malawi using an ASReml program, fitting an individual tree model. The data were from 88 clones assessed at 18, 23, 30, 35, and 40 years of age. Heritability estimates for height, dbh, and volume were moderate to high ranging from 0.19 to 0.54, from 0.14 to 0.53, and from 0.20 to 0.59, respectively, suggesting a strong genetic control of the traits at the individual level, among families, and within families. The genetic and phenotypic correlations between the growth traits were significantly high and ranged from 0.69 to 0.97 and from 0.60 to 0.95, respectively. This suggests the possibility of indirect selection in trait with direct selection in another trait. The predicted genetic gains showed that the optimal rotational age of the *Pinus kesiya* clonal seed orchard is 30 years; therefore, it is recommended to establish a new *Pinus kesiya* clonal seed orchard. However, selective harvest of clones with high breeding values in the old seed orchard should be considered so that the best parents in the old orchard can continue to contribute until the new orchard is well established.

1. Introduction

Pinus kesiya Royle ex Gordon occurs naturally in Himalaya region (Asian): Burma, China, India, Laos, Philippines, Thailand, Tibet, and Vietnam [1]. This species particularly grows well at altitudes from 600 to 1800 m above sea level [2]. The trees can reach heights of 30–35 or 45 m tall with straight, cylindrical trunk [3]. *Pinus kesiya* is a major exotic plantation species in Malawi and other Southern African countries. Its success as an exotic is due to its fast growth rate and wide adaptability. With the increasing demand for wood products globally [4], maximizing wood production on available land resources is of major importance. The high growth rate of *Pinus kesiya*, the variation evident in natural stands and plantations in Malawi, and the need to improve timber quality and production led to the establishment of a breeding programme in Malawi in the 1970s [5]. The breeding programme included phenotypic mass selection in *Pinus kesiya* stands and use of the material for seed production in clonal seed orchards.

Seed orchards are plantations created for production of genetically improved seeds to create commercial forest crops [6]. The genetic quality of the seeds depends on the genetic superiority of the plus trees, their relationships, their combination ability, and the rate of pollen contamination, among other factors [7]. The major constraint to the efficient breeding of *Pinus kesiya* in Malawi has been the lack of genetic parameter information to guide decisions on the most appropriate breeding strategy and, more generally, to monitor genetic progress.

Genetic parameters estimates available for *Pinus kesiya* are mainly from studies in Brazil. Heritability estimates for diameter at breast height (dbh) and height from these studies were high [1]. There appear to be no estimates of genetic parameters for height and dbh in *Pinus kesiya* grown in Malawi, which is an issue of concern as fast growing tree crops are likely to exhibit different genetic parameters than slower ones [8]. According to Díaz et al. [9], genetic parameters may differ among regions. This lack of genetic parameter estimates for these economically important traits has potentially adverse consequences for realizing genetic progress in *Pinus kesiya* in Malawi.

TABLE 1: Characteristics of the data sets.

Age (years)	Trait	Mean	SD	CV%	Number of trees
18	Height (m)	12.9	1.23	9.2	1869
	dbh (cm)	24.4	2.01	8.6	1869
	Volume (m^3)	0.283	0.09	23.5	1869
23	Height (m)	22.5	1.95	8.6	1840
	dbh (cm)	28.2	3.26	7.3	1840
	Volume (m^3)	0.659	0.15	22.4	1840
30	Height (m)	26.4	1.93	8.9	1743
	dbh (cm)	36.7	2.41	7.5	1743
	Volume (m^3)	1.298	0.34	20.2	1743
35	Height (m)	28.3	1.56	9.1	1731
	dbh (cm)	37.1	2.53	8.0	1731
	Volume (m^3)	1.409	0.67	21.8	1731
40	Height (m)	28.5	1.62	8.4	1619
	dbh (cm)	37.4	2.49	7.8	1619
	Volume (m^3)	1.463	0.69	23.4	1619

SD is the standard deviation; CV is the coefficients of variation.

Many models have been proposed for estimation of genetic parameters of quantitative traits in sort of mixed mating system. The model proposed by [10] is the most complete because it considers use of unbalanced data and allows more accurate prediction of genetic values. The optimum estimation/prediction procedure of genetic values is Reml/Blup, that is, the estimation of the components of variance by restricted maximum likelihood (Reml) and the prediction of genetic values by the best linear unbiased prediction (Blup).

The aim of the study was to estimate genetic parameters: variance components, heritability, genetic and phenotypic correlations, and genetic gains for height, dbh, and volume traits for *Pinus kesiya* clonal seed orchard in Malawi using Reml and prediction of additive genetic and genotypic values by Blup.

2. Materials and Methods

2.1. Study Site. The study was conducted in Malawi located in Southern Africa in the tropical savannah region at Mapale, Dedza (14°21′ S, 34°19′ E, and 1690 m above sea level). Mapale receives from 1200 mm to 1800 mm rainfall per annum, with annual temperature ranging from 7°C to 25°C. It is situated about 85 km southeast of the capital Lilongwe.

2.2. Plant Material and Seed Orchard. The study was carried out with 88 clones of *Pinus kesiya* seed orchard, which was established in 1972. The clones were selected phenotypically for growth from plantations in Kenya, Malawi, Zambia, and Zimbabwe. The seed orchard was established in a 10 × 10 triple lattice, five trees per plot, and planted following a randomized complete block design in four replications. The trees were planted at a space of 6 × 6 m. At the ages of 18, 23, 30, 35, and 40 years after planting, data were collected for the following traits: total height (distance along the axis of the stem of the tree from the ground to the uppermost point), dbh, and true volume. Total height was measured using a Suunto clinometer with standard, while dbh was measured at 1.3 m above the ground for each standing tree using a calliper. Tree volume was calculated from the tree dbh and height using a tree volume function [5].

2.3. Statistical Analysis. Data obtained were subjected to Kolmogorov-Smirnov D and normal probability plot tests using Statistical Analysis of Systems software version 9.1.3 [11]. This was done in order to check the normality of the data. The characteristics of the data sets for the traits analysed are given in Table 1. Estimation of variance components, heritability, predicted breeding values (EBVs), genetic and phenotypic correlations, and genetic gains was undertaken with the statistical software ASReml [12] using the following individual tree model:

$$Y = Xb + Za + Wc + e, \tag{1}$$

where Y is vector observation; b, a, c, and e are the data vectors of fixed effects (block means), of additive genetic effects (random), of plot effects (random common environment effects of the plots), and of the random errors, respectively; X, Z, and W are known matrices of incidences, formed by the values 0 and 1 which associate the incognita b, a, and c, respectively, with the data vector Y. Approximate standard errors of statistics were obtained by Taylor expansion within the ASReml programme.

3. Results and Discussion

The overall means, standard deviation, and coefficient of variation for height, dbh, and volume and the number of trees at each age are shown in Table 1. The coefficient of variation at all ages was relatively low for height and dbh, ranging from 8.4% to 9.2% and from 7.3% to 8.6%, respectively,

TABLE 2: Variance components and heritability (standard errors) for individual height, dbh and volume traits at different ages.

Age (years)	Trait	σ_A^2	σ_P^2	σ_e^2	σ_w^2	h^2 (s.e)	G_s (%)
18	Height (h)	174	14	81	269	0.48 (0.01)	13.5
	dbh	258	45	324	627	0.26 (0.01)	15.7
	Volume	22	2	11	35	0.46 (0.01)	18.9
23	Height (h)	179	12	80	265	0.50 (0.02)	18.5
	dbh	267	44	311	614	0.28 (0.03)	21.1
	Volume	23	2	11	35	0.48 (0.04)	23.4
30	Height (h)	180	12	80	264	0.50 (0.02)	19.7
	dbh	275	43	309	610	0.29 (0.02)	22.3
	Volume	24	2	11	34	0.51 (0.03)	24.5
35	Height (h)	118	13	71	265	0.34 (0.03)	13.3
	dbh	191	36	323	624	0.19 (0.02)	15.8
	Volume	16	2	10	33	0.36 (0.03)	17.6
40	Height (h)	65	10	70	255	0.19 (0.02)	10.6
	dbh	147	35	320	621	0.15 (0.01)	11.2
	Volume	9	2	10	32	0.20 (0.03)	11.9

σ_A^2: additive genetic variance, σ_P^2: genetic variance among families, σ_w^2: genetic variance within families, σ_e^2: residual variance, h^2: heritability, G_s: genetic gain.

TABLE 3: Variance components and heritability (standard errors) among families for height, dbh and volume traits at different ages.

Age (years)	Trait	σ_A^2	σ_P^2	σ_e^2	σ_w^2	h^2 (s.e)	G_s (%)
18	Height (h)	130	25	174	329	0.51 (0.01)	11.2
	dbh	200	31	207	438	0.51 (0.01)	12.5
	Volume	19	2	8	29	0.59 (0.01)	14.8
23	Height (h)	135	26	168	320	0.53 (0.03)	14.3
	dbh	206	32	200	431	0.53 (0.04)	16.4
	Volume	20	2	8	28	0.59 (0.04)	17.6
30	Height (h)	137	27	166	318	0.54 (0.05)	15.6
	dbh	210	32	200	429	0.53 (0.06)	16.8
	Volume	21	2	8	27	0.59 (0.04)	17.9
35	Height (h)	123	21	185	349	0.45 (0.03)	11.9
	dbh	195	23	216	444	0.43 (0.02)	12.1
	Volume	13	2	15	39	0.47 (0.05)	12.6
40	Height (h)	114	16	241	385	0.33 (0.02)	10.1
	dbh	172	19	299	492	0.32 (0.04)	10.9
	Volume	11	2	24	42	0.38 (0.05)	11.3

σ_A^2: additive genetic variance, σ_P^2: genetic variance among families, σ_w^2: genetic variance within families, σ_e^2: residual variance, h^2: heritability, G_s: genetic gain.

indicating that reliable estimates can be obtained from the variance analyses. The coefficient of variation for volume was moderate, ranging from 20.2% to 23.5%. Higher CV value for volume is expected, when comparing to height and dbh parameters, as volume is estimated from these two variables, combining the experimental errors of both of them [13].

3.1. Variance Components and Heritability Estimates. Variance components and heritability values are given in Tables 2, 3, and 4. The results indicate that additive variances for all traits in all the three levels (individual, among familie and within families) peaked at 18 years and continued to increase with age up to 23 years and almost remained constant up to 30 years of age and then decreased with age. A similar trend for heritability was also observed. All heritabilities were relatively high, suggesting a strong genetic control of the traits at the individual level, among families, and within families. These results suggest that important genetic progress can be achieved using a simple individual selection in the orchard or a combined selection among and within families. The values observed in this study at the age of 18 years are in agreement with those reported by [1]. However, they are higher than those reported by [14] working with *Pinus caribaea hondurensis* from Isla de Guanaja, confirming the promising genetic control of the traits as well as the high potential of the population for selection.

This study and that of [15] differ in the age of maximum heritability; the estimates were maximum at the same mean

TABLE 4: Variance components and heritability (standard errors) within families for height, dbh and volume traits at different ages.

Age (years)	Trait	σ_A^2	σ_p^2	σ_e^2	σ_w^2	h^2 (s.e)	G_s (%)
18	Height (h)	156	12	113	281	0.51 (0.01)	17.8
	dbh	184	48	385	617	0.22 (0.01)	18.5
	Volume	21	4	15	40	0.43 (0.01)	22.7
23	Height (h)	158	10	102	278	0.53 (0.03)	24.6
	dbh	196	36	376	614	0.26 (0.03)	26.7
	Volume	22	4	14	39	0.46 (0.04)	28.9
30	Height (h)	159	10	99	276	0.54 (0.05)	24.9
	dbh	210	32	371	610	0.29 (0.04)	28.3
	Volume	22	4	13	38	0.47 (0.04)	29.6
35	Height (h)	140	17	108	298	0.41 (0.04)	20.7
	dbh	180	53	380	658	0.19 (0.03)	21.4
	Volume	20	6	15	41	0.33 (0.03)	23.2
40	Height (h)	117	23	113	301	0.31 (0.03)	11.2
	dbh	175	79	381	681	0.14 (0.05)	11.6
	Volume	16	7	16	45	0.20 (0.03)	12.5

σ_A^2: additive genetic variance, σ_p^2: genetic variance among families, σ_w^2: genetic variance within families, σ_e^2: residual variance, h^2: heritability, G_s: genetic gain.

TABLE 5: Genetic (below diagonal) and phenotypic (above diagonal) correlations and their standard errors in parenthesis for height, dbh and volume traits at different ages.

Age (years)	Trait	Height (h)	dbh	Volume
18	Height (h)		0.68 (0.05)	0.74 (0.04)
	dbh	0.79 (0.04)		0.95 (0.01)
	Volume	0.86 (0.03)	0.97 (0.01)	
23	Height (h)		0.67 (0.05)	0.72 (0.04)
	dbh	0.76 (0.05)		0.93 (0.02)
	Volume	0.85 (0.04)	0.95 (0.01)	
30	Height (h)		0.65 (0.05)	0.69 (0.05)
	dbh	0.75 (0.05)		0.92 (0.02)
	Volume	0.82 (0.03)	0.94 (0.01)	
35	Height (h)		0.62 (0.06)	0.68 (0.05)
	dbh	0.71 (0.06)		0.91 (0.02)
	Volume	0.81 (0.04)	0.92 (0.02)	
40	Height (h)		0.60 (0.05)	0.63 (0.07)
	dbh	0.69 (0.07)		0.90 (0.03)
	Volume	0.78 (0.05)	0.91 (0.03)	

height, suggesting a possible link between mean height and heritability estimate. This is consistent with findings of [16]. The change in heritability in long rotation crops such as trees is not surprising since genes involved in growth may change with age [17], and these changes may be related to different growth phases [18]. In animals, this change in heritability with age was also attributed to the fact that the trait may change genetically with age [19] and is probably related to different growth phases as reported for trees. These growth phases might be due to changing influences of maternal effects in animals and to a lesser extent in trees and to nursery or competition effects in trees. Changes in heritability with age here may also be attributed to management practices.

3.2. Genetic and Phenotypic Correlations. The genetic and phenotypic correlations for height, dbh, and volume are given in Table 5. Age-age genetic and phenotypic correlations between traits were high, ranging from 0.69 to 0.97 and from 0.60 to 0.95, respectively. As the age interval increased, genetic and phenotypic correlations for all traits decreased. Genetic correlations were generally higher than phenotypic correlations. This suggests the possibility of indirect selection in trait with direct selection in another trait. As previously discussed, volume was the indicative trait for selection, but this trait was associated with a high experimental error. Thus, as dbh and volume presented very high genetic and pheno-typic correlations (0.97 and 0.95, resp.) and dbh is an easily measurable trait, the selection can be based on this specific trait, resulting in indirect gains in volume. This means that selection of plus trees, aiming to maximise the genetic gains in volume, must be based on the dbh because of the high additive genetic correlation and low standard deviation between dbh trait and volume. These results are in agreement with those in the literature [13–15].

3.3. Genetic Gains. Estimates of genetic gains for selection of 50% of families and 50% of trees within families for height, dbh, and volume for *Pinus kesiya* at different ages are given in Tables 2, 3, and 4. The genetic gains were high for all studied traits. These results suggest that the growth improvement through individual selection in the orchard or a combined selection among and within families is possible. The results show that the genetic gains increased with an increased age up to the age of 30 years and decreased with an increased age after the age of 30 years for all the traits. This indicates that the rotational age of the *Pinus kesiya* clonal seed orchard is 30 years of age. According to Andersson et al. [20], Eriksson et al. [21], and Prescher [22], genetic gain is one of the important factors to consider when considering optimal active life span of a seed orchard. Prescher [22] explained that as long as the genetic gain is increasing, the seed orchard can produce genetically better reproductive material. Therefore, this paper recommends an establishment of a new *Pinus kesiya* clonal seed orchard. However, selective harvest of clones with high breeding values in the old seed orchard should be considered so that the best parents in the old orchard can continue to contribute until the new orchard is well established.

4. Conclusion

The evaluated *Pinus kesiya* clonal seed orchard presented potential for improvement in view of high heritability estimates for the traits' height, dbh, and volume. The correlation among growth traits was significantly high and the accuracy of the predicted genotypic values was also of high magnitude, confirming the reliability of the genetic gain estimates. The predicted genetic gains showed that the optimal rotational age of the *Pinus kesiya* clonal seed orchard is 30 years of age; therefore, it is recommended to establish a new *Pinus kesiya* clonal seed orchard. However, selective harvest of clones with high breeding values in the old seed orchard should be considered so that the best parents in the old orchard can continue to contribute until the new orchard is well established.

Acknowledgments

The authors thank Mr. Edward Moyo and his colleagues of Forestry Research Institute of Malawi (FRIM) Centre for providing them with data that were used in this study.

References

[1] R. F. Missio, A. M. Silva, L. A. S. Dias, M. L. T. Moraes, and M. D. V. Resende, "Estimates of genetic parameters and prediction of additive genetic values in *Pinus kesiya* progenies," *Crop Breeding and Applied Biotechnology*, vol. 5, pp. 394–401, 2005.

[2] N. Nyunaï, *Pinus Kesiya Royle Ex Gordon*, PROTA, Wageningen, The Netherlands, 2008.

[3] M. Ertekin, *Clonal Variations in Flowering, Cone Production and Seed Characteristics of Black Pine (Pinus nigra Arnold. Subsp. pallasiana (Lamb.) Holmboe) Seed Orchard originated from Yenice-Bakraz [Ph.D. thesis]*, Zonguldak Karaelmas University, Zonguldak, Turkey, 2006.

[4] D. P. Gwaze, J. A. Wolliams, P. J. Kanowski, and F. E. Bridgwater, "Interactions of genotype with site for height and stem straightness in *Pinus taeda* in Zimbabwe," *Silvae Genetica*, vol. 50, no. 3-4, pp. 135–140, 2001.

[5] C. L. Ingram and N. W. S. Chipompha, *The Silvicultural Guide Book of Malawi*, FRIM, Malawi, 2nd edition, 1987.

[6] D. Gömöry, R. Bruchánik, and R. Longauer, "Fertility variation and flowering asynchrony in *Pinus sylvestris*: consequences for the genetic structure of progeny in seed orchards," *Forest Ecology and Management*, vol. 174, no. 1–3, pp. 117–126, 2003.

[7] R. Díaz and E. Merlo, "Genetic variation in reproductive traits in a clonal seed orchard of *Prunus avium* in northern Spain," *Silvae Genetica*, vol. 57, no. 3, pp. 110–118, 2008.

[8] C. T. Chao and D. E. Parfitt, "Genetic analyses of phenological traits of pistachio (*Pistacia vera* L.)," *Euphytica*, vol. 129, no. 3, pp. 345–349, 2003.

[9] R. Díaz, R. Zas, and J. Fernández-López, "Genetic variation of Prunus avium in susceptibility to cherry leaf spot (*Blumeriella jaapii*) in spatially heterogeneous infected seed orchards," *Annals of Forest Science*, vol. 64, no. 1, pp. 21–30, 2007.

[10] K. Ritland, "Extensions of models for the estimation of mating systems using n independent loci," *Heredity*, vol. 88, no. 4, pp. 221–228, 2002.

[11] *SAS 9.1.3 Qualification ToolS User's Guide*, SAS Institute, Cary, NC, USA, 2004.

[12] A. R. Gilmour, B. J. Gogel, B. R. Cullis, and R. Thompson, *ASReml User Guide Release 3.0*, VSN International, Hemel Hempstead, UK, 2009.

[13] E. V. Tambarussi, A. M. Sebbenn, M. L. T. de Moraes, L. Zimback, E. C. Palomino, and E. S. Mori, "Estimative of genetic parameters in progeny test of *Pinus caribaea* Morelet var. *hondurensis* Barret & Golfari by quantitative traits and microsatellite markers," *Bragantia*, vol. 69, no. 1, pp. 39–47, 2010.

[14] P. T. B. Sampaio, M. D. V. Resende, and A. J. Araújo, "Estimativas de parâmetros genéticos e métodos de seleção para o melhoramento genético de *Pinus caribaea* var. *hondurensis*," *Pesquisa Agropecuária Brasileira*, vol. 35, no. 11, pp. 2243–22253, 2000.

[15] R. F. Missio, L. A. S. Dias, M. L. T. Moraes, and M. D. V. Resende, "Selection of *Pinus caribaea* var. *bahamensis* progenies based on the predicted genetic value," *Crop Breeding and Applied Biotechnology*, vol. 4, pp. 399–3407, 2004.

[16] D. I. Matziris, "Genetic variation and realized genetic gain from Aleppo pine tree improvement," *Silvae Genética*, vol. 49, no. 1, pp. 5–10, 2000.

[17] V. P. G. Moura and W. S. Dvorak, "Provenance and family variation of *Pinus caribaea* var. *hondurensis* from Guatemala and Honduras, grown in Brazil, Colombia and Venezuela," *Pesquisa Agropecuária Brasileira*, vol. 36, no. 2, pp. 225–234, 2001.

[18] R. B. da Costa, M. D. V. de Resende, A. J. de Araujo, P. de Souza Gonçalves, and A. R. Higa, "Selection and genetic gain in rubber tree (*Hevea*) populations using a mixed mating system," *Genetics and Molecular Biology*, vol. 23, no. 3, pp. 671–679, 2000.

[19] E. Missanjo, V. Imbayarwo-Chikosi, and T. Halimani, "Estimation of genetic and phenotypic parameters for production traits

and somatic cell count for jersey dairy cattle in zimbabwe," *ISRN Veterinary Science*, vol. 2013, Article ID 470585, 5 pages, 2013.

[20] B. Andersson, B. Elfving, T. Persson, T. Ericsson, and J. Kroon, "Characteristics and development of improved *Pinus sylvestris* in northern Sweden," *Canadian Journal of Forest Research*, vol. 37, no. 1, pp. 84–92, 2007.

[21] G. Eriksson, I. Ekberg, and D. Clapham, *An Introduction To Forest Genetics*, Uppsala, Sweden, 2006.

[22] F. Prescher, *Seed orchards—genetic considerations on function, management and seed procurement [Ph.D. thesis]*, Swedish University of Agricultural Sciences, Umeå, Sweden, 2007.

10

Using the Contingent Grouping Method to Value Forest Attributes

Pere Riera,[1] Joan Mogas,[2] and Raul Brey[3]

[1] *Department of Applied Economics, Autonomous University of Barcelona, 08193 Bellaterra, Spain*
[2] *Department of Economics, Rovira i Virgili University, Avinguda Universitat 1, 43204 Reus, Spain*
[3] *Department of Economics, Pablo de Olavide University, Ctrretera de Utrera km 1, 41013 Sevilla, Spain*

Correspondence should be addressed to Joan Mogas; joan.mogas@urv.cat

Academic Editors: P. Newton and M. Pensa

This paper presents the first application of a recently proposed stated preference valuation method called contingent grouping. The method is an alternative to other choice modeling methods such as contingent choice or contingent ranking. It was applied to an afforestation program in the northeast of Spain. The attributes included (and the marginal values estimated per individual) were allowing picnicking in the new forests (€2.47), sequestering 1000 tons of CO_2 (€0.04), delaying the loss of land productivity by 100 years, due to erosion in the new forests area (€0.783), and allowing four-wheel driving (€6.5), which is perceived as a welfare loss.

1. Introduction

Forest ecosystems generate a wide variety of goods and services not only for the forest owners but also for society at large. They provide a number of public goods, like enjoyment from recreational opportunities, nontimber products (e.g., mushrooms, berries, or aromatic herbs), carbon sequestration, erosion prevention and biodiversity preservation, among others. In order to make sound decisions for the whole society, forest planning and management ought to take into account the value of forests for both the landowner and the other affected persons. The field of economics helps in this process by being able to estimate the value, in monetary units, of the forest at stake. Their estimation could constitute a significant source of information for further forest policy design and the development of financial instruments.

Forest valuation is often undertaken from choice modeling techniques. They involve surveying people and asking them to state their preferences among a set of alternatives characterized by attributes fixed at different levels [1]. These preferences may be stated, for example, selecting the most preferred alternative from a choice set (named choice experiment; see, e.g., Louviere et al. [2]) or ranking the alternatives included in the choice set (named contingent

ranking; see, e.g., Chapman and Staelin [3]) according to their preferences. The different choice modeling variants, like the aforementioned contingent choice and contingent ranking, are able to obtain separate social values for different forest goods and services.

Recently, Brey et al. [4] proposed a variant named contingent grouping (CG). It requests individuals to classify alternatives included in a choice set as "better than" or "worse than" a status quo or reference situation. The purpose of this paper is to illustrate an application of CG in order to determine how Catalan people choose among potential afforestation programs described by six attributes. The marginal economic values of several forest functions are estimated using a survey with a CG elicitation format undertaken in Catalonia, a region in the northeast of Spain. Following Hensher and Greene [5] and Hensher et al. [6], the WTP values are presented in three different ways: a point estimates, the confidence interval, and the whole distribution. The latter can help to better understand the results obtained.

The structure of this paper is as follows. Section 2 introduces the CG method. Section 3 describes the application. Section 4 reports and discusses the results. Finally, Section 5 summarizes the main conclusions.

2. Methodology

Within the different valuation methods at hand, stated preference methods are most often used when valuing changes in forests or forest-related goods and services. The contingent valuation method (CVM) tends to be applied when a holistic approach is required, whereas attribute-based valuation methods (ABVM) can isolate the value of different forest attributes.

Typically, CVM analyzes the individual tradeoff between the provision of a good and a payment. The elicitation question can take different formats [7]. For instance, in the closed-ended dichotomous choice variant [8, 9], people may be asked whether they would accept the provision of a nonmarket good at a given price (usually called bid) that varies among subsamples. The allowable answer is therefore closed and dichotomous: yes or no, although a fraction of the population may not know or may not the answer. From the yes or no responses to the different bid amounts, the researcher estimates a response probability function, out of which the mean or median of the individual maximum willingness to pay (WTP) is computed as economic value. Some CVM applications to forests are, for example, [10–16].

An ABVM contains a set, or several sets, of alternatives (choice set) defined by options with different attribute levels, varying across the sample. Usually, the choice sets include a status quo or "business-as-usual" (BAU) option. Individuals express their preferences for the alternatives making some kind of choices, like picking the most preferred-called choice experiment or contingent choice (CC), or ranking the alternatives of the choice set from best to worse, or vice versa, labeled contingent ranking (CR). If one of the attributes is the money that a person would have to pay to secure the change, it is possible to generate estimates of the marginal value of changes in each attribute. In a sense, these methods are able to provide more information than CVM, but at the expense of a more demanding statistical treatment, and maybe a higher burden on the individuals.

CC seems to be relatively easier for respondents to answer than CR, since it resembles the kind of choices that individuals face in actual markets. From a statistical point of view, though, CC provides less information per choice set and individual than CR [1]. The increased complexity of CR may lead some individuals to rank arbitrarily or to engage in strategies of ranking different from utility maximization, which could lead to unreliable welfare measures [17–19]. CC exercises have been applied to forest ecosystems [11, 20–23]. Examples of CR applications in the field of forest economics can be found [24, 25].

A new variant of the ABVM family for modeling preferences for goods, where goods are described in terms of their attributes and levels, is the contingent grouping method (CG) [4, 26]. This variant requests individuals to classify alternatives included in the choice set as "better than" or "worse than" a status quo or reference alternative. This elicitation method provides welfare measures conforming to standard consumer theory. Contingent grouping improves upon CC by generally collecting more information without the cognitive demands of CR [4]. In a way it can be regarded

as an appealing alternative to CC and CR, when balancing the amount of statistical information and burden on respondents.

The CG method can be introduced as follows. Consider an individual i, $i \in \{1, \ldots, N\}$, facing a set of five alternatives, as will be used in the empirical application. Denote these alternatives as a, b, c, d, and q, where q represents the status quo or BAU alternative. The utility provided by an alternative $j \in \{a, b, c, d, q\}$ can be expressed with a deterministic and a stochastic part, as in

$$U_{ij} = \beta_i' x_{ij} + \varepsilon_{ij}, \qquad (1)$$

where x_{ij} is a vector of observable variables describing the alternative j for individual i, β_i is a vector of unobserved coefficients representing the tastes of individual i, and ε_{ij} is an unobserved stochastic component independent of x_{ij} and β_i.

Next, consider that the individual is requested to classify the alternatives as "better than" or "worse than" the status quo alternative. Under the assumption of rational behavior, the individual will group the different alternatives according to whether they provide more or less utility than q. Therefore, there are 16 possible complete grouping results (see Table 1). In other words, assuming that the probability for an individual to select an alternative j from the choice set is given by the probability that its utility is greater than the sum of the utilities of all the other alternatives included in the choice set and that the utilities of these alternatives are independent (i.e., we have independent RUM models [27]), information about the individual preferences underlying the sixteen different groupings can be obtained as shown in Table 1.

Depending on the BAU position in the choice set in terms of the utility provided to the individual, the information obtained differs. Thus, a BAU dummy variable is to be included in the utility function model to obtain consistent estimates. This variable informs of where option q stands in relation to the other alternatives in the preferences of the individual. Using conditional logit models to estimate these probabilities, Brey et al. [4] showed by numerical simulation that CG generally provides better welfare measure estimates than CC but worse than CR.

3. Application

The case study application took place in Catalonia, a region in the northeast of Spain. About 40% is occupied by forests. Having a Mediterranean climate in most of the region, Aleppo Pine (*Pinus halepensis*) is the dominant species; coniferous trees account for 73% of the total forest area. Deciduous trees occupy 14% of the forest land, being Holm oaks (*Quercus ilex*) the most abundant species. Three quarters of the forest are in private hands, whereas the rest is publicly own, mostly municipal [28]. However, commercial forests are only marginal, with 2% of the agrarian production in Catalonia [29]. The main reason is the low profitability of its timber. Nevertheless, forests provide the Catalan society with many goods and services that seem to increase due to the economic development in the last decades [30].

TABLE 1: Grouping and information about the individual's preferences.

Groupings*	Information about the individual's preferences**
(i) $a, b, c, d > q$	(i′) $P(U_a \mid U_a, U_q)P(U_b \mid U_b, U_q)P(U_c \mid U_c, U_q)P(U_d \mid U_d, U_q)$
(ii) $a, b, c > q > d$	(ii′) $P(U_a \mid U_a, U_d, U_q)P(U_b \mid U_b, U_d, U_q)P(U_c \mid U_c, U_d, U_q)P(U_q \mid U_d, U_q)$
(iii) $a, b, d > q > c$	(iii′) $P(U_a \mid U_a, U_c, U_q)P(U_b \mid U_b, U_c, U_q)P(U_d \mid U_c, U_d, U_q)P(U_q \mid U_c, U_q)$
(iv) $a, c, d > q > b$	(iv′) $P(U_a \mid U_a, U_b, U_q)P(U_c \mid U_b, U_c, U_q)P(U_d \mid U_b, U_d, U_q)P(U_q \mid U_b, U_q)$
(v) $b, c, d > q > a$	(v′) $P(U_b \mid U_a, U_b, U_q)P(U_c \mid U_a, U_c, U_q)P(U_d \mid U_a, U_d, U_q)P(U_q \mid U_a, U_q)$
(vi) $a, b > q > c, d$	(vi′) $P(U_a \mid U_a, U_c, U_d, U_q)P(U_b \mid U_b, U_c, U_d, U_q)P(U_q \mid U_c, U_d, U_q)$
(vii) $a, c > q > b, d$	(vii′) $P(U_a \mid U_a, U_b, U_d, U_q)P(U_c \mid U_b, U_c, U_d, U_q)P(U_q \mid U_b, U_d, U_q)$
(viii) $a, d > q > b, c$	(viii′) $P(U_a \mid U_a, U_b, U_c, U_q)P(U_d \mid U_b, U_c, U_d, U_q)P(U_q \mid U_b, U_c, U_q)$
(ix) $b, c > q > a, d$	(ix′) $P(U_b \mid U_a, U_b, U_d, U_q)P(U_c \mid U_a, U_c, U_d, U_q)P(U_q \mid U_a, U_d, U_q)$
(x) $b, d > q > a, c$	(x′) $P(U_b \mid U_a, U_b, U_c, U_q)P(U_d \mid U_a, U_c, U_d, U_q)P(U_q \mid U_a, U_c, U_q)$
(xi) $c, d > q > a, b$	(xi′) $P(U_c \mid U_a, U_b, U_c, U_q)P(U_d \mid U_a, U_b, U_d, U_q)P(U_q \mid U_a, U_b, U_q)$
(xii) $a > q > b, c, d$	(xii′) $P(U_a \mid U_a, U_b, U_c, U_d, U_q)P(U_q \mid U_b, U_c, U_d, U_q)$
(xiii) $b > q > a, c, d$	(xiii′) $P(U_b \mid U_a, U_b, U_c, U_d, U_q)P(U_q \mid U_a, U_c, U_d, U_q)$
(xiv) $c > q > a, b, d$	(xiv′) $P(U_c \mid U_a, U_b, U_c, U_d, U_q)P(U_q \mid U_a, U_b, U_d, U_q)$
(xv) $d > q > a, b, c$	(xv′) $P(U_d \mid U_a, U_b, U_c, U_d, U_q)P(U_q \mid U_a, U_b, U_c, U_q)$
(xvi) $q > a, b, c, d$	(xvi′) $P(U_q \mid U_a, U_b, U_c, U_d, U_q)$

* " $>$ " reads "preferred to."
** For example, $P(q \mid a, b, c, d, q)$ denotes the probability of selecting the alternative q from the choice set composed by the alternatives a, b, c, d, and q.

Fire is probably the principal disturbance of the Catalan forest. It plays a relevant role in determining the landscape structure and plant community composition, but also in the amount of carbon sequestration and soil erosion.

An afforestation program was proposed in the valuation exercise, increasing the forest surface from 40% to 50% of Catalonia. The expansion would come at the expense of marginal agricultural land. The proposal was in line with the government's policy. Social and economic changes that occur within developed societies lead up to a situation in which large pieces of agricultural land are being left abandoned as many rural areas become depopulated. Afforestation may be an attractive way of managing fallow lands. Numerous European countries implement subsidized afforestation programs, which provide land owners with financial support for afforestation and management of the planted forest [31, 32].

3.1. CG Design. The program contemplated five different nonmonetary attributes, allowing for picnics in the new forests, four-wheel drive access, picking mushrooms, different amounts of CO_2 sequestration, and deferring erosion over time and a required payment. The attributes and levels are summarized in Table 2. The selection of these attributes and levels come from a combination of expert opinions, focus groups, and pilot tests.

A total of 512 combinations were obtained out of the levels and attributes ($2^3 \times 4^3$). A fractional factorial design was then applied to obtain 16 alternatives that were included in four choice sets of four alternatives each.

3.2. Questionnaire. The first part of the questionnaire described the most relevant aspects of the current forest situation in Catalonia and the afforestation program, with its likely effects. The central part contained the description of the market institution and the value elicitation questions. Each individual was faced with one of the predetermined choice sets and was asked to classify the alternatives as better or worse than the BAU option. The final part of the questionnaire collected several debriefing questions and the request of personal and economic data from the individuals.

The questionnaire went through different focus group sessions and two pilot tests. Focus group participants found the program to be very credible and its description in the questionnaire is understandable for the general public. More than 90% of surveyed people perceived the afforestation program as an overall positive or very positive initiative. Figure 1 reproduces the map included in the questionnaire showing the forested areas before and after the proposed afforestation.

3.3. Data Collection. The questionnaire was administered face-to-face in personal home interviews. A stratified random sample of 800 individuals was selected in Catalonia, from which 732 valid responses were obtained. The strata used were age, gender, and size of the town of residence. The interviews involved 25 randomly selected locations and administered to the population of 18 years of age or older proportionally to the population of each location. In each location, the questionnaires were distributed using random survey routes,

TABLE 2: Attributes and levels used in the choice sets.

Attribute	Description	Levels
Picnic	Picnicking allowed in the new forests (BAU* = No)	Yes No
Drive	Driving by car allowed through the new forests would be allowed (BAU = No)	Yes No
Mushrooms	Picking mushrooms allowed in the new forests (BAU = No)	Yes No
CO_2	CO_2 sequestered annually by the new forests. Equivalent to the pollution produced annually by a city of... (BAU = 0)	300.000 people 400.000 people 500.000 people 600.000 people
Erosion	Erosion risk (If not afforested, land would become unproductive....)** (BAU = 100)	After 100 years After 300 years After 500 years After 700 years
Payment	The afforestation cost per person and year (BAU = 0)	6 euros 12 euros 18 euros 24 euros

*BAU: business-as-usual alternative. **If afforested, erosion would be prevented indefinitely.

with the sample being stratified to include ten respondents selected in terms of gender and age. No significant problems were appreciated in applying the survey. The response rate (91.5%) has been considerably high, suggesting that the cognitive burden of the grouping task may be not too strong.

4. Results and Discussion

The parameters were estimated by maximizing the log-likelihood function $\sum_{i=1}^{N} w_i \ln P_i$, where P_i denotes the i'–xvi' expression for the corresponding event associated to individual i, w_i are the weights associated to the individual, and N is the number of sampled individuals. Weights were included in the log-likelihood function to give the same importance to each individual. Expressions i'–xvi' were estimated using random parameter logit models [5, 33]. Thus, CG is formulated here as a random parameter logit model with one or repeated choices with stable tastes from the sampled individuals.

The estimated coefficients are shown in Table 3. The estimations reported here were programmed in NLOGIT 3 [34], version of August 2005. The program code can be supplied by the authors.

The random variables were determined combining two approaches. One was the classical procedure based on estimating different models and using the likelihood ratio tests to select from them. The other procedure was based on

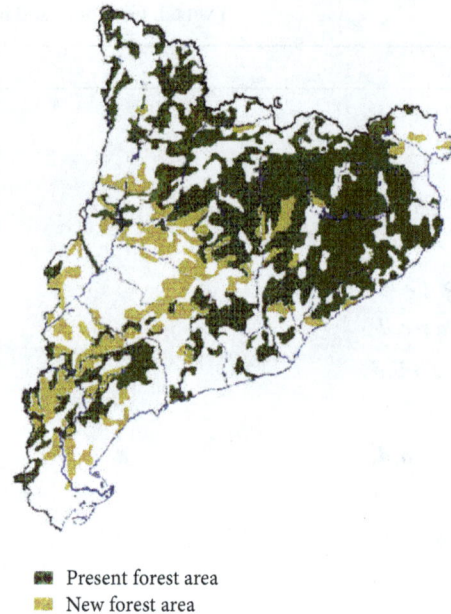

FIGURE 1: Forest areas in Catalonia earlier and the proposed expansions.

Present forest area
New forest area

the inclusion of artificial variables, as suggested by McFadden and Train [35]. The application of both approaches determined variables CO_2 and Payment as random. The estimated spreads for these variables were different from zero at 5% significance level. Out of a number of different distributions tested, the triangular showed the best fit. A triangular density function has the shape of a triangle [36]. One of the main advantages of this distribution is that it has relatively short tails, avoiding the strong influence of outliers on the mean estimation [5]. On the other hand, it may provide difficult-to-explain sign changes for some variables. This could be the case, for example, with variable Payment, where the coefficient is always expected to be negative. To avoid this problem, Hensher and Greene [5] propose making the standard deviation or spread of this distribution a function of its mean, thus ensuring that the distribution provides only positive or negative values. However, it was not necessary to introduce this constraint since the estimated values of the spread for the random parameters were significantly lower than their mean.

All the estimated coefficients, except for Mushrooms, are significant at 10% significance level, and all but Mushrooms and Picnic are significant at 5%. The model contains multiplicative interactions of the status quo dummy variable with some dummies representing different age intervals, since respondents seemed to be more or less prone to select the "business as usual" alternative depending on their age. These dummy variables Age_1, Age_2, and Age_3 are set to 1 for individuals between 30 and 44 years of age, 45 and 64, or more than 60-year-old, respectively. The interval age between 18 and 29-year-old is then considered as the reference interval. The results indicate that the probability of choosing the status quo, everything else held constant, increases with the age of the respondents.

TABLE 3: Model estimations.

Nonrandom parameters	Coefficient	t-statistic
Picnic	0.238	1.924
Drive	−0.626	−4.965
Mushrooms	−0.190	−1.325
Erosion	−0.075	−2.633
SQ	−1.972	−5.812
SQ*Age$_1$	3.416	9.764
SQ*Age$_2$	3.637	11.129
SQ*Age$_3$	4.112	10.322
Random parameters: mean		
CO$_2$	0.263	2.929
Payment	−0.117	−5.663
Random parameters: spread		
CO$_2$	3.436×10^{-4}	3.380
Payment	1.016×10^{-4}	3.602
Observations	732	
log-likelihood	−628.105	
log-likelihood at 0	−821.4110	

The model was estimated using the Halton sequences with 500 replications. In the results reported in this paper, the variables CO$_2$ and Erosion were divided by 10^5 and 10^2, respectively, to facilitate estimation.

4.1. Marginal WTP Estimates. Following Hensher and Greene [5] and Hensher et al. [6], the mean WTP, the confidence interval, and its entire distribution were estimated. The simplest approach estimates coefficients as fixed points. Assuming a linear utility model, the WTP is estimated as the negative of the ratio between the mean coefficients of the corresponding attribute and the coefficient of the Payment attribute. Figure 2 shows the WTP estimates.

To estimate the sampling variance, the Krinsky-Robb procedure was adopted [37]. This procedure uses random draws from the estimated asymptotic normal distribution of parameter estimates to calculate different WTP estimates. Figure 2 shows WTP point estimate and the 95% confidence intervals calculated with 1000 replications.

The estimation of the whole marginal WTP distribution is also based on the Krinsky-Robb procedure. For each replication, marginal WTP values were obtained considering the triangular distribution of the random variables according to the following expressions:

$$\frac{\beta_S}{\beta_{\text{COST·MEAN}} + \beta_{\text{COST·SPREAD}}t_1}, \quad (2)$$

for attributes Picnic, Drive, Mushrooms, and Erosion (S = {Picnic, Drive, Mushrooms, Erosion}), and

$$\frac{\beta_{\text{CO}_2\text{·MEAN}} + \beta_{\text{CO}_2\text{·SPREAD}}t_2}{\beta_{\text{COST·MEAN}} + \beta_{\text{COST·SPREAD}}t_3}, \quad (3)$$

for attribute CO$_2$, where t_1, t_2, and t_3 denote random variables following a triangular distribution with centre at zero and spread of one.

The values obtained from the previous expressions can be used to apply some nonparametric density estimators, allowing the description of the distribution of the marginal WTP. This is done without assuming any underlying analytical distribution. Two different estimators of the density function will be reported here: a histogram, which constitutes the simplest representation, and the kernel density estimator, which can be seen as a generalization and improvement over histograms [38]. Figure 2 shows the histograms and the Kernel density estimates constructed from 5000 simulated observations of the marginal WTP, for the five attributes.

The values of 2.47, −1.978, and −6.505 of Picnic, Mushrooms, and Drive, respectively, correspond to the maximum amount of euros (at 2005 values) that on average an individual would pay annually for a discrete change, from not being allowed to picnic, pick mushrooms, or drive cars in the new forests, to being allowed to do those recreational activities. The marginal CO$_2$ WTP of 2.735 euros reflects the value that a new forest provides to society by sequestering a quantity equivalent to the emissions of CO$_2$ that, on average, a Catalan city of 100.000 inhabitants generates annually in production and consumption activities, that is, approximately equivalent to 68000 tone of CO$_2$ per year [39]. Moreover, in the case of the CO$_2$ sequestered annually by the new forests, the distribution of the marginal WTP has presented a long tail indicating that a considerable part of the population is willing to pay amounts significantly larger than the estimated mean. Finally, 0.78 euros is the marginal WTP for delaying the loss of land productivity for a hundred years, or in other words, not avoiding erosion for that 100 year period would imply a land productivity cost of 0.783 euros per person and year.

Respondents valued positively the availability of new forests where picnicking is allowed, whereas they seem to be more indifferent toward the possibility of picking mushrooms. The nonsignificance of the Mushrooms coefficient might be due to the fact that picking mushrooms is a leisure activity requiring some specialized knowledge, and thus the number of people involved in this activity being more reduced than in other leisure activities in the forests.

The marginal WTP of Drive indicates that people tend to assess negatively this attribute. Furthermore, the estimated density function never reaches positive values, with a relatively long tail to the left. Thus, the four-wheel driving would not be part of the benefits that individuals perceive from a forest area. Allowing driving may be interpreted by individuals as a source of pollution, confronting with some of the environmental values associated to forests. Four-wheel driving may bring to forests one thing that many respondents might be trying to escape from.

Results also suggest that people tend to assess positively the environmental attributes considered, reinforcing the importance of forest environmental services to society.

5. Conclusions

This paper has illustrated the use of a recently proposed stated preference variant, the contingent grouping (CG) method, to analyze the way in which individuals regard some attributes

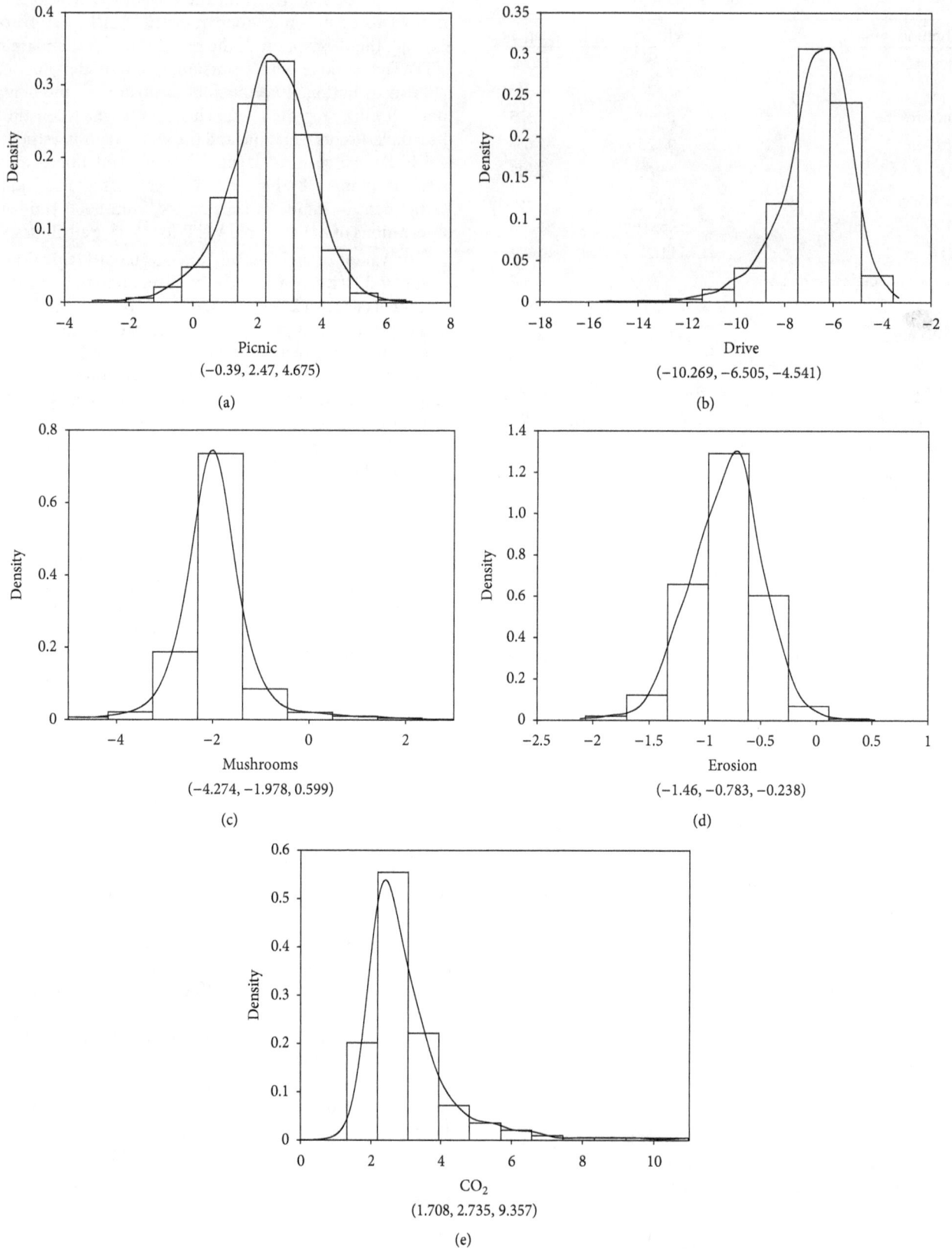

FIGURE 2: Point estimate, confidence intervals, Kernel's density functions, and histograms of the marginal willingness to pay (WTP) estimates for the five attributes (in € of 2005). NB: The first and last values in brackets represent the 95% confidence interval limits, and the values in the middle represent the corresponding point estimates.

representing both recreational and environmental functions of forests in Catalonia, Spain. CG requests individuals to classify alternatives as "better than" or "worse than" a status quo or reference alternative. This elicitation format has only recently been proposed and its feasibility in the empirical ground has to be tested. The results obtained seem to be in line with expectations regarding the sign for the attributes and no particular problems were detected in the CG application, when exposed to respondents. The response rate (91.5%) has been considerably high, suggesting that the cognitive burden of the grouping task may be not too strong.

In this paper, the CG estimations have been formulated as a random parameter logit model with one or repeated choices with stable tastes from the sampled individuals. The WTP estimates have been presented in three different ways: point estimates, confidence intervals, and their entire estimated distribution. On average, in 2005 values, a Catalan over 18-year-old would pay 2.47 euros per year for being allowed to picnic in the new forests and 2.73 euros per year for the CO_2 sequestered annually by the new forests, which is equivalent to the emissions from a Catalan city of 100.000 inhabitants. The relatively large tail of the estimated distribution seems to indicate that a portion of the population has large WTP for this attribute. Respondents value negatively the possibility that driving through the new forests is allowed; this implying an individual cost of 6.5 euros per year. Likewise, they would be willing to pay 0.783 euros per year to delay one 100-year period of the loss of land productivity caused by erosion.

The results may be relevant for policy design in the afforestation practices in Catalonia. For example, the WTP estimates that could be used by forest planners in a social cost-benefit analysis to assess different variants of an afforestation program are the best to implement. Likewise, policy makes facing mushroom picking legislation ought to be aware of the social effects of such regulations. Management plans for public forests can benefit from the estimates if social welfare is taken into consideration. The results could also be used in forest damage assessments and the estimation of compensations.

CG does not seem to have notably different problems from other choice modeling variants. However, more applications would be needed to further test the applicability of CG.

Acknowledgments

The authors greatly acknowledge the partial financial support from the Ministry of the Environment of the Spanish and Catalan governments, and from the Forest Technological Center of Catalonia through its MEDFOREX program and the institutional support of the Ministerio de Educación y Cultura of the Spanish government (grant ECO2010-17728).

References

[1] N. Hanley, S. Mourato, and R. E. Wright, "Choice modelling approaches: a superior alternative for environmental valuation?" *Journal of Economic Surveys*, vol. 15, no. 3, pp. 435–462, 2001.

[2] J. J. Louviere, D. A. Hensher, and J. D. Swait, *Stated Choice Method. Analysis and Application*, Cambridge University Press, Cambridge, UK, 2000.

[3] R. G. Chapman and R. Staelin, "Exploiting rank ordered choice set data within the stochastic utility model," *Journal of Marketing Research*, vol. 19, no. 3, pp. 288–301, 1982.

[4] R. Brey, O. Bergland, and P. Riera, "A contingent grouping approach for stated preferences," *Resource and Energy Economics*, vol. 33, no. 3, pp. 745–755, 2011.

[5] D. A. Hensher and W. H. Greene, "The mixed logit model: the state of practice," *Transportation*, vol. 30, no. 2, pp. 133–176, 2003.

[6] D. A. Hensher, J. M. Rose, and W. H. Greene, *Applied Choice Analysis. A Primer*, Cambridge University Press, New York, NY, USA, 2005.

[7] R. C. Mitchell and R. T. Carson, *Using Surveys to Value Public Goods: The Contingent Valuation Method*, Resource for the Future, Washington, DC, USA, 1989.

[8] R. C. Bishop and T. A. Heberlein, "Measuring values of extramarket goods: are indirect measures biased?" *American Journal of Agricultural Economics*, vol. 61, no. 5, pp. 926–930, 1979.

[9] W. M. Hanemann, "Welfare evaluations in contingent valuation experiments with discrete responses," *American Journal of Agricultural Economics*, vol. 66, no. 3, pp. 332–341, 1984.

[10] R. A. Kramer and D. E. Mercer, "Valuing a global environmental good: U.S. residents willingness to pay to protect tropical rain forests," *Land Economics*, vol. 73, no. 2, pp. 196–210, 1997.

[11] E. Lehtonen, J. Kuuluvainen, E. Pouta, M. Rekola, and C. Z. Li, "Non-market benefits of forest conservation in southern Finland," *Environmental Science and Policy*, vol. 6, no. 3, pp. 195–204, 2003.

[12] M. Lockwood, J. Loomis, and T. DeLacy, "A contingent valuation survey and benefit-cost analysis of forest preservation in East Gippsland, Australia," *Journal of Environmental Management*, vol. 38, no. 3, pp. 233–243, 1993.

[13] M. Lockwood, J. Loomis, and T. DeLacy, "The relative unimportance of a nonmarket willingness to pay for timber harvesting," *Ecological Economics*, vol. 9, no. 2, pp. 145–152, 1994.

[14] J. B. Loomis and A. Gonzalez-Caban, "A willingness-to-pay function for protecting acres of spotted owl habitat from fire," *Ecological Economics*, vol. 25, no. 3, pp. 315–322, 1998.

[15] M. Rekola and E. Pouta, "Public preferences for uncertain regeneration cuttings: a contingent valuation experiment involving Finnish private forests," *Forest Policy and Economics*, vol. 7, no. 4, pp. 635–649, 2005.

[16] R. Scarpa, S. M. Chilton, W. G. Hutchinson, and J. Buongiorno, "Valuing the recreational benefits from the creation of nature reserves in Irish forests," *Ecological Economics*, vol. 33, no. 2, pp. 237–250, 2000.

[17] M. Ben-Akiva, T. Morikawa, and F. Shiroishi, "Analysis of the reliability of preference ranking data," *Journal of Business Research*, vol. 24, no. 2, pp. 149–164, 1991.

[18] V. Foster and S. Mourato, "Testing for consistency in contingent ranking experiments," *Journal of Environmental Economics and Management*, vol. 44, no. 2, pp. 309–328, 2002.

[19] J. A. Hausman and P. A. Ruud, "Specifying and testing econometric models for rank-ordered data," *Journal of Econometrics*, vol. 34, no. 1-2, pp. 83–104, 1987.

[20] W. Adamowicz, J. Swait, P. Boxall, J. Louviere, and M. Williams, "Perceptions versus objective measures of environmental quality in combined revealed and stated preference models of

environmental valuation," *Journal of Environmental Economics and Management*, vol. 32, no. 1, pp. 65–84, 1997.

[21] P. C. Boxall and B. MacNab, "Exploring the preferences of wildlife recreationists for features of boreal forest management: a choice experiment approach," *Canadian Journal of Forest Research*, vol. 30, no. 12, pp. 1931–1941, 2000.

[22] P. Horne, P. C. Boxall, and W. L. Adamowicz, "Multiple-use management of forest recreation sites: a spatially explicit choice experiment," *Forest Ecology and Management*, vol. 207, no. 1-2, pp. 189–199, 2005.

[23] J. Rolfe, J. Bennett, and J. Louviere, "Choice modelling and its potential application to tropical rainforest preservation," *Ecological Economics*, vol. 35, no. 2, pp. 289–302, 2000.

[24] G. D. Garrod and K. G. Willis, "The non-use benefits of enhancing forest biodiversity: a contingent ranking study," *Ecological Economics*, vol. 21, no. 1, pp. 45–61, 1997.

[25] J. Siikamaki and D. Layton, "Pooled models for contingent valuation and contingent raking data: valuing benefits from biodiversity conservation," Working Paper, Department of Agricultural and Resource Economics, University of California, Davis, Calif, USA, 2001.

[26] P. Riera, M. Giergiczny, J. Peñuelas, and P. A. Mahieu, "A choice modelling case study on climate change involving two-way interactions," *Journal of Forest Economics*, vol. 18, no. 4, pp. 345–354, 2012.

[27] R. D. Luce and P. Suppes, "Preference, utility and subjective probabiblity," in *Handbook of Mathematical Psychology*, R. D. Luce, R. R. Bush, and E. Galanter, Eds., J. Wiley and Sons, New York, NY, USA, 1965.

[28] Departament d'Agricultura, *RamaderIa I Pesca, DARP*, Pla General de Política Forestal. Generalitat de Catalunya, Barcelona, Spain, 1994.

[29] C. García, *Estimació de les macromagnituds agràries de les comarques de Catalunya, 1993*, Serveis de Publicacions, Universitat de Lleida, Lleida, Spain, 1997.

[30] M. Merlo and E. Rojas, "Policy instruments for promoting positive externalities of Mediterranean forests," in *Proceedings of the European Forest Institute, Annual Conference*, Chartreuse Ittingen, Switzerland, 1999.

[31] Council Regulation (EC) No 2080/92 Community aid scheme for forestry measures in agriculture, 30 June 1992.

[32] Council Regulation (EC) No 1257/1999 Support for rural development from the European Agricultural Guidance and Guarantee Fund (EAGGF) and amending and repealing certain Regulations, 17 May 1999.

[33] K. Train, *Discrete Choice Methods with Simulation*, Cambridge University Press, New York, NY, USA, 2003.

[34] W. H. Greene, *NLogit Version 3. 0 Reference Guide*, Econometric Software, Plainview, NY, USA, 2002.

[35] D. McFadden and K. Train, "Mixed MNL models for discrete response," *Journal of Applied Econometrics*, vol. 15, no. 5, pp. 447–470, 2000.

[36] M. Evans, N. Hastings, and B. Peacock, *Statistical Distributions*, John Wiley and Sons, New York, NY, USA, 3rd edition, 2000.

[37] I. Krinsky and A. L. Robb, "On approximating the statistical properties of elasticities," *The Review of Economics and Statistics*, vol. 68, no. 4, pp. 715–719, 1986.

[38] B. W. Silverman, *Density Estimation for Statistics and Data Analysis*, Chapman and Hall, New York, NY, USA, 1986.

[39] Departament de Medi Ambient, *Les emissions a l'atmosfera a Catalunya. Una aproximació quantitativa. Quaderns de Medi Ambient*, 5. Generalitat de Catalunya, Barcelona, Spain, 1996.

11

Poverty Alleviation through Optimizing the Marketing of *Garcinia kola* and *Irvingia gabonensis* in Ondo State, Nigeria

A. D. Agbelade[1] and J. C. Onyekwelu[2]

[1] Department of Forestry, Wildlife and Fisheries Management, Ekiti State University, P.M.B. 5363, Ado Ekiti, Ekiti State 300001, Nigeria
[2] Department of Forestry and Wood Technology, Federal University of Technology, P.M.B. 704, Akure, Ondo State 340001, Nigeria

Correspondence should be addressed to J. C. Onyekwelu; onyekwelujc@yahoo.co.uk

Academic Editors: M. Kanashiro and P. Newton

The paper examines poverty alleviation through optimizing the marketing of *Garcinia kola* and *Irvingia gabonensis* in Ondo State, Nigeria. Data for this study were collected using structured questionnaire. Two categories of pretested structured questionnaires were used to obtain information from the respondents (farmers and the marketers of the species). Data analysis was done using descriptive analysis, and Student t-test was used to compare the income generated by the producers and the marketers of the fruits of the tree species. In addition, analysis of variance (ANOVA), arranged in randomized complete block design, was employed to test the significance of price variable across the three market structures (i.e., farm gate price, rural market price, and urban market price). Marketing of forest fruits species is a profitable enterprise with an average profit of ₦19,123.37 per marketer per month. The analysis of variance for the two forest fruit species indicated that *Irvingia gabonensis* generated the highest annual income in rainforest ecosystem while *Garcinia kola* generated the highest annual income in derived savanna ecosystem. Major constraints militating against these forest fruit species are poor market access and infrastructure development. The paper recommended among other things that domestication and interventions of these forest fruit species should be encouraged for proper management and sustainability.

1. Introduction

Nontimber forest products (NTFPs) as part of forestry sector in any economy have always been supportive for many rural dwellers that live within and around the forests estates. In many rural communities, the people depend solely on farming and marketing of NTFPs in order to generate income, boost their economic lives, improve their nutritional intakes and sustain their livelihood. However, the socioeconomic, nutritional, cultural factors, are importance values of NTFPs, especially to rural communities that depend on them [1–3], and were only brought to limelight in recent time. The awareness of the benefits of NTFPs has been on the increase due to the roles it play within the microlevel of the economy and high potential of the products to contribute to the livelihood of the people. In most part of developing countries employment opportunities from traditional industries are declining, people within forest reservation areas look for alternative

sources of income and often turn to the collection of these products from the nearby forest [4].

Garcinia kola (Bitter kola) fruits are harvested annually between July and October, which makes it a highly seasonal product. Bitter kola fruits are smooth and elliptically shaped, with yellow pulp and brown seed coat. *Garcinia kola* has economic value across West African countries where the seeds are commonly chewed and used for traditional ceremonies and medicines. It is highly valued for its perceived medicinal attributes, and the fact that consumption of large quantities does not cause indigestion (as cola nuts do) makes it a highly desired product [5]. The bark when soaked into water can be used as a treatment for intestinal worms and to cure stomach pain. The edible part of *Garcinia kola* fruit aids digestion when eaten raw. The potential utilization of *Garcinia kola* as hop substitutes in beer brewing has been reported [6]. The fruit constitutes an integral part of the rural livelihood of the people, and it boosts their economic status within the rural setting. The potentials of some NTFPs like *Garcinia kola* (Bitter kola) in rural livelihood make it imperative to create awareness on the uses of the fruit as well as its economic importance.

Irvingia gabonensis fruit is harvested annually between April and June of every year. The species, commonly called Bush mango, belongs to the family of Irvingiaceae. The fruit is similar to a small domesticated mango. It is generally green but becomes yellowish when ripe, and the ripe fruit contains a lot of fibers which is good for the body. Commercially, *Irvingia gabonensis* products are highly valued in Nigeria, Cameroon, and Côte d'Ivoire. Market for the species but also not only exists within its natural range, it's also widely traded outside it natural range [3]. Although market for the species is mostly rural, it is also sold in urban centres, where it attracts higher prices. Apart from local and regional market, there is a growing international market for the products of the species. The kernel (Ogbono) is marketed in USA and Europe where about 100,000 potential consumers are found [7, 8]. In Nigeria, annual demand was estimated at 80,000 tonnes [9]. Marketing of Ogbono has the prospect of providing a considerable income generating opportunity for rural people. Currently, a cup of Ogbono costs ₦158 (about US$1) in Nigeria.

2. Methodology

The study was conducted in two ecological zones (Rainforest and Derived Savanna) of Ondo State, Nigeria. Ondo State is one of the thirty-six states in Nigeria, located within the South-Western part of the country. The state lies between latitudes 5°45′ and 7°52′N and longitudes 4°20′ and 6°5′E.

2.1. Data Collection and Analysis. Data for this study were collected using two sets of structured questionnaires. Market information from traders who market the products of the species in selected rural and urban markets was collected, which indicated the rate of supply, demand, prices, and income generation on the forest fruit species. Farmers with *G. kola* and *I. gabonensis* trees on their farms, home gardens, or fallow field were selected and the questionnaire administered to them. Total of 60 questionnaires were administered to farmers while 120 questionnaires were administered to traders. This was complemented with focus group discussion as well as participatory observation. Data collected from the field were analysed using descriptive statistical tools such as mean, frequency, and percentages.

2.2. Linear Statistical Model. In addition, analysis of variance (ANOVA) arranged in randomized complete block design was also employed to test for significant difference parameters (e.g., price, etc.) from the three market structure (farm gate price, rural market price, and urban market price) in the two ecological zones.

The linear statistical model for randomized complete block design (RCBD) is

$$Y_{ijk} = \mu + B_i + T_j + \Sigma_{ijk}, \tag{1}$$

where Y_{ijk} is individual observation for the treatment in the block, μ is general mean, B_i is ecological zones (rainforest and derived savanna ecological zones), T_j is market structure (farm gate, rural and urban markets), and Σ_{ijk} is experimental error.

3. Result

3.1. Age Distribution of the Respondent

3.1.1. Market Assessment. The age range of traders of the two forest fruits tree species in rainforest and derived savanna ecosystems is between 21 and 60 years (Table 1). However, there were indications that majority of the traders are middle aged, especially those involved in the sale of *I. gabonensis* products. For example, results in Table 1 show that between 80 and 93% of traders of *I. gabonensis* in the rainforest ecosystem are between 31 and 50 years old while between 55% and 70% of traders of *G. kola* are between the same age ranges. A good percentage of traders (45% and 20% in rural and urban markets, resp.) of *G. kola* in the rainforest ecosystem are over 50 years old. In the derived savanna ecosystem, the traders are between the age of 31 and 60 years old, except for *I. gabonensis* in urban markets (Table 1). Similar to the results for rainforest, majority of *I. gabonensis* traders are within the 31–50 years age bracket. Over 45% of the traders of *G. kola* in both urban and rural markets in the derived savanna are over 50 years old (Table 1).

3.1.2. On-Farm Assessment. The age range of farmers of these forest food tree species in rainforest and derived savanna ecosystems is between 31 and over 60 years (Figure 1). The results in Figure 1 indicated that elderly people are mostly involved in the production of forest fruit species. In both ecological zones, between 70 and 90% of the farmers are 50 years and above (Figure 1). The results indicated that between 33.3% and 48.4% of the farmers are above 60 years old which indicated that the farmers of forest fruit species are elderly people. The result also shows that middle-aged people are not fully involved in the production of forest fruit species, as only about 4.2% and 9.7% of the farmer's are aged between 31 and 40 years in the two ecological zones of Ondo State (Figure 1).

3.2. Gender of the Respondent

3.2.1. Market Assessment. The result reveals that the female folks are more involved in the marketing of the forest food tree species as shown in Tables 2(a) and 2(b), 10.5% and 89.5% of traders of *Irvingia gabonensis* in rural markets in rainforest ecosystem are males and females, respectively, and all traders (100%) of the species in urban markets in rainforest are female (Table 2(a)). The same trend was observed for *Garcinia kola* trade, where about 9.1% males and 90.9% females were observed to be involved in the sale of the species in rural markets while only female's (100%) are involved in the sales of the fruit in urban. Results (Table 2(b)) revealed that only females are involved in marketing all these forest fruits

TABLE 1: Age distribution of respondent (traders) for market assessment in the two ecological zones (%).

Ecological zone	Age range	Irvingia gabonensis		Garcinia Kola	
		Rural market	Urban market	Rural market	Urban market
Rainforest	21–30 yrs	5.3	6.7	0.0	0.0
	31–40 yrs	36.8	60.0	18.2	40.0
	41–50 yrs	47.4	33.3	36.4	30.0
	51–60 yrs	10.5	0.0	46.4	20.0
	>60 yrs	0.0	0.0	0.0	10.0
Derived Savanna	21–30 yrs	0.0	6.2	0.0	0.0
	31–40 yrs	43.8	12.6	15.4	13.3
	41–50 yrs	50.0	56.2	38.4	40.0
	51–60 yrs	6.2	25.0	23.1	13.3
	>60 yrs	0.0	0.0	23.1	33.4

TABLE 2: (a) Gender and religious status of respondent for market assessment in rainforest ecosystem. (b) Gender and religious status of respondent for market assessment in derived savanna ecosystem.

(a)

	Irvingia gabonensis		Garcinia kola	
	Rural market	Urban market	Rural market	Urban market
Gender				
Male	10.5	0	9.1	0
Female	89.5	100	90.9	100
Religion				
Christianity	100	86.7	100	100
Traditional	0	0	0	0
Islam	0	13.3	0	0

(b)

	Irvingia gabonensis		Garcinia kola	
	Rural market	Urban market	Rural market	Urban market
Gender				
Male	0	0	0	0
Female	100	100	100	100
Religion				
Christianity	56.2	87.5	38.5	60.0
Traditional	0	0	0	6.7
Islam	43.8	12.5	61.5	33.3

species in rural and urban markets in the derived savanna ecosystems.

3.2.2. On-Farm Assessment. The results of on-farm assessment indicated that more males are involved in the farming of the tree species (domestication) than the female across the sampled communities in both ecosystems. Figure 2 shows that between 76.2% and 92.3% males are involved in the

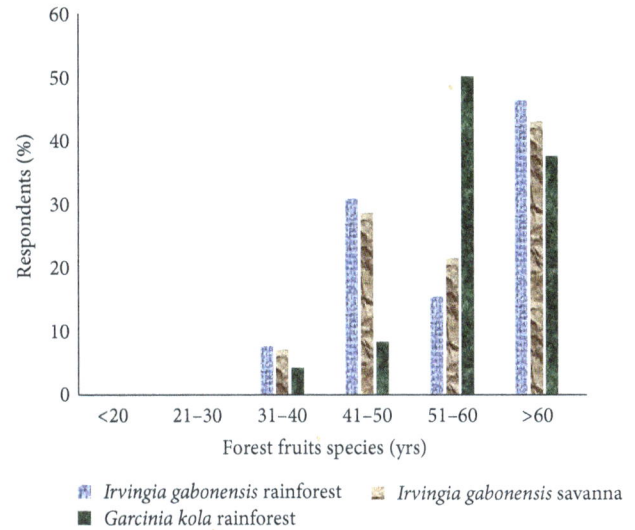

FIGURE 1: Age distribution of respondent for on-farm assessment in the two ecological zones.

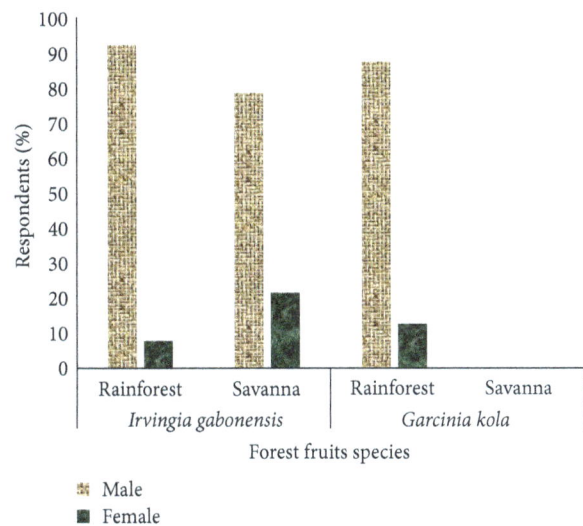

FIGURE 2: Gender distribution of on-farm respondent in the two ecological zones.

conservation and/or planting of trees of *I. gabonensis* and *G. kola* in the rainforest and derived savanna ecosystems while the percentage of females involved in domestication of the species ranged between 7.7 and 23.8%. No farmer is involved in the planting/conservation of *G. kola* in derived savanna (Figure 2) because the ecosystem is outside the natural range of the species.

3.3. Income Generation (in Naira) from Sale of the Species. Annual income generated from sale of the species ranged from <₦50,000.00 to >₦200,000.00 ($316.50 to $1265.80) in the derived savanna ecosystem and >₦50,000.00 to >₦200,000.00 ($316.50 to $1265.80) in the rainforest ecosystem as shown in Table 3. Generally, higher income is generated from the sale of the fruits of the three species by

TABLE 3: Income generated (in Naira) from sale of forest fruit species.

	Irvingia gabonensis		Garcinia kola	
	Rural market	Urban market	Rural market	Urban market
Rainforest				
<50,000	0.0	0.0	0.0	0.0
50,000–100,000	0.0	0.0	0.0	10.0
100,001–150,000	36.8	33.3	27.2	0.0
150,001–200,000	31.6	13.3	45.5	20.0
>200,000	31.6	53.4	27.3	70.0
Derived savanna				
<50,000	37.5	0.0	38.5	0.0
50,000–100,000	12.5	0.0	0.0	0.0
100,001–150,000	18.8	18.8	15.4	33.3
150,001–200,000	31.2	25.0	46.1	40.0
>200,000	0.0	56.2	0.0	26.7

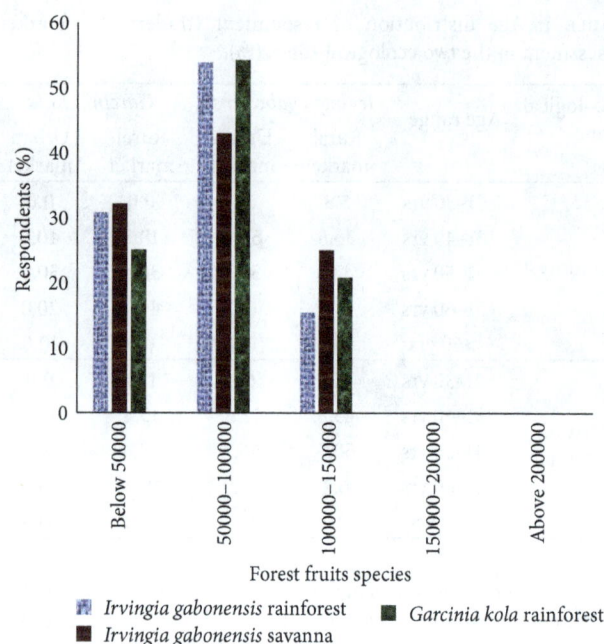

FIGURE 3: Total annual income generated (in Naira) from forest fruits.

the marketer in urban markets than rural markets. For example while between 53.4 and 70.0% of traders in urban markets in the rainforest ecosystem generated over ₦200,000.00 ($1265.80) per annum, only between 27.3 and 31.6% of traders in rural markets generated as much in the rainforest ecosystem. In the derived savanna ecosystem, no trader in the rural markets generated higher income than ₦200,000.00 ($1265.80) per annum. In the rainforest ecosystems majority of urban market traders earned between ₦150,000.00 and >₦200,000.00 ($949.40 and $1265.80) while in rural markets, majority earned between ₦100,000.00 and ₦200,000.00 ($632.91 to $1265.80). In the derived savanna majority of urban market traders earned between ₦100,000.00 and >₦200,000.00 ($632.91 and $1265.80) while in rural markets, majority earned between <₦50,000.00 and ₦200,000.00 ($316.50 to $1265.80) as shown in Table 3. The result indicated that a high percentage (up to 37.5%) of traders in the derived savanna ecosystem generated low annual income of ₦50,000 ($316.50) for *Irvingia gabonensis*, while 38.5 generated low annual income of ₦50,000 ($316.50) for *Garcinia kola*.

3.3.1. On-Farm Assessment. The producers (farmers) of the forest fruit species generated less income than the traders of the forest fruit species across the two ecological zones of Ondo state. The result of income generation shown in Figure 3 indicated that between 32.1% and 30.8% of the farmers in the two ecological zones generated annual income of less than ₦50,000 ($316.50) for *Irvingia gabonensis*, while 25% earn the same amount for *Garcinia kola* in rainforest ecosystem. Higher percentage of farmer (between 25% and 54.2%) had annual income generation of between ₦50,000 to ₦150,000 ($316.50 to $949.40) from the sale of the fruits in the two ecological zones as shown in (Figure 3). No farmer generates any income from the sales of *Garcinia kola* in derived savanna (Figure 3) because the ecosystem is outside the natural range of the species.

3.4. Comparison of Income Generated from Marketing of Forest Fruits Species. The results of analysis of variance indicated that there are significant differences between the average annual incomes generated by respondents from the sale of the fruits of the three species within the two ecological zones of the study as well as between the various market types as shown in Tables 4(a) and 4(b). The results showed a significant difference in income generated from the sale of the fruits of the species as one moves from farm gate (on farm) through rural markets to urban markets. The analysis indicated that income generated was significantly highest in urban markets in the two ecological zones, and it was found to be significantly higher than income obtained from rural markets and farm gates. The difference in income generated was more noticeable in rainforest than in derived savanna ecosystem. For example, while average annual income generated at urban markets in the rainforest ecosystem was ₦191,667.00, ($1213.10), and it was only ₦66,667.00 ($422.00) at farm gate, a difference of over 300% (Table 4(a)). The difference in annual income generation at urban markets ₦123,330.00 ($780.60) and farm gates ₦75,000.00 ($474.70) in derived savanna ecosystem was about 200%. Annual income generated from rural markets in the two ecological zones was significantly higher than that of on-farm (farm gate) income generation, which is the least and less than ₦100,000.00 Naira ($632.91 USD) per annum within the two ecological zones (Table 4(a)). The results also show higher-income generation in rainforest ecosystem than derived savanna for the market analysis and reverse being the case for on-farm analysis which reveals that derived savanna had the higher income than the rainforest ecosystem. The analysis of variance for the two forest fruit species within the two ecological zones is presented in Table 4(b). *Irvingia gabonensis* generated the

TABLE 4: (a) Results of analysis of variance for mean annual income generated (in Naira) and assessed vertically in the two ecological zones and the three market types. (b) Results of analysis of variance for annual income (in Naira) generated from the sales of the two forest fruit species assessed vertically in the two ecological zones.

(a)

Ecological zones	On farm	Rural market	Urban market	
Rainforest	66,667	123,333	191,667	(P > 0.05)
Derived Savanna	75,000	111,667	123,330	

Significantly different from each other at P > 0.05 level of significance.

(b)

Ecological zones	Garcinia kola	Irvingia gabonensis	
Rainforest	133,330	191,667	(P > 0.05)
Derived Savanna	135,000	98,333	

Significantly different from each other at P > 0.05 level of significance.

highest annual income in rainforest ecosystem ₦191,667.00 ($1213.10) of Ondo state while *Garcinia kola* generated the highest annual income in derived savanna ecosystem ₦135,000.00 ($854.43). *Irvingia gabonensis* generated significantly higher income in rainforest than derived savanna ecosystem, and *Garcinia kola* generated significantly higher income in derived savanna ecosystem than rainforest.

4. Discussion

Most of the respondents (54%) are females involved in the sales of forest fruits species, indicating that forest fruits marketing is a female dominated enterprise, and male's are the major collectors and also involved in the production. Majority of the marketers (72%) are within the active labour age range of 20–50 years. Majority of the marketers used "size and sweetness" as the standard measure of selling. Prices of forest fruits are arrived through bargaining power of the sellers and buyers as attested to by 70% of the respondents. About 54% of the respondents claimed that forest fruits are not always available throughout the year due to their seasonal nature and perishability. The most common method of informing buyers of forest fruits is through open display as attested to by 58% of the respondents. The most common strategy of optimising forest fruits marketing among respondents is by processing before sales as attested to by 38% of the respondents. Forest fruits marketing is a profitable enterprise with an average monthly profit of ₦19,123.97 ($121.04) per marketer per month and in turn can alleviate poverty within the household, and this is support of Adebisi 2004 findings [5]. The results of this study revealed that the production, collection, and marketing of *I. gabonensis* and *G. kola* constitute major economic contribution to the livelihood of the people in the rainforest and derived savanna ecological zones of Ondo State. The use of these two forest fruits species adds crucial

dimension to a diversified livelihood base of the rural populace and thereby reducing poverty. Thus, they act as safety net particularly when there is a shortfall in agricultural production and thus fill the gap of food shortage and reduce malnutrition.

5. Recommendations

Forest fruit species contribute significantly to the people's economy and livelihood. Thus, priority should be given to the conservation of the mother trees to ensure sustainable production of the fruits while effort should be made towards their domestication. Currently in the study area, especially in the rainforest ecosystem, domestication of the forest fruit species is inadequate. It is therefore recommended that forest management strategies and interventions should be designed for domestication of the forest fruit species in the two ecological zones to properly address the management and sustainability of the resource base of the species.

References

[1] E. T. Ayuk, B. Duguma, S. Franzel et al., "Uses, management and economic potential of *Irvingia gabonensis* in the humid lowlands of Cameroon," *Forest Ecology and Management*, vol. 113, no. 1, pp. 1–9, 1999.

[2] J. C. Okafor, "Improving edible species of forest products," *Unasylvia*, vol. 42, no. 165, pp. 17–22, 1991.

[3] J. C. Onyekwelu and B. Stimm, "*Irvingia gabonensis*, (Aubrey-Lecomte ex O. Rorke) Baill," in *Enzyklopädie der Holzgewächse*, P. Schütt, H. Weisgerber, H. J. Schuck, and A. Roloff, Eds., vol. 43 of *Ergänzungslieferung*, p. 14, Ecomed, Munich, Germany, 2006.

[4] A. A. Adepoju and A. S. Salau, "Economic valuation of non-timber forest products," MPRA Paper 2689, 2007, http://mpra.ub.uni-muenchen.de/2689/.

[5] A. A. Adebisi, "A case study of *Garcinia kola* nut production-to-consumption system in J4 area of Omo forest reserve, southwest Nigeria," in *Forest Products, Livelihoods and Conservation. Case Studies of Non-Timber Forest Product Systems*, T. Sunderland and O. Ndoye, Eds., vol. 2, pp. 115–132, CIFOR, Africa, 2004.

[6] A. F. Eleyinmi and R. A. Oloyo, "Pilot scale brewing trials using formulated blends of selected local vegetables as hop substitute," *Journal of Food Science and Technology*, vol. 38, no. 6, pp. 609–611, 2001.

[7] FAO, *Non-Wood Forest Products of Central Africa: Current Research Issues and Prospective for Conservation and Development*, T. C. H. Sunderland, L. E. Clark and P. Vantomme, Ed., FAO, Roma, Italy, 2003.

[8] D. O. Ladipo, J. M. Foundoun, and N. Ganga, "Domestication of the Bush Mango (*Irvingia* spp.): some exploitable intraspecific variations in West and Central Africa," in *Domestication and Commercialization of Nontimber Forest Products for Agroforestry*, R. R. B. Leakey, A. B. Temu, M. Melnyk, and P. Vantomme, Eds., Non Wood Products No. 9, pp. 193–205, FAO, Rome, Italy, 1998.

[9] L. C. Nwoboshi, "Meeting the challenges of deforestation in Nigeria through effective reforestation," in *Proceedings of the 1996 Annual Conference of the Forestry Association of Nigeria*, A. B. Oguntala, Ed., Minna, Nigeria, 2000.

12

Sustaining Cavity-Using Species: Patterns of Cavity Use and Implications to Forest Management

Fred L. Bunnell

Forest Sciences Department, University of British Columbia, 3041-2424 Main Mall, Vancouver, BC, Canada V6T 1Z4

Correspondence should be addressed to Fred L. Bunnell; Fred.Bunnell@ubc.ca

Academic Editors: F. Castedo-Dorado, T. S. Fredericksen, H. Nahrung, and J. F. Negron

Many bird and mammal species rely on cavities in trees to rear their young or roost. Favourable cavity sites are usually created by fungi, so they are more common in older, dying trees that are incompatible with intensive fiber production. Forestry has reduced amounts of such trees to the extent that many cavity-using vertebrates are now designated "at risk." The simple model of cavity use presented helps unite research findings, explain patterns of use, and clarify trade-offs that can, or cannot, be made in snag management. Predictions generated are tested using data from over 300 studies. Implications to forest management are derived from the tests, including the following: ensure sustained provision of dying and dead trees, retain both conifers and hardwoods and a range of size and age classes, sustain a range of decay classes, ensure that some large trees or snags are retained, promote both aggregated and dispersed retention of dead and dying trees, meet dead wood requirements for larger species where intensive fibre production is not emphasized, do not do the same thing everywhere, and limit salvage logging after tree mortality. The paper focuses on species breeding in the Pacific Northwest, but draws on data from throughout those species' ranges.

1. Introduction

Most cavities in trees begin with fungi. Because trees resist decay, it takes time for fungi to soften wood enough that cavity excavation by birds is possible. By that time, trees are often old and beginning to die. Old and dying trees reduce economic efficiencies within managed forests, so for decades we have sought to remove them. Our actions were successful, and cavity sites have been much reduced [1–3]. I focus on the Pacific Northwest of North America (PNW), here defined as Alaska, Yukon Territory, Alberta, British Columbia, Washington, Oregon, Idaho, Montana and northern Nevada, and California. Of the 67 vertebrate species commonly using cavities in the PNW, 20 (30%) are designated "at risk" or "potentially at risk." Where forestry has been practiced longer, the proportion of cavity users among forest-dwelling vertebrates designated "at risk" is higher [4].

I review kinds of cavity use, present a general framework of cavity use in the PNW, review key factors influencing cavity use, and interpret those in terms of management implications. Focus is on primary excavators, but all birds and

mammals commonly using cavities are included. Summary tables and figures highlight regional differences: coastal forests (under maritime influence), subboreal plus boreal forest, and inland (all other forests). Common and scientific names follow the American Ornithologists' Union (birds) and British Columbia Conservation Data Centre (mammals).

2. Kinds of Cavity Use

Two broad groups of cavity users are distinguished: primary excavators, such as woodpeckers, that excavate their own cavities and secondary cavity users, such as bufflehead ducks (*Bucephala albeola*), tree swallows (*Tachycineta bicolor*), and northern flying squirrels (*Glaucomys sabrinus*), that use holes excavated by primary cavity excavators. Secondary users include species that seek a particular, often uncommon, form of natural cavity, such as brown creepers (*Certhia americana*), Vaux's swifts (Chaetura vauxi), and several bats plus opportunistic species that use cavities 50% or more of the time in some areas but not everywhere, such as black bears (Ursus americanus), porcupine (Erethizon dorsatum), and some bat and rodent species. In analyses following, such unconventional and opportunistic users are grouped with other secondary cavity users.

Primary excavators show different abilities to excavate. Two groups are recognized: species that forage by drilling, boring, or hammering into wood or soil and species that probe or glean bark, branches, and leaves to acquire prey. The former group is termed strong excavators. It includes most woodpeckers, sapsuckers, and the northern flicker (*Colaptes auratus*). Strong excavators are typically well-adapted for creating holes in trees and have reinforced skulls and ribs and chisel-like beaks [6]. Weak excavators include chickadees, nuthatches, and those woodpeckers that forage primarily by probing and gleaning, extracting seeds, or capturing insects in flight (e.g., acorn (*Melanerpes formicivorus*), downy (*Picoides pubescens*), Lewis's (*M. lewis*), Nuttall's (*P. nuttalli*), and white-headed (*P. albolarvatus*) woodpeckers). Despite using cavities to nest, weak excavators are less well adapted to excavation than are species that drill into wood to forage, so often use cavities initiated by strong excavators. Strong excavators are generally large birds; weak excavators are mostly smaller species.

The larger species of strong primary excavators can act as keystone species, providing nest, den, and roost sites for other cavity-using species. If their requirements are lacking, secondary cavity users may be lost [7–11]. Similarly, sapsucker foraging activity creates feeding opportunities for many other species. At least 23 bird species, 6 mammal species, and numerous arthropods (9 orders and 22 families) have been reported feeding at sapsucker holes [12–14]. Woodpeckers also can sometimes constrain the abundance of forest "pest" insects [15, 16]. Loss of strong excavators would seriously disrupt forest ecosystems.

In the PNW, 67 vertebrate species use cavities more than 50% of the time, either generally or regionally; more species are opportunistic in their use of tree cavities. A small component of strong, primary excavators creates cavity sites for many more species (Figure 1). There are 22 primary cavity excavators and 45 secondary cavity users relying on hollows or cavities, not all of these excavated by birds. Only 9 species are strong excavators, affirming the role of strong excavators as keystone species. Most secondary users rely on holes excavated by primary excavators. The proportion of nest sites of secondary users excavated by other species ranged from 89 to 100% with one exception [7]; neither of two flammulated owl (*Otus flammeolus*) nest sites was excavated.

Birds (48 species) and bats (11 species) represent 88% of species consistently or commonly using cavities. More bird than mammal species use cavities; mammals using cavities or hollows range in size from bats to grizzly bears (*Ursus arctos*). Other than for some bats and squirrels, mammal use of cavities is more opportunistic than it is for birds. Amphibians and reptiles also use cavities in snags and stumps opportunistically [17–19].

Larger snags provide more room and are longer lived, so provide greater opportunities for cavity use. The number of cavity-using species thus decreases with decreasing diameter of the dominant tree species. In the north, where trees are small (Spruce Willow Birch of Figure 1), the numbers of cavity users is much reduced.

Figure 1 excludes opportunistic cavity users that are included with secondary users in analyses following. Many of these are smaller birds and mammals that do not require cavities but will use them; some larger species (e.g., great horned owl, *Bubo virginianus*, and porcupine) also are opportunistic. In any forest type there also are a few individualist species that seek particular cavities not excavated by birds. Vaux's swifts, for example, nest and roost in hollow snags large enough that they can circle into and out of them [20]. Brown creepers, like some amphibians and reptiles, often nest or seek cover under slabs of loose bark [21]. Some cavity users (American martens, *Martes americana*; fishers, *M. pennanti*; black bears) are too large to rely on cavities excavated by birds. They rely on cavities formed by decay or fire. Such cavities are becoming uncommon, because old, large trees are increasingly uncommon.

3. Biology of Cavity Use and Associated Predictions

Hairy woodpeckers (*Picoides villosus*) need about 20 days to excavate their nest [22], as do yellow-bellied sapsuckers (*Sphyrapicus varius*) [23]. That is 20 days of hammering bill and head against wood in a small space, at rates of 100 to 300 times per minute [24]. There has to be some benefit to this behavior, and there is: cavity-using species benefit from the protection cavities, provide their young from predators, and shelter from weather in the nest, den, or roost. Because of their high-energy demands and sensitivity to temperature, bats often select cavities with favourable microclimates [25, 26]. For birds, the ideal cavity nest is located where the outer wood is hard to prevent predators from tearing open the nest, but the inner wood is soft to allow easy excavation. Soft inner wood may be required, due to the difficulty of chiseling wood in an enclosed space. Heart rots, preferably localized, are thus of prime importance. Generally, all primary excavators seek decayed heartwood [27–31]. Stronger excavators will chisel through sound wood to reach decay. For weak primary cavity nesters, harder outer wood is desirable to reduce predation, but these species are not adapted to excavating hard wood. They must either compromise with softened sapwood, or find existing holes in hard outer wood, such as those at dead branch stubs. The ideal cavity also should be high up a tree to avoid ground-dwelling nest predators, such as weasels (*Mustela*) and rodents.

The bole or branch, where the nest is located, must be wide enough to accommodate the animal and its eggs or young. Size of the cavity can affect productivity; several cavity-using species have been found to be more productive in larger nest boxes [32–35]. Together, these two factors (protection from predators and size) promote a preference for large diameter trees in which the bole is a suitable width at a greater height. When trees are sound, greater size also reduces the danger of the tree or snag breaking. The importance of

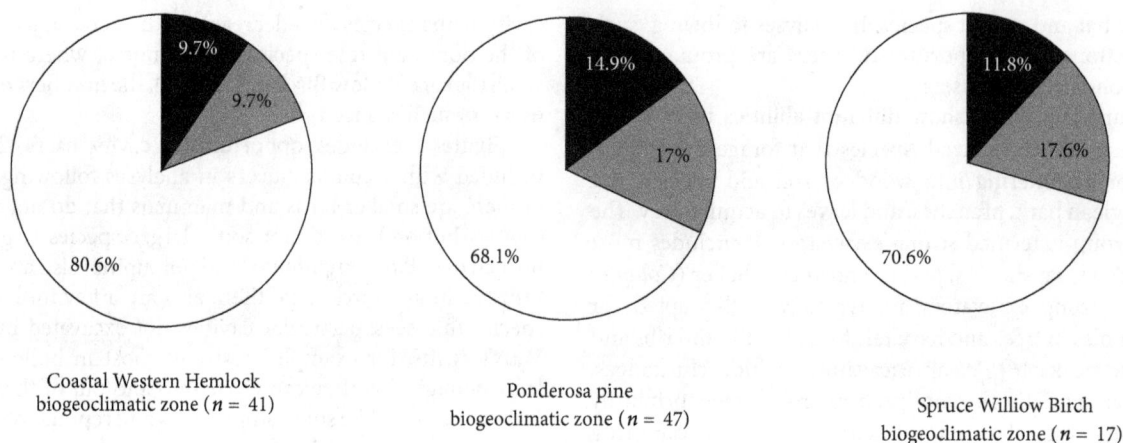

FIGURE 1: Proportions of strong primary excavators (black), weak primary excavators (gray), and secondary cavity users (white) among vertebrates using cavity sites in three biogeoclimatic zones of British Columbia: Coastal Western Hemlock, Ponderosa Pine, and Sitka Willow Birch (*n* = number of cavity-using species). Biogeoclimatic zones are described by [5].

softened heartwood also promotes preference for larger trees, because larger trees are usually older trees that have had time to experience heart rot. The search for decayed wood, appropriate stem diameter at the nest, and height above the ground explain many of the preferences documented for primary excavators.

Unlike primary excavators, excavating a cavity is not part of the breeding behaviour of secondary cavity users. With regard to cavities, secondary cavity users "take what they can get" with certain limitations. The most obvious limitation is that the cavity must be big enough. Mountain bluebirds (*Sialia currucoides)*, for example, rely primarily on nests made by flickers, rather than those of the smaller woodpeckers or sapsuckers [11]. The nest also needs to be near the appropriate foraging habitat, which is not forest for many secondary cavity users (e.g., cavity-nesting ducks, swallows, bluebirds, American kestrels, *Falco sparverius*). Whereas primary excavators usually make or modify a new cavity each year, secondary users will return to the same cavity as long as it remains stable [36]. Because secondary users are relatively abundant compared to the primary excavators, competition for cavities can be intense [37, 38]. Where opportunities exist (e.g., holes at shed branches, stem breakage), secondary users seek more stable, living trees, but many end up using preexcavated holes in dead trees.

The preceding statement on the biology of cavity use provides a simple model or framework to unite research findings, explain patterns of observed use, and clarify trade-offs that can, or cannot, be made in snag management. The framework also generates expectations or predictions that can guide the questions posed in synthesis. Four broad predictions and associated corollaries were tested using attributes commonly employed to describe trees or snags used as cavity sites (tree species, size of tree, decay state, and snag density).

(1) Ability to excavate wood permits greater selectivity, with the range of sites used becoming broader from strong excavators through weak excavators to secondary users. Where possible, secondary users will use live trees.

(2) Strong excavators will select trees with less visible signs of decay than will weak excavators. Intense competition among secondary users will relegate most to use dead trees, where existing cavities are more common, rather than more stable, live trees.

(3) To attain desired protection, selection of tree species will reflect the tree's ability to compartmentalize decay. Deciduous trees more often contain internal rot surrounded by a sound outer shell than do conifers, so will be sought preferentially.

(4) The search for sufficient room and protection from predators will produce selection for the largest suitable cavity site available, as high above ground as possible. Strong excavators will enact selection more effectively.

4. Decay

The energy demands of excavation make softened wood or decay a dominating factor in selection of cavity sites. I review how decay is initiated, then examine patterns of use by primary excavators of cavities and secondary users.

4.1. Natural Disturbance and Decay. How a tree dies influences snag longevity and onset of decay. Research has identified tree species, tree size, decay stage, crown scorch, and stand density as major factors determining snag longevity [39–43]. Natural fire regimes influence the kinds and amounts of dead wood present [44]. The role of fire is apparent within the 12 broad forest types of British Columbia. Within forest types, the proportion of species using cavity sites decreased significantly with increasing fire size and intensity; species richness of cavity users decreased significantly as fire-return interval lengthened, and snags were created less frequently [45, 46]. Generally, snags created by fire fall sooner than do other snags. The species of snag has less effect on snag longevity than the method of creation and tree diameter [42]. Trees of larger diameter remain standing longer regardless of source of mortality, and pines (*Pinus*) stand longer than firs (*Abies*) [42]. *Pinus ponderosa* killed by fire remained standing longer than those killed by bark beetles [47–49]. Fire can encourage beetle attack by weakening the tree. Beetle-killed trees are more attractive to cavity nesters that tend to excavate nest sites in trees on which they have foraged [50].

Wounds influence the suitability of snags by encouraging different patterns of decay. Tops broken by wind or snow encourage nesting by species from black-capped chickadees (*Poecile atricapillus*) to spotted owls (*Strix occidentalis*). In conifers, decay occurs primarily in older live trees with defects in the outer sapwood, such as broken tops. Heart rots can create the hollow trees and hollow logs sought by Vaux's swift, American marten, fisher, and black bears. Large cavities result from a fire or a living tree's defense against fungi. A tree without decay that succumbs to windthrow will never become hollow. In most most temperate hardwoods, heart rots may not require a wound to enter, occur earlier than in conifers, and are common among otherwise healthy, young trees. In some species, sapwood decays at the same time as heartwood (e.g., Douglas-fir, *Pseudotsuga menziesii*, on wet sites), while in others, sapwood may not begin to soften until decades after the tree has died, and the centre has rotted (e.g., cottonwood and poplar, *Populus balsamifera* sbsp).

Because different agents of wounding or mortality create differ forms of decay and snags, local natural disturbances help explain the abundance of particular cavity-using species in an area. Usually, the forms of mortality are too varied to have predictive utility over large regions, but successful attempts to reduce specific forms of mortality can impact the fauna. Fire suppression in lodgepole pine (*Pinus contorta*) helped create the greatest insect mortality recorded in North America, with subsequent major impacts on bird species [51]. Similarly, harvest of potential nest trees before heart rot became prevalent would seriously impact Vaux's swifts. More generally, natural disturbance regimes can provide broad guidelines to the provision of cavity sites by revealing regional patterns in the onset of decay.

4.2. Species Patterns with Decay. Many decay agents are not readily visible in the field, so researchers classify decay of live trees and snags using a set of visual classes. Most researchers in the PNW follow the classification of [16]. Wood in classes 6 through 9 of that classification is too soft to be used by cavity nesters; the five harder classes are used here. Those classes are most simply defined as 1 (live), 2 (declining), 3 (dead, bark intact), 4 (dead, loose bark), and 5 (dead, no bark). Hardness of living trees and their resistance to excavation as determined mechanically is not consistently related to visual signs of decay [52], but broad patterns with visual signs of decay are evident (Figure 2).

Primary Excavators. In North America, primary excavators, both weak and strong, often create 80% or more of their nest

sites, relying on natural cavities for the rest [7, 9, 11, 53, 54]. Prediction 1 that strong excavators are able to select cavities in less decayed wood than can weak excavators is met (Figure 2).

When nesting in trembling aspen (*Populus tremuloides*), the weighted mean decay classes for strong and weak excavators are 1.9 (live) and 3.8 (dead), respectively. Visual estimates indicate more advanced decay when nesting in conifers: 2.8 and 4.8 for strong and weak excavators, respectively. Sapwood of aspen remains firm for years after tree death [67]. Even in forests dominated by conifers, large strong excavators (e.g., yellow-bellied sapsucker, *Sphyrapicus varius*, and pileated woodpecker, *Dryocopus pileatus*) preferentially nest in trembling aspen having decayed heartwood surrounded by sound sapwood [29, 55]. They use live and dead aspen relatively indiscriminately. Most pileated woodpecker nests in aspen are in live trees [68], affirming their ability to excavate live wood and the value of a sound sapwood shell. Conversely, weak excavators, such as boreal chickadee (*Poecile hudsonica*), nuthatches (*Sitta* spp.), downy, and acorn woodpeckers, do not excavate through hard sapwood shells. They seek dead trees or dead portions of living trees [46, 69, 70]. Among secondary cavity users, both birds and mammals tend to use less advanced stages of decay in hardwoods than in conifers (Figure 2). For waterfowl, too few decay classes for nest sites in conifers have been reported to permit comparisons with hardwood use.

More studies report whether a nest tree is living or dead than report decay class (Figure 3). Two points are clear. First, tree death provides nesting opportunities; 17 of 22 species in Figure 3 rely primarily on dead trees as nesting sites (decay class 3 and greater). The large portion of dead trees among cavity sites of all excavators emphasizes the importance of rot in determining nesting opportunities. In montane aspen stands, the number of nest holes was significantly and positively correlated with the percent of trees infected by *Fomes* (*Phellinus*) [71]. In coastal forests of the PNW, the marked preference cavity nesters show for western hemlock (*Tsuga heterophylla*) over western red cedar (*Thuja plicata*) has been attributed to the greater frequency of rot in hemlock [27, 72]. Woodpeckers may nest low in decay-resistant trees where butt rots are more prevalent than stem or top rots [27, 73]. In short, excavators create cavities where decay permits, and those trees are often dead.

A second point is that weak and strong excavators use trees differently. Across all 9 species of strong excavators in Figure 3, the interspecific average use of dead trees is 53.4%. Of 4403 nest trees used by strong excavators, 2229 (51.8%) were dead. Eleven of 13 species of weak excavators relied primarily on dead trees (Figure 3), even when more cavity-prone hardwoods comprised 80–100% of the nest sites. The acorn woodpecker is an exception that nests in live trees but seeks out dead limbs for cavity sites [87]. Excluding the acorn woodpecker, weak excavators located 70% (1831/2617) of their cavities in dead trees; the interspecific average across species was 74.8%. The abundance of suitable dead trees influences both groups, but weak excavators are less able to nest in living trees, and they are the larger group of species (Figure 1). Some small, weak excavators exploit advanced

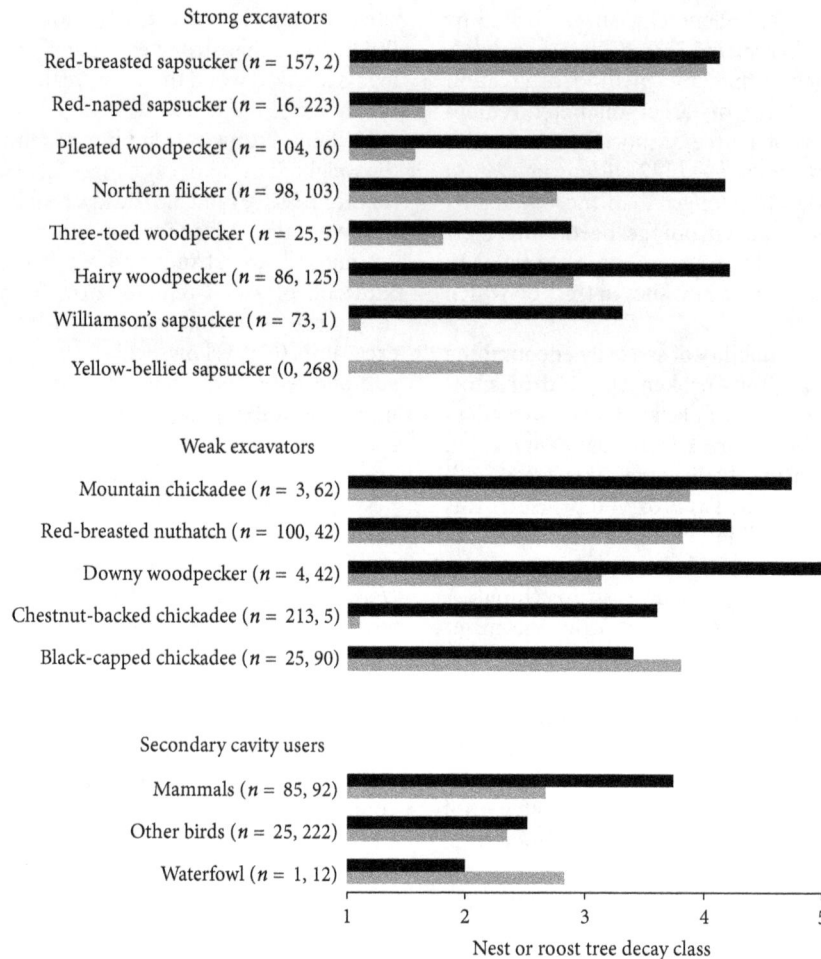

FIGURE 2: Mean decay class of trees used by strong and weak primary cavity excavators and by secondary cavity users in North America. Black bars represent conifers; grey bars are aspen; numbers in parentheses are sample sizes—the first number for conifers, the second for hardwoods. Decay classes of [16]. Data of [28, 55–66].

decay, even though it provides less protection. For example, 39 of 94 cavities excavated by chestnut-backed chickadees (*Poecile rufescens*) were close to the ground in rotting stumps [88].

The period that a snag remains firm enough to provide useful protection is shorter than the life of the snag. Analysis of preference for decay classes by comparing use to availability [46] showed the most strongly preferred decay classes were classes 3 and 4 (recently dead trees). How long a tree remains in these decay classes depends on cause of mortality, tree species, and site. Pileated woodpeckers used cavities in ponderosa pine (*Pinus ponderosa*) for 3–8 years after the trees were killed by fire [39].

Secondary Users. There are many more secondary users of cavities than primary cavity excavators (Figure 1). Dead trees provide most opportunities for secondary cavity users, whether cavities have been excavated by primary excavators or formed otherwise (Figure 4). The search for live trees and greater permanence is apparent as lower means in decay classes in Figure 2 (cavity sites in decay classes 1 and 2

are sought where available), but prior excavations and other natural cavities are more common in trees already dead (Figure 4).

Mammals denning in trees do not excavate cavities, so are more dependent upon natural cavities and rot patterns than are many birds. However, because competition for cavity sites is intense among secondary cavity-nesting birds, there is little difference in relative use of dead wood between them and mammals (Figure 4). The mean percent of dead trees across mammal species is 63.0%, it is 60.4% for secondary cavity-nesting birds. Bats are least able to modify an existing cavity and show the strongest selection for dead trees (74.9% across the seven species). Other, larger mammals are more capable of modifying and enlarging an existing cavity in living trees so are less restricted to dead trees (38.4% across 6 species). These larger species require hollows too large to be created by birds.

Decay can create hollow, living trees, and ultimately, hollow logs through heart rot and loss of heartwood residues [156]. Less than 40% of denning trees of northern flying

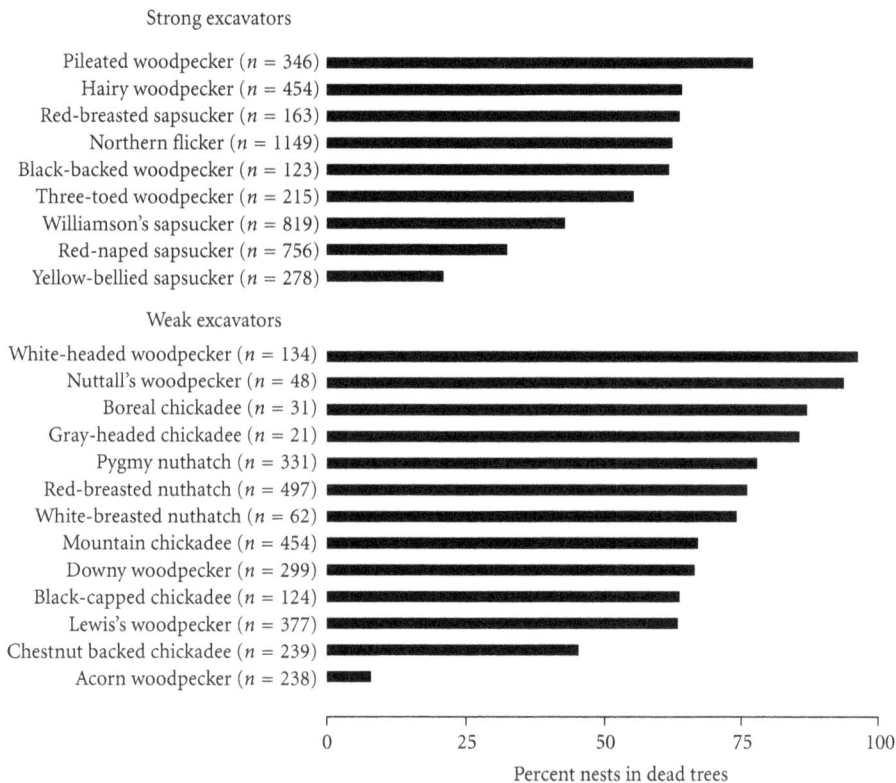

FIGURE 3: Percentage of dead trees used as nest sites by strong excavators and weak excavators of the Pacific Northwest (n = number of nest trees). Data of [8, 11, 28, 53, 55–57, 66, 70, 74–103].

squirrels, American marten, American fisher, and black bears were dead, indicating the importance of sustaining older trees with large rot pockets (Figure 4). For example, most black bear dens recorded from coastal forests of the Pacific Northwest are associated with wooden structures, including trees, logs, and stumps [134, 157]. Den sites in southern, inland forests also are commonly in trees [133]. In dry inland forest, 20 of 23 den sites of radio-fitted black bears were associated with trees. Ten of these were entered through the treetop into hollow centres created by fungi [158]. Even coastal grizzly bears sometimes use hollows in living conifers or stumps [159].

Some birds also use hollow trees. All 60 Vaux's swift nests and roosts located in northeastern Oregon were in live or dead grand firs (*Abies grandis*) with hollow interiors [114]. Amphibians and reptiles use cavities rarely and opportunistically [18, 160]. Such use tends to occur in snags of decay class 3 and 4, when bark is loose and sloughing [17]. Their use of dying and dead trees is thus similar to that of the brown creeper, whose nests are almost always between the trunk and a loose piece of bark on a large, typically dead or dying tree [161].

Predictions 1 and 2 are met. Because strong excavators have more options, their cavity sites are more closely associated with more stable and preferable sites in live and recently dead trees (Figure 2). Decay classes used by weak excavators tend to be more advanced. The mean decay class of trees used

by secondary users is relatively low (Figure 2), indicating a preference for living trees, but most cavity sites are found in dead trees (Figure 4).

5. Tree Species

The ideal cavity is in a stable, living tree. How a tree compartmentalizes fungal attack determines the likelihood of an ideal cavity. Broadleaved, deciduous tree species commonly isolate decay-softened wood inside a hard shell. When assessed by visual decay classes, broadleaved trees harbouring internal decay more often appear healthy (e.g., fully leaved) than do decayed conifers. Broadleaved tree species are conventionally termed "hardwoods." For a cavity-excavating bird, however, they are "softer" or better sites for excavation than are conifers (conventionally termed "softwoods"). The intact, outer sapwood shell provides stability and structure, with decayed wood a short distance beyond the shell. Conifers, however, often remain standing longer after death than do most hardwoods (large cottonwoods may be an exception). We thus expect (prediction 3) selection to be for live hardwoods when available and towards dead conifers when hardwoods are not available, with foraging preferences sometimes dominating. That expectation is met. Cavity nesters often chose hardwoods for 80–95% of their nest sites even where hardwoods comprised only 5–15% of

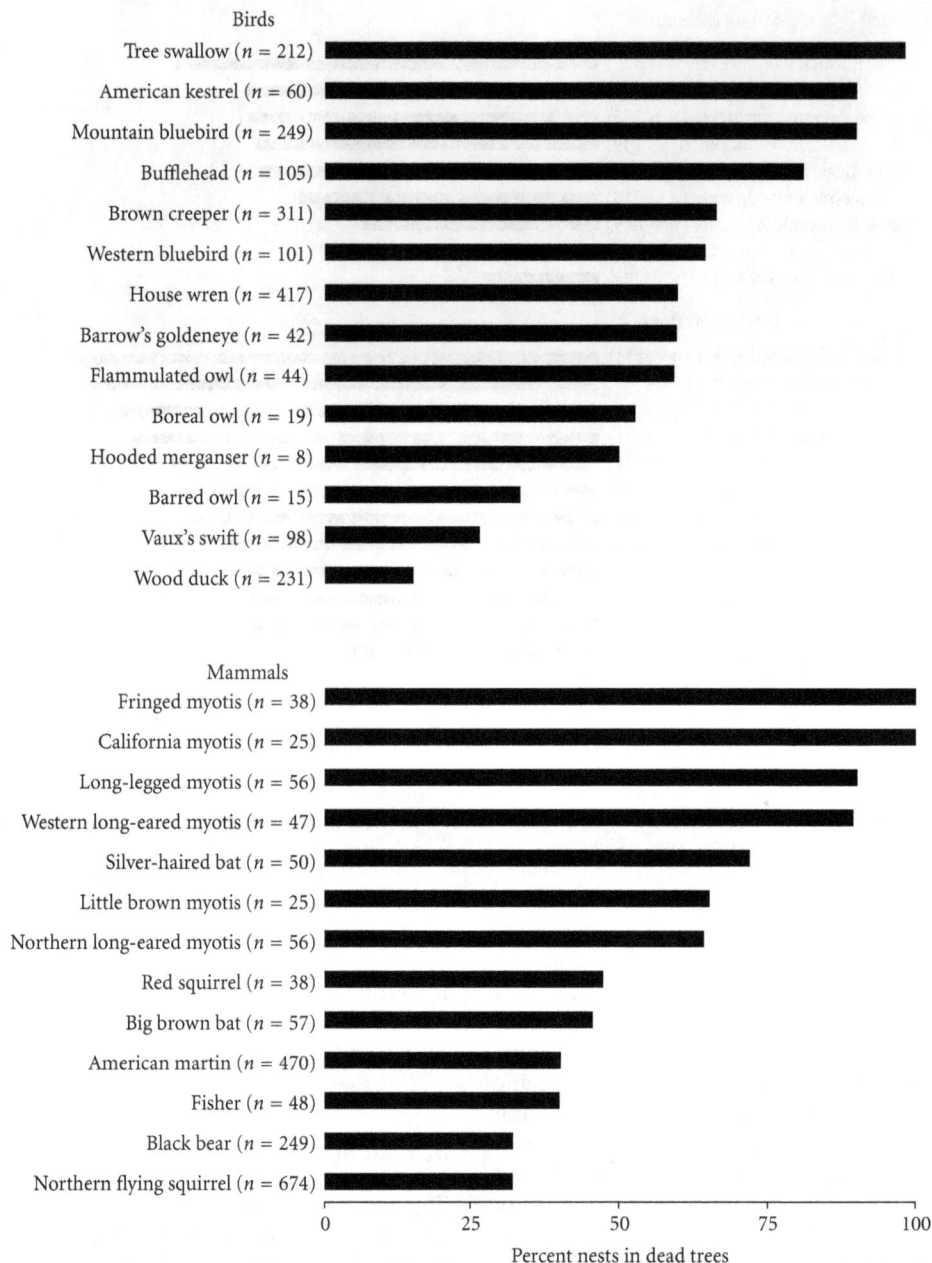

FIGURE 4: Proportion of dead trees used by secondary cavity users of the Pacific Northwest (n = number of nest trees). Data of [25, 60, 66, 70, 72, 75, 77, 94, 95, 98, 104–155].

the available tree stems [29, 53, 55, 162]. For bird species in Figure 2, conifer nest trees tended to exhibit more advanced decay by visual classes than did hardwood nest trees. Most conifers were dead.

Weak excavators are more often limited to dead trees (Figure 3), so availability of living hardwoods has greater influence on cavity site selection by strong excavators. Across 9 species of strong excavators, the percent of dead trees used declined significantly, as the percent of hardwood trees used increased (P < 0.001) [163]. The relationship accounts for much variability among the 46 studies aggregated in Figure 5. For example, in a large sample for pileated woodpeckers

(n = 105 nests) all available nest trees were conifers, and 99% of the nests were in dead trees [57]. Conversely, where hardwoods were available but scarce (<10% of stems), 26 of 27 nest trees were in trembling aspen, more than 85% of them living [29]. Similarly, northern flicker selected snags for 33% of their nest sites when nesting in hardwoods (n = 261), increasing to 65% when nesting in conifers (n = 141) [77].

Selection for specific tree species as cavity sites has been evaluated for studies reporting both use and availability [163]. In the PNW, tree species selected disproportionately to their availability include trembling aspen, western larch (*Larix occidentalis*), and ponderosa pine. Trembling aspen is

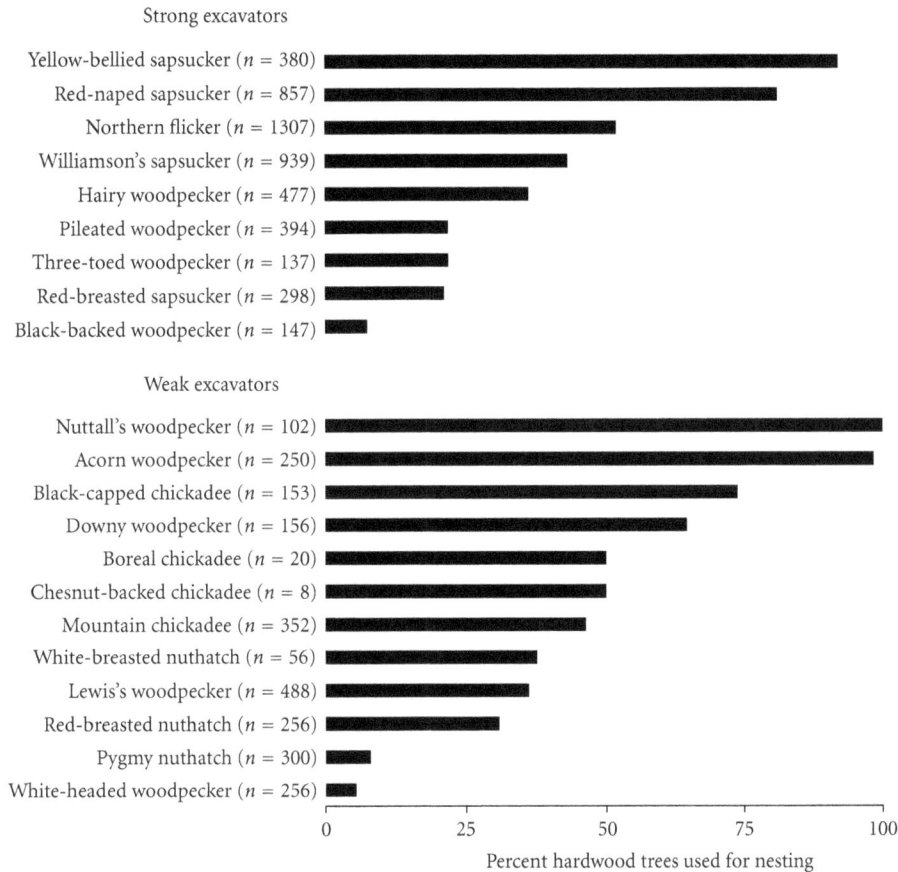

FIGURE 5: Percent use of hardwood by primary cavity excavators of the Pacific Northwest (n = number of nest trees). Data of [8, 11, 28, 53, 57, 58, 61–64, 70, 77, 79, 81–85, 87–93, 95–101, 103, 110, 164–185].

especially favoured. Several bat species also prefer to roost in aspen, probably because aspen is cavity prone [25, 60, 147]. In coastal forests, where aspen is lacking, primary excavators favour bigleaf maple (*Acer macrophyllum*) and red alder (*Alnus rubra*) disproportionately. Generally, lodgepole pine and the true firs (*Abies*) appear least favoured as cavity nesting sites [163]. However, a bird must feed itself before it can nest. Where black-backed (*Picoides arcticus*) and American three-toed (*Picoides dorsalis*) woodpeckers concentrate their foraging in beetle-ridden lodgepole pine, they also concentrate their nesting in lodgepole pine [164, 186]. In the absence of recent fires and beetle-infested trees, three-toed woodpeckers' primary habitat is mature or old-growth coniferous forests with an abundance of insect-infested snags or dying trees [187]. In old-growth Engelmann spruce (*Picea engelmannii*)-Subalpine fir (*Abies lasiocarpa*) forests all nests of three-toed woodpeckers were in Subalpine fir [165].

Although hardwoods are sought when available, use of hardwoods often is flexible. In the southern interior of British Columbia, all 243 nests of primary excavators found occurred in hardwoods, although hardwoods covered only 5% of the study area dominated by Douglas-fir [29]. In western Oregon, where hardwoods were scarce, the identical bird species nested mainly in Douglas-fir [188]. Flammulated owls show similar differences over their range. Cavities in oaks

(*Quercus* spp.) and trembling aspen were used where they were prevalent [189–191]; elsewhere in its range, the owl nests in ponderosa pine and Douglas-fir [192–194]. Similarly, where large, hollow hardwoods were not available, all cavities used by Vaux's swift (n = 58) were in large grand fir [114].

Flexibility is not universal and is constrained by foraging requirements. Nesting of some weak excavators (e.g., white-headed woodpecker; pygmy nuthatch, *Sitta pygmaea*) is largely restricted to conifers (Figure 5). Throughout its range, only 14 of 256 white-headed woodpecker nests have been reported from hardwoods. The restriction is related to a diet high in pine seeds [8, 93]. The pygmy nuthatch shows a similar strong and almost exclusive preference for long-needled pine forests [195]. Restricted diets limit the acorn woodpecker and some sapsuckers (once trees leaf out) largely to hardwoods and other woodpeckers, such as black-backed and three-toed woodpeckers, to conifers (Figure 5). Weaker excavators, such as Lewis's and white-headed woodpeckers, that nest and forage in coniferous trees, are potentially threatened by forest practices, because the conifers in managed forests may not become old and decayed enough to permit excavation. Both Lewis's and white-headed woodpeckers are designated "at risk" in the PNW.

The general model predicts that selection of tree species should reflect the tree's ability to compartmentalize decay, frequently leading to selection of hardwoods. Foraging needs modify selection, but selection for cavity sites still reflects trees' species-specific decay patterns. In the PNW, that generally means hardwood species such as aspen, cottonwood, and birch (*Betula*) are preferred. Among hardwoods, primary excavators select decay-prone species over more decay-resistant species [196]. Fungi commonly invade live aspen, softening the heartwood, while the sapwood remains unaffected. Douglas-fir may be more abundant in the area but often decays from the outside in; that is, decay softens the sapwood before it affects the heartwood [197, 198]. By the time the heartwood is sufficiently decayed to allow excavation, outer layers of wood are sloughing. Smaller, weaker excavators, such as chickadees and nuthatches, sometimes have no choice but to use decay-softened sapwood [72, 88, 199]. Where hardwood species are not common, birds excavate more dead conifers. Rot patterns change both regionally and with the mix of tree species, impeding ready transfer of findings among areas. In coastal forests, western hemlock often harbours heart rot [200, 201], while the more decay resistant western red cedar can contain butt rots [27]. It appears to be the relative amounts of these fungi that shift nesting preferences between western hemlock and western red cedar. Similarly, where the root rot *Armilleria sinapina* creates butt rot in aspen, bat roosts may be as low as 10 cm above ground [202].

Preferences for hardwoods, coupled with regional differences in relative abundance of hardwoods, produce regional differences in use of hardwoods. These are illustrated for birds in Figure 5 and mammals in Table 1. The trend in declining proportions of hardwoods used by mammals as den or roost sites from boreal through inland to coastal forests reflects the relative abundance of hardwoods. Black bears are an exception; boreal hardwoods do not attain suitable diameters for bear dens. Without context, the data of Figure 5 do not make a strong case for selection of hardwood cavity sites by birds. Mean percentages of hardwoods as nest trees are 41.8% and 50.1% for strong and weak excavators, respectively. The context is that these species are using from 40 to 50% hardwoods as nest sites in forests where prevailing climate strongly favours conifers over hardwoods [230], and the large majority of the trees are conifer [163, 231]. Prediction 3, hardwood nest sites are sought preferentially, is affirmed.

6. Size of Tree

The general model states that decay is the major influence on nest site selection, and susceptibility to decay is a major factor in tree species selection (Section 5). Decay, when "walled off" by living wood, provides the initial cavity site and protection. Size confers additional protection (height above ground) and room. The broad prediction 4 of the model embraces more specific corollaries about selection for size.

   (i) Birds seek larger diameters (partly by seeking decay or older, larger trees; partly by seeking greater cavity height above ground).

   (ii) Selection for taller heights will be less evident than for larger diameter (due to stem taper and many nest trees and snags having broken tops).

   (iii) Larger bird species will tend to select larger, older nest trees (partly to seek more room, but primarily because they are better able to excavate so can pursue opportunities for greater protection in taller, larger trees).

   (iv) For the same bird species, diameters sought in nest trees will be larger in conifers than in hardwoods (largely because conifers usually require longer periods to attain rot, during which they attain greater size).

   (v) Relations between size of nest tree and size of species will be less apparent among secondary cavity users (because they "take what they can get").

   (vi) No matter the animal's size, most den trees of mammals will be large, especially in conifers (because they rely greatly on natural cavities; trees are older, thus larger; height above ground may be less important for mammals because they are better able to defend themselves than are birds).

6.1. Diameter. I found 194 samples across 19 bird species that related nest tree diameter at breast height (dbh) to size of bird when nesting in either hardwood or coniferous trees. Larger diameter trees or snags are clearly preferred by cavity-nesting birds. In 30 of 31 comparisons, diameters of trees and snags used as nest trees were greater than the mean diameter of the available pool [232]. The same bird species does tend to use larger diameters when nesting in conifers than in hardwoods (Figure 6). For example, mean dbh of nest trees of pileated woodpeckers nesting in inland conifers was 79.3 cm, but only 54.2 cm in inland hardwoods; comparable values from coastal sites were 77.9 and 44.5. cm. Across all strong excavators, mean nest tree diameters in conifers were 82.2 cm on the coast and 53.7 cm inland. Comparable values when nesting in hardwoods were 45.9 and 36.8 cm. Values for weak excavators nesting in conifers were 86.8 cm on the coast and 55. 3 cm inland; when nesting in hardwoods, values were 45.4 cm and 38.4 cm.

This pattern follows from patterns of rot. In the PNW, hardwoods generally rot at younger ages and smaller sizes than do conifers [197, 233]. Conifers, in particular, grow faster on the coast and attain larger sizes before rot has developed. The tendency for birds to select much larger conifers on the coast than in the interior affirms the pattern. Regional differences in size and growth rates are less for hardwoods, so interregional differences in size of hardwood trees used are less. Size of nest tree selected differs little between strong and weak excavators on the coast or inland.

Larger birds seeking larger trees is apparent only for inland forests (Figure 6). In coastal forests, there is little tendency for larger strong excavators to seek larger trees, because even the smallest of them sometimes excavate in very large trees. The latter observation suggests that coastal conifers attain large sizes before even small pockets of rot

TABLE 1: Regional use of standing hardwood trees as den, nest, and roost sites by mammals. Data of [25, 59–61, 128–130, 139, 140, 143, 144, 148–153, 203–229].

	% Hardwoods (number of sites)		
	Boreal	Inland	Coast
All bats	80.0 (307)	16.9 (1307)	10.8 (301)
Northern flying squirrel	41.7 (224)	25.3 (185)	9.1 (993)
Fisher	100 (27)	93.2 (80)	11.1 (33)
Black bear	0 (116)	64.2 (545)	0 (302)

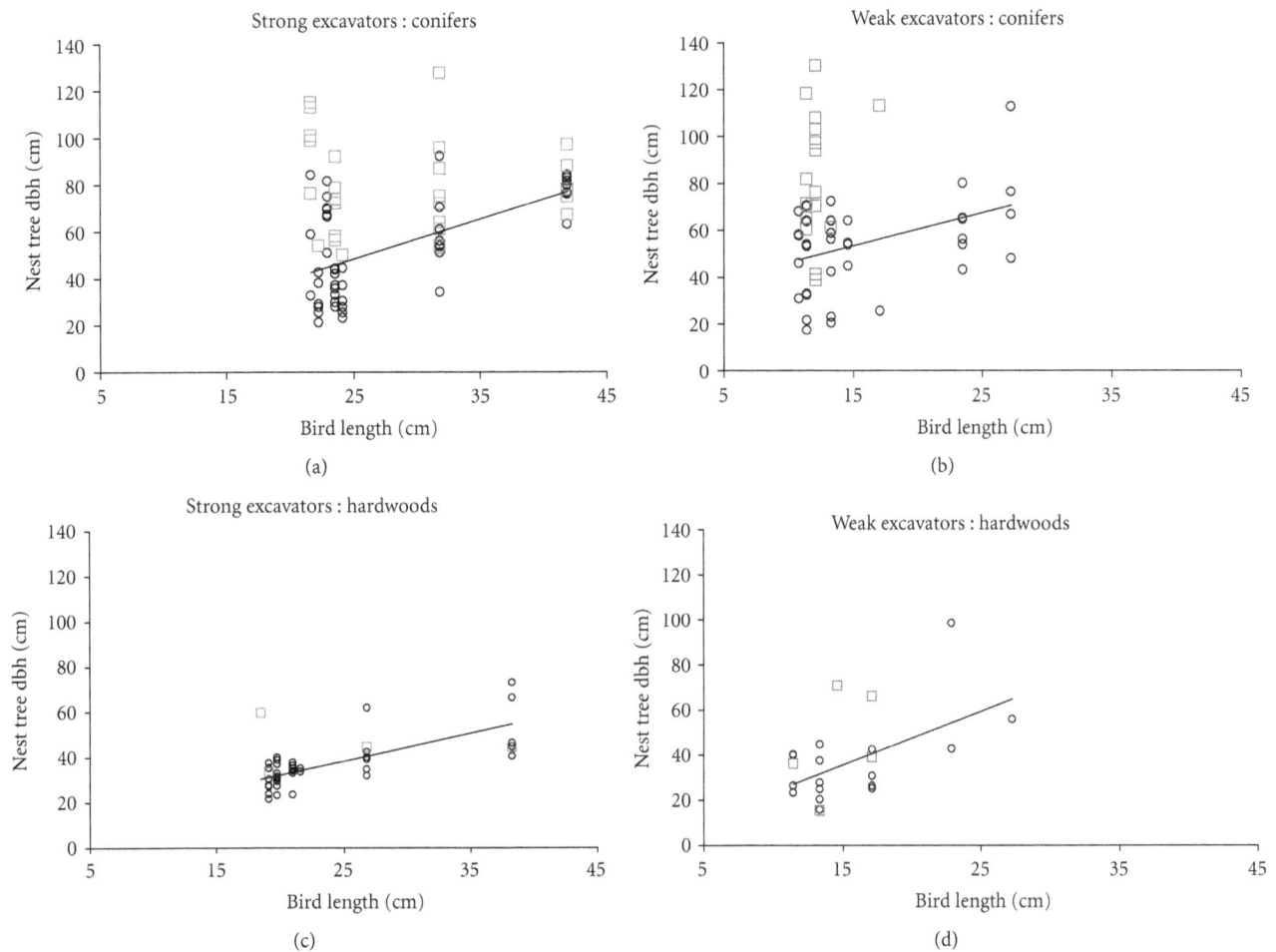

(a)

(b)

(c)

(d)

FIGURE 6: Mean diameter (dbh) of primary excavator nest trees or snags as a function of bird length (bill to tail length) in inland (O) and coastal forests (□) of the Pacific Northwest. Regression lines for inland forests are shown; there were no significant regressions for coastal forests. Bird lengths from [234]. Data of [8, 11, 53–55, 57, 61–64, 68–70, 72, 74, 77, 79, 80, 82, 83, 86, 87, 91–94, 96–99, 103, 118, 162, 164–172, 188, 199, 235–251].

develop, again affirming the role of rot in nest tree selection. Both strong and weak excavators select larger conifers when nesting in conifers than when nesting in hardwoods. The tendency is stronger in small, weak excavators that require well-developed decay (Figure 6).

There is little relation between body size and secondary cavity nesters in conifers, but not in hardwoods (Figure 7). Even relatively small secondary nesters select large trees when nesting in conifers, there apparently is no need to do that in hardwoods. Nest tree diameters must be large

enough to accommodate a cavity with room for an adult bird and nestlings, but sizes in conifers usually exceed that requirement. The selection of trees much larger than the size of cavity required reflects not only pursuit of height above ground, but age and the size at which heart rot develops. That occurs at younger ages and smaller sizes in hardwoods. Collated diameters of conifer nest trees of tree swallows ranged from 18 to 78 cm. Flammulated owls are only slightly larger than a sparrow, but nested in ponderosa pine averaging 57.7 cm dbh on southern aspects and 71.7 cm on northern

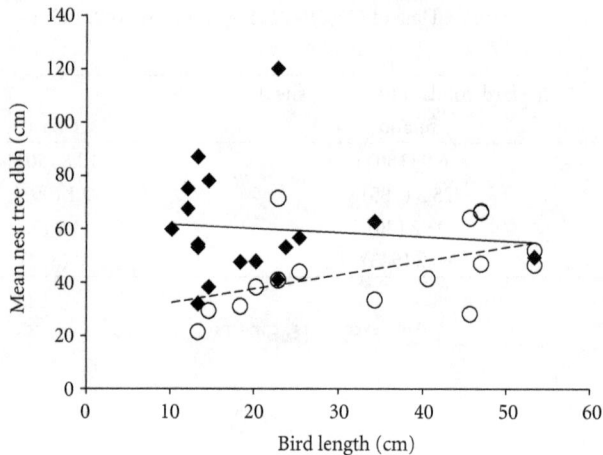

FIGURE 7: Mean diameter (dbh) of nest trees or snags used by
secondary cavity nesters as a function of bird length (bill to tail
length) in conifers (◆) and hardwoods (○) of the Pacific Northwest.
Linear regression relationships are shown by a solid line for conifers
(nonsignificant) and dotted line for hardwoods (*P* < 0.01). Bird
lengths from [234]. Data of [11, 55, 61, 94, 97, 98, 104, 107, 115,
116, 120, 122, 155, 253–257].

aspects (data of [194]). The difference reflects greater rates
of growth on north aspects, thus size at the age when rot
appears. Pygmy nuthatches, flammulated owls, white-headed
woodpeckers, and pileated woodpeckers all nest primarily in
large ponderosa pine or Douglas-fir in at least some regions
[57, 92, 192, 252], showing little relationship between size of
bird and diameter of tree during nest site selection in conifers
(Figures 6 and 7). I found few recorded diameters of conifer
nest trees for larger species, such as waterfowl and the barred
owl (*Strix varia*). Northern hawk owl (*Surnia ulula*), barred
owl, bufflehead, hooded merganser (*Lophodytes cucullatus*),
Barrow's golden eye (*Bucephala islandica*), and wood duck
(*Bucephala islandica*) nests were recorded primarily in hardwood species.

Milling efficiency constrains the sizes of trees grown
in managed forests. Optimal milling efficiency varies with
mill configuration, but for most sawmills in the PNW the
preferred maximum log diameter is from <45 to 50 cm
[258]. That has been further lowered where salvage of trees
killed by mountain pine beetles dominates the harvest. It
has been suggested for inland forests that trees must be at
least 23 cm dbh to provide a nesting site [236]. Most reported
values exceed 23 cm dbh (Figure 6), but a few smaller mean
values have been reported for black-capped and mountain
chickadees (*Poecile gambeli*), red-breasted nuthatches (*Sitta
canadensis*), and downy woodpeckers (*Picoides pubescens*),
especially when nesting in hardwoods (Figure 6). Mean
diameters of conifer nest trees in inland forests consistently exceeded 50 cm dbh for several species, including red-
breasted (*Sphyrapicus ruber*), yellow-bellied and Williamson's
sapsuckers (*Sphyrapicus thyroideus*), and pileated and white-
headed woodpeckers. In all but one study, data for the
northern flicker [97] and Lewis's woodpecker [245] also
exceeded 50 cm dbh.

Of these species, only the white-headed and Lewis's
woodpeckers are weak primary excavators, and several play

keystone roles in particular regions. Three are candidates for
designation or are designated "at risk" in portions of the PNW
(Lewis's, pileated and white-headed woodpeckers), and one
has two subspecies designated (Williamson's sapsucker). In
coastal forests, some small birds (chestnut-backed chickadee,
red-breasted nuthatch) consistently selected nest trees with
a mean dbh >50 cm, indicating the strong role of age and
decay in nest site selection by weak excavators (the chestnut-
backed chickadee frequently nests in stumps when it cannot
find large trees with rot). Habitat maintenance and milling
efficiency are incompatible for some species.

Preferred roost trees also can exceed the maximum
diameter for milling efficiency. Unlike nests, which often are
excavated anew each year, roosts usually are sought in natural
cavities. The average diameter of 123 pileated woodpecker
roost trees was 71 cm [242]. Most of these were grand fir
extensively decayed by Indian paint fungus (*Echinodotium
tinctorium*). Pygmy nuthatches use communal roosts housing
up to 100 birds during winter [259]. In Arizona, these roost
sites averaged 73.2 cm dbh [252]. Similarly, the hollow trees
or snags used by Vaux's swift must be large enough to allow
the swift to fly up and down within the tree or snag. The mean
dbh of 18 roost trees of Vaux's swift was 77 cm [260].

Mammals do select large trees for cavity sites (Table 2).
Emphasis here is on implications to forestry, so only studies
for which ≥40% of roosting or denning sites were in tree
cavities, or bark fissures are summarized in Table 2. In some
areas, species such as the red squirrel (*Tamiasciurus hudson-
icus*) and northern flying squirrel primarily use platforms or
mistletoe brooms; some bat species (e.g., pallid bat, *Antrozous
pallidus*, and fringed myotis, *Myotis thysanodes*) use a variety
of substrates for roosting. For cavity or roost sites, even
small bats use larger trees. Most studies of tree-using bats
have found that species selected for larger diameters and
greater heights than those of random trees [261]. Data of
Table 2 extend that finding. One result is that diameters and
heights selected vary with region, generally increasing from
subboreal and boreal forests through inland forest, attaining
their greatest values in the coastal forests where trees are
largest (Table 2). The pattern reflects the age, thus size, at
which common tree species develop heart rot or deeply
furrowed bark.

Differences in regional availability of large trees are most
clear for the largest species in Table 2, black bear. Along the
coast, from Vancouver Island south through Oregon, >95%
of black bear dens were in large dead or dying trees and
downed wood (Table 2 and [307]). In inland forests with
trees of smaller stature, 69% of black bear dens were in
wooden structures (Table 2). Mean dbh of hollow, black bear
den trees is >100 cm in both coastal and inland forests. In
boreal forests, only 12.3% of dens were in wooden structures.
That value underestimates the significance of large trees to
black bears. Of the 89 earthen den sites in boreal forest,
41% were under trees or stumps, 23% were under logs, and
only 36% were directly into soil [300]. Grizzly or brown
bears also use hollowed trees, though infrequently, in British
Columbia [159] and Europe [308]. Most brown bear dens in
Scandinavia were reported from ant hills and stumps [309].

Table 2: Characteristics of trees and snags used as denning and roosting sites by mammals, weighted means by species. Data of [25, 55, 59, 60, 123–130, 134–144, 146, 148–150, 152, 153, 162, 203–222, 262–306].

Species	ST[a]	Reg[b]	N	Percent use of substrate type						Diameter (cm dbh)				Height (m)	
				Wood[c]	Tree	Dead	HW[d]	Con[e]	n	HW	n	Both	n	Ht	n
Big brown bat *Eptisicus fuscus*	Ro	B	27	100	100	11.0	100		—	35.8	27		—	25.6	27
		I	145	99.2	99.2	60.1	37.1	31.4	50		—	59.6	71	20.0	129
California myotis *Myotis californicus*	Ro	I	49	100	100	100	0	48.7	41		—		—	28.0	39
Fringed Myotis *Myotis thysanodes*	Ro	I	51	62.5	62.5	100	0	43.3	9		—		—	6.9	9
		C	23	100	100	100	0	120.8	23		—		—	40.5	23
Keen's myotis *Myotis keeni*	Ro	C	94	97.5	91.5	50.0	0	95.0	86		—		—		—
Little brown bat *Myotis lucifugus*	Ro	B	97	97.9	96.9	50	54.3	24.1	13	40.6	32	32.9	48	21.7	33
		I	19	52.6	47.4	83.3	0	62.2	5		—	52.0	3	21.2	7
		C	8[f]	62.5	50.0	25.0	0	192.5	4		—		—	30.25	4
Long-eared myotis *Myotis evotis*	Ro	B	61	80.6	69.0	100	0		—		—		—		—
		I	96	45.8	40.6	100	0	52.0	14		—		—	32.5	14
		C	73	100	41.1	70.0	20.0	93.0	20		—		—	34.0	20
Long-legged myotis *Myotis volans*	Ro	I	268	84.0	84.0	96.4	2.2	42.6	15		—	64.0	192	27.4	192
		C	82	97.6	97.6	90.0	0	100.1	80		—		—	39.7	67
Northern myotis *Myotis septentrionalis*	Ro	B	156	100	100	19.9	68.6		—		—	40.0	156	17.8	44
		I	333	100	100	75.4	54.7	55.1	28	45.8	115	45.7	136	18.9	226
Pallid bat *Antrozous pallidus*	Ro	I	13	84.6	84.6	63.6	0	107.0	8		—		—	28.5	8
Silver-haired bat *Lasionycteris noctivagans*	Ro	B	11	100	100	—	100		—	42.5	11		—	22.1	11
		I	153	100	98	65[b]	44.0	42.3	54	52.3	65		—	19.8	135
		C	1	100	100	0	0	225.0	1		—		—	26.0	1
Yuma myotis *Myotis yumanensis*	Ro	C	20	90	90	11.1	55.6		—		—	115.2	18	19.6	18
All bats	Ro	B	307	99.5	99.2	29.4	80.0	24.1	13	39.6	70	37.4	230	22.1	141
		I	1150	84.8	82.9	83.5	16.9	53.8	225	48.1	225	55.3	392	22.5	796
		C	301	92.5	82.0	49.4	10.8	137.7	214		—	115.8	18	31.7	133
Northern flying squirrel *Glaucomys sabrinus*	N	B	224	100	99.5	30.5	41.7	31.5	79	33.4	3	39.5	141	19.4	141
		I	185	96.7	96.7	22.4	25.3	37.7	4	45.6	16	29.2	136	13.9	154
		C	993	98.8	97.9	41.3	9.1	73.0	173	35.8	12	76.7	792	36.9	161
Red squirrel *Tamiasciurus hudsonicus*	N	I	98	100	100	57.1	63.7		—		—	39.0	98	14.1	98
American marten *Martes americana*	D	B	13	100	76.2	16.7	0	>40	4		—	60.9	7		
	Re	B	85	93.9	75.9	16.7	18.0	>50	10		—				
	D	I	219	90.2	40.1	79.2	1.0	83.0	12	79.0	7	66.3	40	23.0	12
	Re	I	1099	80.1	46.3	68.6	0.4	91.6	318		—	57.9	42	15.9	300
	D	C	29	96.6	89.3	44.0	0	76.3	2		—	98.0	22		
	Re	C	379	87.6	69.1	30.9	0		—		—	87.5	262		

TABLE 2: Continued.

Species	ST[a]	Reg[b]	N	Percent use of substrate type				Diameter (cm dbh)						Height (m)	
				Wood[c]	Tree	Dead	HW[d]	Con[e]	n	HW	n	Both	n	Ht	n
Fisher	D	B	27	100	100	0	100			54.7	27				
Martes pennanti	Re	B	65	39.1	33.1		80								
	D	I	80	100	97.1	46.9	93.2	55.9	6	103	5				
	Re	I	748	93.2	87.0	27.6	23.2	75.7	194	81.8	152	76.9	88	33.6	50
	D	C	33	100	81.8	44.4	11.1	89.0	1			77.5	26	21.4	26
	Re	C	1251	99.0	91.2	22.0	13.4	104.6	116	89.7	102			16.6	54
Black bear	D	B	116	12.3	30.0	0	0								
Ursus americanus	D	I	545	69.0	49.2	22.6	64.2	161.3	64	117.5	41	163.3	3	19.7	90
	D	C	302	95.8	52.2	11.4	0	155.3	192					43.4	29

[a] Site type: D: den, N: nest, Re: rest, and Ro: roost; [b] region: B: boreal, I: inland (south of boreal), and C: coast; [c] all wooden sites including logs and stumps; [d] hardwood species; [e] conifer species; [d] may include Yuma myotis; [e] living or dead not consistently specified; [f] this value is a minimum.

Other than for the largest species, there is little relation between the size of tree selected and the size of the mammal. Bats use larger trees (Table 2), because they often use either natural hollows or cavities excavated by woodpeckers. The difference between hardwoods and conifers noted for birds applies (Figure 6). Diameters of bat roost trees on inland sites averaged 53.8 cm dbh (n = 225) for conifers and 48.1 cm for hardwoods (n = 225; Table 2). Six mammal taxa in Table 2 that use trees or snags >50 cm are listed as "sensitive" or "at risk" in the PNW (Keen's myotis, northern myotis, black bear (subspecies *Kermodei*), grizzly bear, fisher, and American marten).

Shape of the diameter distribution is more informative than the mean. Cumulative frequency distributions (CFDs) of snags tend to have long tails as small numbers of scattered large snags gradually accumulate (Figure 8). These tails raise mean values. That pattern is particularly evident in Figures 8(a) and 8(c). In coastal forests, active (used) snags were larger (height and diameter) than inactive or unused snags in old-growth areas (P < 0.05); snags in 70- to 100-year-old second growth were smaller than either active or inactive old-growth snags (P < 0.01) [27]. Only 14.7% of nest sites were located in snags <46 cm dbh (Figure 8(a)). Figure 8(b) illustrates the CFD of diameters of aspen nest trees over a range of about 15 to 55 cm dbh. Steep parts of the curves indicate the most sought diameters of nest trees in the region and are clearly evident in Figures 8(b) and 8(c). Most data on snag use are collected from unmanaged forests. Figure 8(d) compares nest trees selected by red-breasted nuthatches in managed and unmanaged forests of different tree species. The similarity of snag sizes selected suggests some governing relationship within the two disparate forest types.

Three points are apparent. First, birds show a central tendency (steep portion of the CFD) when selecting nest tree diameters, even across different forest types (see also [28]). Second, because of the long tails in many distributions, this tendency is better reflected by medians (which typically are smaller than means), but means are more often reported. Third, minimum nest tree diameter is a poor management target, because it ignores the central tendency. A more appropriate management target would be to provide trees larger than the median diameter at which heart rot commonly accrues in that forest type.

6.2. Height. The fact that dbh of trees used for nesting may be much larger than the size of cavity needed by birds reflects the benefits of nesting high in trees to achieve greater protection from mammalian predators [113, 311]. In living trees, diameter and height usually are closely related. That is not true of snags because many have broken tops. Nonetheless, we expect cavity-nesting birds to seek out taller nest trees, both live and dead, from those available.

Taller snags generally are used disproportionately to their availability, but the mean height of nest trees exceeds that of the mean available only about half the time [232]. That occurs because many snags with broken tops are used. For example, in interior Douglas-fir, 14 of 20 nests of red-breasted nuthatches were within 2 meters of the top of decay class 4 Douglas-fir snags with broken tops; 21 of 22 nests in Englemann spruce-Subalpine fir forest were in similar class 4 Subalpine fir snags, broken off by heart rot and wind [232].

On the coast, mean reported nest heights averaged 16.9 m, 14.10 m, and 10.1 m for strong primary excavators, weak primary excavators, and secondary users, respectively. Comparable values from inland sites were 9.4, 8.9, and 7.4 m (e.g., Figure 9). Small, weak excavators often use soft substrate, commonly found in short, broken-topped snags or stumps. That tendency is most apparent among chickadees and accounts for some of the short nest tree heights in Figure 9(b). Larger excavators, both weak and strong, also occasionally use older, broken-topped snags or stumps (e.g., northern flickers, downy, black-backed, three-toed, and white-headed woodpeckers in Figures 9(a) and 9(b)). Secondary users most often nest close to the ground (Figure 9(c)). Reported tree or snag heights are naturally taller than nest heights. On the coast they were 25.2 m, 16.5 m, and 18.1 m for strong primary excavators, weak primary excavators, and secondary users, respectively. Comparable values for inland sites were 19.5 m, 12.5 m, and 16.1 m. The relatively tall tree heights for some secondary users reflects the use of cavities in larger taller trees by larger species such as kestrels and owls.

Among strong excavators, larger bird species locate their nests higher above ground (P < 0.05), and there is a tendency for larger birds to locate nests in taller trees and snags (P < 0.05), but there is great variability due to broken tops (Figure 9(a)). Across all species, however, nest height shows little relation with size of the bird (r^2 < 0.10). A general lack of relationship between bird size and nest height is expected, because stem taper means that larger cavities sought by larger birds cannot be as near the top of the tree as those of smaller birds. Management targets can be based on diameters alone.

7. Density and Distribution

There is little clarity on meaningful measures of density and distribution of trees for cavity nesters. Several issues, including scale, obscure appropriate metrics. For example, the area over which to measure density is ill defined and almost certainly a function of species-specific territory sizes, which themselves are a function of foraging opportunities; many cavity-using species use trees or snags for foraging as well as nesting (forage and nest trees have different attributes, but both are necessary); primary and secondary users of cavities usually forage differently; there are tradeoffs in the relative use of hardwoods and conifers, but researchers sometimes report only conifers; and numbers of snags in the immediate area of the nest tree likely overestimate requirements, because snags tend to occur in clumped patterns that do not extrapolate over large areas.

Predictions of expected responses are thus limited.

(i) Responses to density will be inconsistent and better exposed by extremes in simpler managed stands than in highly variable natural or near-natural stands.

(ii) Cavity nester density will be asymptotic against nest tree density and limited by factors other than snag density as snag density increases.

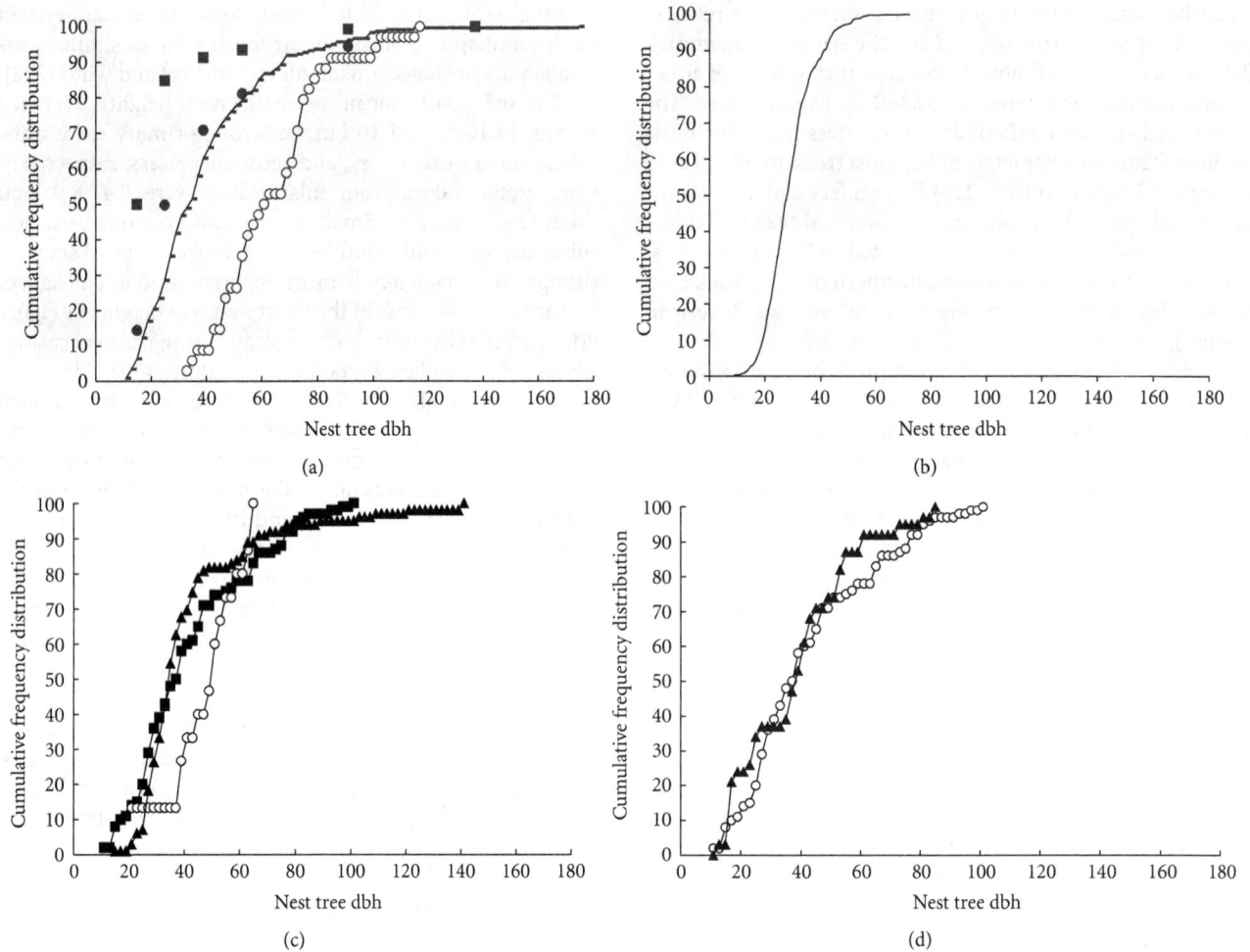

FIGURE 8: Cumulative frequency distributions (percent) of snag and nest tree measurements: (a) dbh of old growth snags without cavities (inactive, dotted line), old growth snags with cavities (active ○), and snags in younger stands, 70 (■) and 100 (●) years of age (all inactive; data of [27, 310] for the Coastal Western Hemlock zone); (b) dbh of aspen nest trees in the Cariboo (data of K. Martin); (c) dbh of nest trees used by three species in unmanaged Interior Cedar-Hemlock: red-breasted nuthatch (■), northern flicker (○), and red-naped sapsucker, *Sphyrapicus nuchalis* (▲) (data of C. Steeger and M. Machmer for the Interior Cedar-Hemlock zone); (d) dbh of nest trees used by red-breasted nuthatches in managed (▲) and unmanaged (○) stands (from [232]: Figure 2).

(iii) There will be little response to snag density in hardwood stands because a high proportion of living hardwoods contains heart rot.

(iv) Clumped distributions of nest trees (including snags) will better sustain primary excavators, because they are more likely to include foraging requirements.

(v) Many secondary users will favour dispersed retention because they forage in more open areas.

Density. Density of cavity-nesting birds is inconsistently related to measures of snag density. Generally, studies surveying natural stands with many snags obtain weak relations (e.g., 27, 61, and 316), while those including managed stands, in which some or all snags were removed, find stronger relations (e.g., [54, 70, 324]). In managed forests where snag density was high, there was no relation between cavity nester and snag density [325]. Including simpler managed stands

in analysis is more revealing of snag effects than using only natural stands. Within stands ranging from unmanaged to scattered seed trees, density of snags 25–50 cm dbh was a strong predictor of red-breasted nuthatch and chickadee densities; $r^2 = 0.94$ and 0.83, respectively (mountain and chestnut-backed chickadees combined) [324]. Such relationships may indicate consistent nest site limitation or better foraging opportunities in stands with more snags (e.g., older stands with more insects versus thrifty managed stands).

Figure 10 illustrates relationships for stands dominated by conifers. Most variation in cavity nester response to snag density is due to sizes of the snags, so estimates of density were limited to larger snags that the birds use preferentially: ≥50 cm dbh for coastal forests and ≥30 cm for inland forests. Presence of an asymptote in density was evaluated by fitting a Michaelis-Menten relationship to data illustrated in Figure 10(a). That invokes the assumption that the rate of response (cavity nester density) is a function of the concentration of

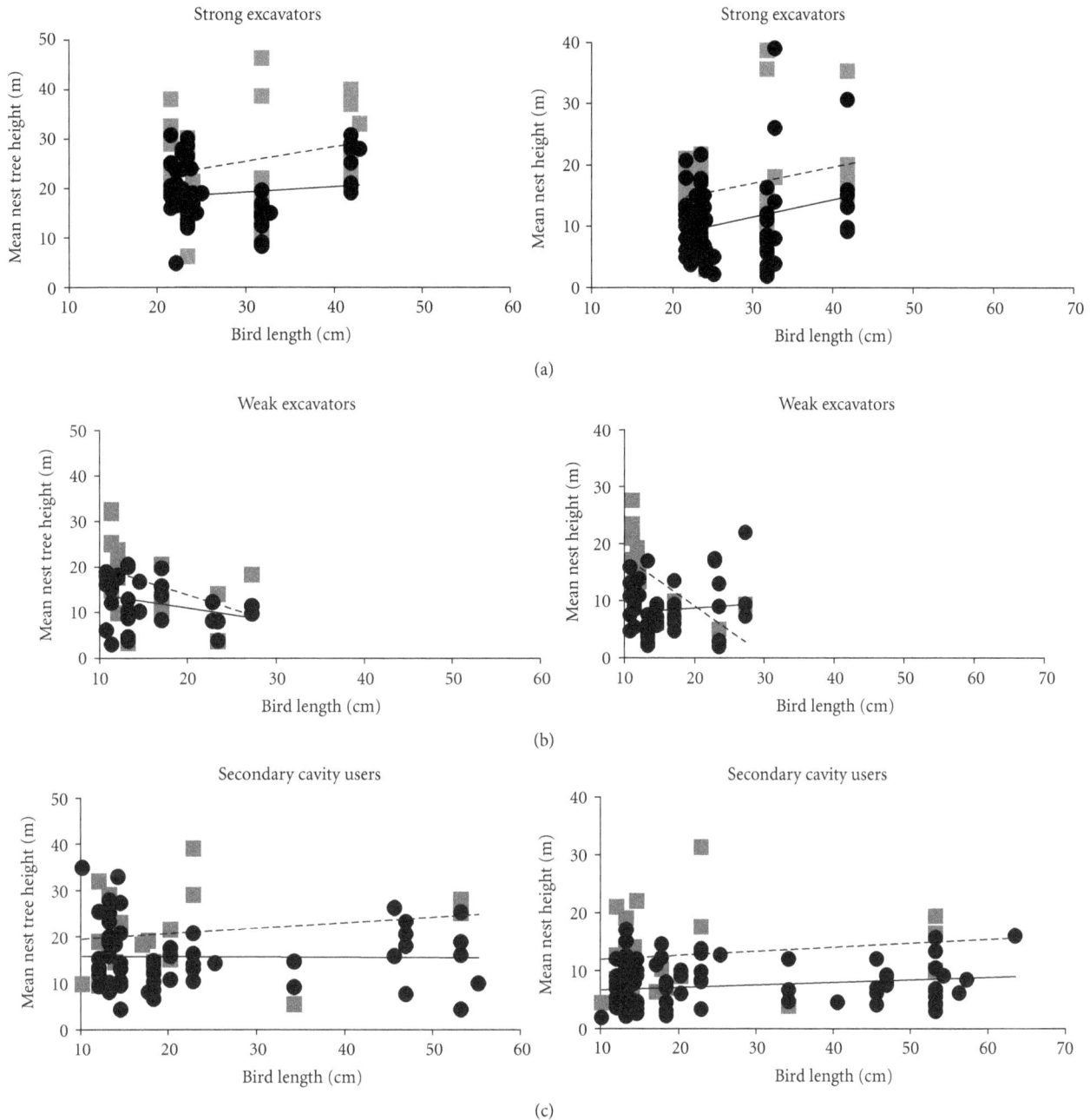

FIGURE 9: Mean height of primary excavator nest trees or snags and nests as a function of bird length (bill to tail) in inland (●) and coastal (■) forests of the Pacific Northwest. All bird lengths from [234]. Regression lines are dotted for coastal forests and solid for inland forests. Data of [7, 8, 11, 29, 53, 55–58, 65, 69, 70, 72, 74, 75, 78–82, 84, 85, 87, 91–94, 98, 103, 108, 112, 169, 173–175, 188, 199, 238, 241, 242, 249, 312–323].

substrate (snags/ha). Response of the seven studies combined is asymptotic, with little additional increase in density of cavity nesters beyond about 3 large snags per ha (Figure 10(a)). That shape is expected among species limited by resources such as food, space, or nest sites. Birds present at 0.0 snags per ha indicate that some species can nest in snags <30 cm dbh (some of them hardwoods, unreported in data on conifers).

The response shown in Figure 10(a) was consistently expressed: each individual study shows an initial increase

in cavity-nester density with increasing snag density. That merely affirms large snags become limiting at densities of ≤1 per ha. The fact that the relation fits a Michaelis-Menten relationship implies that the response is consistent with limitation by substrate abundance. Small differences would shift the estimated half-saturation value between 2 and 3 large snags per ha. Relationships for some smaller individual bird species were more linear (Figures 10(b) and 10(c)). Territorial behaviour is only weakly expressed in the chestnut-backed chickadee but is evident in mountain chickadees. It is possible

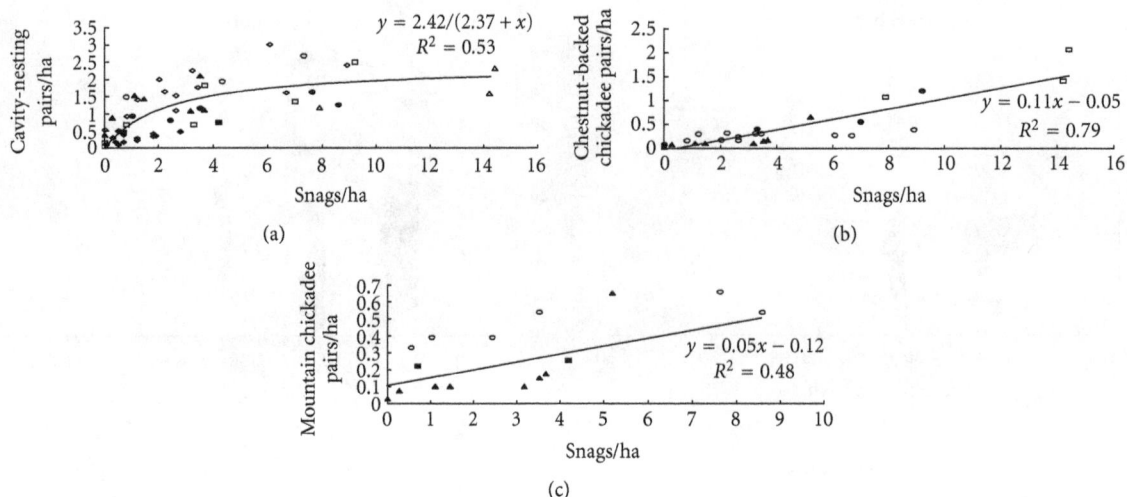

FIGURE 10: Density of cavity-nesting bird pairs versus snag density in primarily coniferous forests of the Pacific Northwest. Hollow symbols represent coastal forests; solid symbols are inland forest types. (a) All cavity-nesting species.[O] = [326]; snags >50 cm dbh]; [▲] = [118]; snags >30.5 cm dbh); □ =[327]; snags >48 cm dbh]; [■] = [328]; snags >31 cm dbh]; [◊] = [329]; snags >51 cm dbh]; [Δ]= [238]; snags >50 cm dbh]; [●]= [70]; >38 cm dbh]; [◊] = [330]; snags >50 cm dbh]. Solid line is the fitted Michaelis-Menten relationship of the form cavity nesters per ha with asymptote = 2.42 and half saturation constant = 2.37 [adapted from [233]]. (b) Chestnut-backed chickadee pairs [▲] = [118]; snags >30.5 cm dbh]; [●] = [327]; snags >48 cm dbh]. [O] = [329]; snags >51 cm dbh]; □ = [238]; snags >50 cm dbh]. (c) Mountain chickadee pairs: [▲] = [118]; [■] = [328]; [O] = [70].

that competition for cavity sites is more strongly expressed within the entire cavity-nesting fauna, than within individual small species. However, data of Figure 10(c) also suggest that other variables may influence the response; at the same snag densities, data of [70] are consistently higher than those of [118]. During a mountain pine beetle epidemic, densities of both red-breasted nuthatches and mountain chickadees increased dramatically and linearly as beetles killed more conifers and provided more food. Once most of the conifers were killed, numbers of conifer snags continued to increase, but the numbers of beetles diminished and so did the nuthatches and chickadees [51]. Although the density of conifer snags increased greatly, the birds continued to nest primarily in aspen; food was a more influential variable than snag density.

Relations of Figure 10 are helpful to the extent that they reveal that low snag densities are limiting and suggest that about 2.5 large snags per ha support 50% of natural densities of cavity nesters. Current snag-retention guidelines for most North American forest types fall between 1 and 10 large snags/ha, converging around 6 to 7 trees/ha [331]. These guidelines derive primarily from work in the Blue Mountains of Oregon and are expected to sustain the full complement of cavity nesters ([16]: Appendix 22], [332]). Because of the great variation in the data, they differ little from the relationship of Figure 10.

There is a problem however. Although small and large snags are both important to cavity nesters, nesting densities usually correlate better with large snag densities than with densities of all snags. For example, in the coastal rain forest of British Columbia densities of all snags did not differ between old-growth sites and 70-year-old second growth

(34.8 and 33.9 snags/ha, respectively [27]. Because many snags were larger in old growth, the proportion of snags used was about 12 times greater (16.0%, compared to 1.3% in second growth). Median diameter was about 18 cm dbh over all second-growth snags and 62 cm dbh for nest trees in old growth. Relations derived from the Blue Mountains and in Figure 10 address nesting, so focus on larger snags. Woodpeckers, however, tend to forage on several smaller snags rather than the single larger snag in which they nest. Smaller snags that represent foraging sites, but are inadequate for nesting, were present in all studies included in Figure 10 and undoubtedly influence the apparent response to snag density. Their abundance often is not reported. The number of small snags required, and the preferable distribution of snags remain obscure. The former is troubling because small snags are necessary foraging sites for many species. The latter can be evaluated by more integrative measures of retention (see below).

Data are few, but snag density appears less important in hardwood stands. In southern hardwood stands of Florida, cavity nester density increased with density of hardwood snags, though not as dramatically as in pine forests [333]; there was no increase in cavity nesters with a doubling in hardwood snags in mesophyll forest in Kentucky [334]. In mixed forests of hardwood and conifer, five of six cavity-nesting species increased dramatically during a mountain pine beetle epidemic and greater foraging opportunities [51]. Although the number of conifer snags increased dramatically, the birds continued to locate 95% of their nests in living or recently dead aspen. Densities of the five cavity nesters responding increased 5- to 6-fold with little change in density of hardwood snags. Black-capped chickadees, which did not change in density, forage primarily on insects found on living trees and showed no response to increasing numbers of beetles or conifer snag density until most trees were dead (when chickadees declined in abundance).

Distribution. Distribution has both spatial and temporal dimensions. From the broadest to finest scale, questions we ask about distribution of a resource are ordered. How do we sustain a continuous supply? Should actions be focused at the landscape level or within stands? How should within-stand efforts be distributed? How much is necessary? What kinds (e.g., sizes)?

Kinds and amounts (how much) were addressed above. Here we address the three broader questions noted. The broadest scale of distribution relates to planning for the sustained contribution of dying and dead wood. We know that if a 200-year-old tree is required 200 years from now, it must be planted now and subsequently reserved from harvest. Models that project the creation and duration of snags reveal projected shortfalls large enough to significantly impact cavity nesters [2, 3, 335]. For example, projections for managed Douglas-fir forest suggest 0.1 large snags per ha per decade [3], well below required levels of Figure 10. These models typically address the temporal scale and indicate that in managed stands snags are not being created or replenished fast enough to ensure a continuous supply. The solution appears simple—leave more trees to become snags. Solution becomes more complicated when we include the spatial scale. There often are landscape-level reserves or set asides (e.g., steep slopes, riparian areas, and reserves for a particular species) that contribute snags now and in the future. Trees and snags also may be retained within stands.

The challenge is then to determine what should be retained within stands, given landscape-level reserves and the natural rates of snag creation or fall, and how should within-stand retention be distributed? The kinds of bookkeeping and projection required are relatively simple, but we confront several challenges: (1) we usually do not have good estimates of rates of key processes (e.g., snag creation or tree mortality rates, snag fall down rates, and progression of decay or period when the snag is useful), (2) dead wood is a more transient habitat that many others and organisms are adapted to particular transient stages of decay (Figure 2 and [307]), (3) the diverse forms of tree mortality ensure stochasticity and a great deal of variation when amounts in natural stands are sampled [307, 336–338], (4) species using dead wood respond variably and may seek different sites for nesting and foraging, and, perhaps most challenging, (5) scale, which makes amount and distribution nearly inseparable. The same amount can be patchily or uniformly distributed. Patchily distributed amounts may have too little dead wood at local scales and what is local varies among species.

It is clear that many managed forests produce too few large nest trees, but once we choose to retain trees, the questions of where that retention should occur and what form it should take remain. Tolerance levels can help address the natural variation in measures of amount [339]. The issue of distribution remains. Retaining trees as landscape reserves or within harvest blocks have different costs. We need to know whether they also have different consequences. The shape of response curves relating cavity-nester abundance to proportion of forest retained permits evaluation of tradeoffs between stand- and landscape-level retention. The response of cavity nester abundance could be directly proportional to the proportion of forest retained, above or below proportional, or reveal potential thresholds (Figure 11).

The nature of the response curve has implications to the distribution of retention:

(i) Directly proportional to the amount of cover retained implies that within-stand and landscape-level retention contribute equally. The total abundance of the species in a landscape depends only on the total retention level, whether this is retained within stands or in landscape reserves or in a combination of the two.

(ii) Above proportional to the amount of cover retained indicates that within-stand retention during harvest is creating greater opportunities for the species within each stand. For example, in Figure 11 the >proportional response indicates that retaining 20% of trees in a cut block maintains a species at 40% of its abundance in uncut forest. That supports using retention in cut blocks at the stand level. For a given total amount of retention, stand-level retention maintains a greater abundance of the species across the landscape than would alternative strategies, such as retaining the same total amount in large reserves and clearcutting the harvested stands (all else being equal).

(iii) Total abundance less than proportional indicates that landscape-level retention is favoured because retention is less useful when "diluted" within individual cut blocks. For example, in Figure 11 the <proportional response indicates that 20% retention maintains the species at only 5% of its abundance in uncut forest. To sustain species showing such a response, it is better to allocate a given retention level to larger reserves, rather than stand-level retention (again, all else being equal).

The most helpful response to management would be a marked threshold response, that is, a sudden decline in a species' abundance when amount of a habitat variable drops below a particular value [341–343]. If well-defined thresholds were evident, they would help guide choices of target levels for retention [344]. However, the many sources of variability in natural systems means thresholds are more likely to occur as softer, sigmoidal-type relationships (Figure 11, [341, 343]). Whereas a hard threshold reveals a clear target value, a sigmoidal response or "softer" threshold suggests a mixed strategy of stand- and landscape-level retention, or of higher and lower retention levels in harvested areas. The practical point is whether the relationship has a point of inflection, where abundance changes from more to less rapidly increasing, as retention levels increase. For example, with 20% overall allocation to retention, the sigmoidal relationship in Figure 11 implies the best strategy would be to retain 35% in about half the stands with minimal retention in the rest.

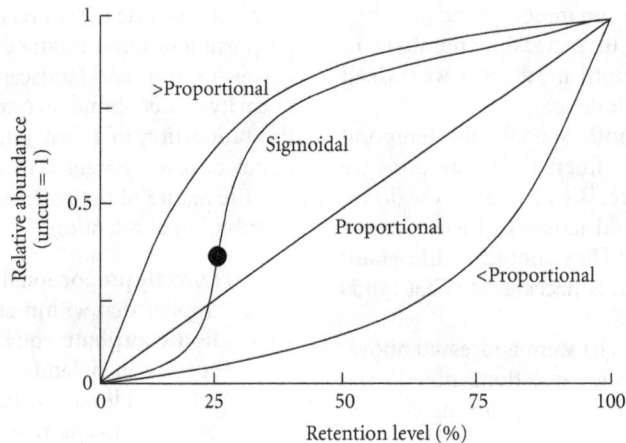

FIGURE 11: Examples of response curves with abundances greater than proportional, less than proportional, or proportional to retention levels, showing a sigmoidal or "soft threshold" relationship with retention levels. The dot on the sigmoidal curve shows the point of inflection. Adapted from [340].

In fact, most conceivable shapes of response appear in nature [345]. Few generalities are apparent among the 15 cavity-nesting species for which meta-analysis was possible: only 1 species (brown creeper in hardwoods) showed a response less than proportional to retention; 5 species showed neutral responses (these sometimes limited to particular seasons or regions); the majority profits from some within-stand retention. For some species, retention had no apparent effect until it reached levels from 15 to 20% of the original volume; others profited from some opening of the canopy. The combined response of 15 species plus 95% confidence intervals is illustrated in Figure 12.

In Figure 12, the "community" is a composite of studies of 15 species [345], each standardized as 100% abundance in uncut forest. It shows that even small amounts of harvest eliminate some species, but over a wide range of retention most species are present, with a few species occurring much more commonly at high retention levels (openings <20% canopy cover). These few species—black-capped chickadee, mountain chickadee, downy woodpecker and red-breasted sapsucker—produce the wide confidence limits in abundance at higher retention levels. Clearly, some cavity nesters profit from opening of the canopy (abundance greater than that in uncut forest) and, other than for a few species, little is gained in abundance beyond 30% retention (Figure 12(a)). Collating data from several sources ensures wide scatter at the lower end, so confidence limits do not converge to near zero at zero retention. For example, northern flicker forages on the ground and is the only cavity nester for which confidence intervals do not contract towards zero abundance at zero retention at the stand level.

Even though desired amounts of retention vary across species, it is clear that all cavity nesters profit from some degree of retention. Retention silvicultural systems (e.g., green-tree retention and variable retention) contribute permanent retention—the trees will never be harvested. Other partial harvesting systems, such as regular or irregular shelterwoods, selection cuts, strip cuts, and group removals, often

leave high levels of retention, but the trees are not intended to be retained permanently. Whatever the approach, trees retained within a treatment unit can be distributed in groups or as scattered individuals (dispersed retention) or as a mix of both.

Review of studies that permitted comparisons of forms of retention revealed three major patterns [346]. First, although dispersed retention of snags (high levels of removal) increased abundance and richness of secondary cavity nesters beyond that found in mature and old-growth forests, abundance of primary excavators was reduced (see also [347]). This response accounts for the increase in similarity of combined species at the highest levels of retention in Figure 12(b). Group retention better sustains primary excavators. In both logged and managed stands, nests of primary excavators often are concentrated in dense patches of snags or broken-topped trees for both strong and weak excavators [70, 72, 74, 244]. Woodpeckers also used snags in groups, even in an area where snags were uniformly distributed by salvage logging [186]. Likewise, there was no difference in nesting levels between clustered and scattered snags that had been created by topping, suggesting that some natural phenomenon was acting [315]. Trees, snags, and down wood typically occur in clumps [348]. It is unclear whether primary excavators seek dense patches of nest trees for some associated value or if it merely reflects the patchy way in which trees are killed by insects or disease.

Second, secondary cavity nesters were more abundant in areas experiencing some timber removal than in mature or old-growth forests. That increase likely reflects the fact that many secondary nesters forage more effectively in openings and accounts for the increase in abundance at midlevels of retention in the combined response of Figure 12(a). Third, in most instances the abundance of primary cavity nesters was little affected by partial harvesting, and in some instances increased in abundance. That may reflect the fact that several primary excavators also favor small openings and edges (e.g., hairy woodpeckers [68], northern flicker [77], three-toed

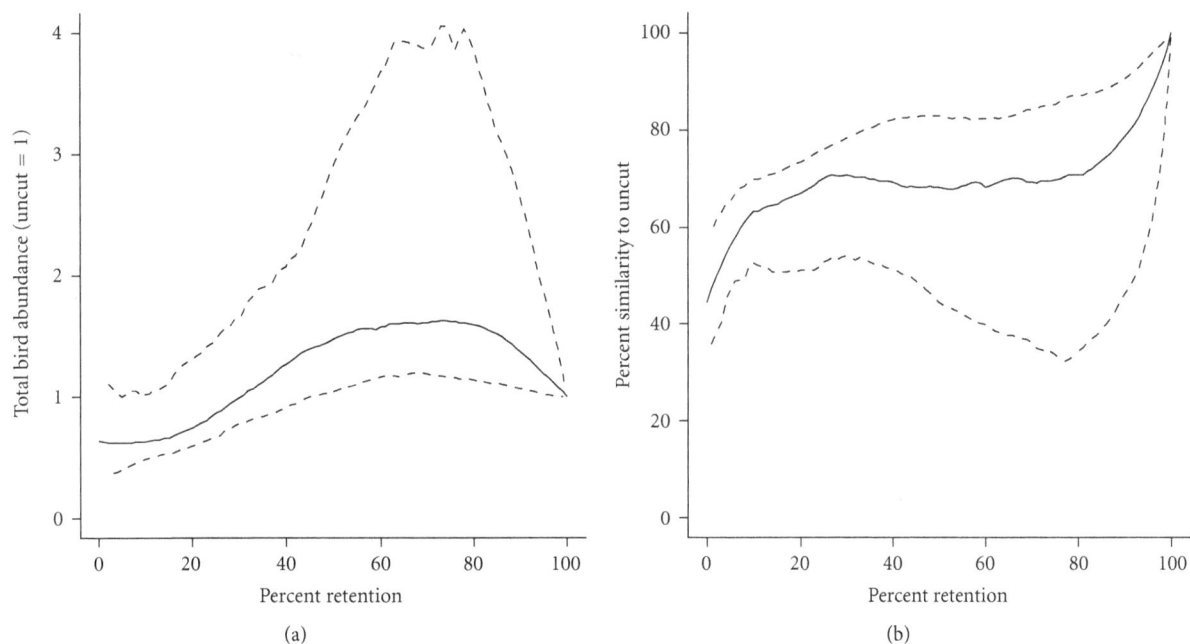

FIGURE 12: Combined response of 15 cavity-nesting species to percent forest retained. (a) Total abundance. (b) Similarity of the bird community in the retained forest to uncut controls. Methods of [340].

woodpecker [165], and perhaps downy woodpecker [349]). In central British Columbia, preferred nest sites for three-toed woodpeckers were within 20 meters of an edge [165].

In summary, it is apparent that snags can be limiting and that from 2 to 3 appropriately sized snags per ha appear to sustain about 50% of the natural cavity-nesting community locally (Figure 10). Individual species, however, show a variety of responses (Figure 12), with broad generalities apparent only between primary and secondary nesters. It is much less clear how much of the forest should have what snag densities. Sustained provision of habitat will differ among forest types, and the rates at which snags are produced, decay and fall. Estimating total snags required must recognize that smaller snags are used as foraging sites, and foraging sites are sometimes more limiting than cavity sites [350, 351]. More dead or dying wood than required for nesting is needed to sustain all cavity-nesting species. Moreover, through provision of perching, foraging, and hawking sites, snags of all sizes tend to increase richness and abundance of birds other than cavity nesters [352, 353].

8. Management Implications

In forests of the PNW, most studies find significant correlations between snag density and the abundance of primary excavators (Figure 10 and review of [354]). Similar relations are evident for secondary cavity users [54]. Because we prefer that trees in managed forests do not die or become rotten, management is designed to reduce snag densities below natural levels [2, 3, 335]. One result is that a large portion of cavity-using species is listed "at risk," or "potentially at risk,"

globally [9] and in the PNW [46]. Over the past two decades, greater effort has been made to provide snags in managed forests (e.g., [355, 356]). Data collated here can help to guide those efforts. Comprehensive regional repositories for dead wood data and species using dead wood are available, such as DecAid for Washington and Oregon [357].

Major challenges to extracting management guidelines from available data are the fact that amounts and distribution of trees and snags are naturally interrelated, and that concern for old growth, particularly in the PNW, has focused research on older forests. The latter condition means that many studies report use based on a wide range of tree sizes; the results are "natural" but potentially inflate diameters required by the species. That is evident in regional differences in nest tree diameters (Table 1). The patterns of cavity use by birds and mammals extracted by this review cannot evade those challenges but still yields general guidelines to assist efforts at sustaining cavity-using species.

Implications following were derived from review of more than 350 studies of 67 cavity-using species occurring in the PNW. The regional focus was intended to restrict the number of species to a naturally occurring assemblage; data on use by a given species were incorporated no matter where the study occurred, including Eurasia. Varied locations act as independent samples of apparent use and preference. Even within the PNW, there are clear regional differences in patterns of nest or roost tree selection (Tables 1 and 2), much of it due to the relative abundance of hardwood and conifer species that are used differently (Figures 2, 5, 6, 7, and 9). Documenting biological reasons for regional differences within a general model aids extrapolation beyond the PNW.

8.1. Ensure Sustained Provision of Dying and Dead Trees.
Many species depend on dying and dead trees and eventually logs (Figures 3 and 4). Snags come from living trees. If a species will need a 150-year-old tree that is required 150 years from now, that tree must begin growth now. Planning to sustain cavity sites must consider not just what is there now, but how that will be sustained. A practical way of providing sustained diversity of structure, including dead trees, is patchwise or group retention ([355]; e.g., Figure 12).

8.2. Retain Trees and Snags of Both Hardwoods and Favoured Conifer Species. Because hardwoods generally are preferred cavity sites (Figure 5) and provide favourable substrate at a younger age than do conifers, it is tempting to rely on hardwoods to meet cavity users' needs. Although they are favored nesting sites, we cannot rely solely on the more decay-prone hardwoods. Hardwoods are uncommon in many forests of the PNW and the varied needs of forest organisms include well-decayed snags, large hollow snags, and snags with loose slabs of bark. Hardwood species will not accommodate all these needs, nor will any one species of conifer. Conifers are longer-lived and provide a longer-lasting source of cavities than do hardwoods [358, 359], so are more likely to sustain snags late into rotations. Moreover, conifer snags are required by species foraging on bark beetles or wood-boring beetles and conifer logs last longer than do hardwood logs. Favoured conifers include western larch, Douglas-fir, and ponderosa pine. Monocultures of less preferred species, such as lodgepole pine, should be avoided.

8.3. Sustain a Range of Decay Classes of Potential Nest Trees. A range of decay classes in standing trees is necessary to sustain biodiversity (fungi, arthropods, birds, and large mammals; e.g., Figure 2). The same is true for species using down wood [307, 360]. Regular inputs from harvest residues are important to maintain the total amount of down wood, but that input tends to be concentrated in a narrow range of decay classes. Retention helps sustain recruitment and ensure continual provision of different decay stages in standing and down wood. Given the patterns of succession among species on and within logs, gaps in recruitment should not greatly exceed 25 years [307]. Organisms in standing trees also use a range of decay classes (Figure 2) and similar recruitment intervals appear appropriate for sustaining standing decay. Some recruitment happens naturally within landscape-level reserves or set-asides (steep slopes, riparian areas, and reserves for particular species) and where retention is practiced. Sustained recruitment over large areas involves planning to avoid large gaps in size and decay stages as well as practices such as retention. Staggering harvest stages among adjacent stands helps sustain a variety of decay classes.

8.4. Retain a Range of Size and Age Classes. Large trees and snags provide nesting or denning sites longer than do small snags [42, 361]. However, smaller snags provide foraging sites, and many more foraging sites are needed than nesting sites. Although larger diameters usually are selected by vertebrates as cavity sites, smaller trees and logs

are used. Where safety considerations encourage falling older snags during harvest, managers should ensure that snags can develop through the rotation. The desirability of a range of decay classes is well documented for bryophytes, insects, terrestrial breeding salamanders, and birds [1, 307]. Well-decayed snags present greater safety risks and are more easily retained in patches. Unless reserve patches are very large, recruitment of well-decayed snags must occur outside of reserve patches. Snags may never become well decayed if operational guidelines require snag falling. Either no-work zones are required during subsequent entries, or silvicultural systems that do not require frequent entries should be employed in at least some areas. Retaining declining live trees, or recently-dead snags, ensures provision of later stages of decay.

8.5. Ensure That Some Large Trees or Snags Are Retained. Some large trees should be retained during both thinning and harvesting operations. Although individual species use a wide range of tree or snag sizes, they tend to select larger ones when available (Figures 6 and 7). In the PNW, studies of vertebrate-forest relations have been concentrated where trees are larger and more valuable, so existing data overestimate requirements where trees are smaller. Although data reveal many instances, particularly in coastal forests, where birds have selected larger trees, most species can be accommodated in managed forests by conifer cavity trees 50 cm dbh; hardwood trees can be smaller, and a dbh >30 cm accommodates most bird species in less productive, inland forest types [232]. However, those values appear too small for some bird species, including four designated "at risk" (Figure 6). Likewise, some mammals, including 6 taxa "at risk," select trees or snags >50 centimeters in diameter or use down wood from 50 to 150 cm in diameter [125, 134, 362]. A major implication is that stands managed intensively for fibre production often will not retain all cavity users. That implies some portion of the landscape should be left unmanaged. There is no unequivocal estimate of how much. While single target diameters are the simpler management target, tolerance intervals accommodate natural variation and the potential bias of large amounts of data collected from old growth forests [339]. Results of Figure 9 affirm that diameter is a sufficient guideline in assessing trees or snag suitability; height merits less attention.

8.6. Provide Both Aggregated and Dispersed Retention of Living and Dead Trees. Figure 10(a) suggests that relatively little is gained by providing more than 3 large snags per ha, but individual species may respond positively to greater snag densities (Figures 10(b) and 10(c)). All forms of response in Figure 11 were observed, indicating that there is no single best approach. When targets are the cavity-nesting community, rather than a single species, about 30% retention maintains about 70% of community members (Figure 12(b)). The importance of variety in the distribution of retention applies equally to species reliant on downed wood [307]. If management for cavity users is to be effective at larger scales, approaches must be developed over a scale appropriate to the range of species affected and recruitment rates of snags, rather than individual land ownership.

8.7. Meet Dead Wood Requirements for Larger Species in Areas Where the Emphasis Is Not on Intensive Fiber Production. There are economic and ecological advantages to zoning the intensity of fiber production [363, 364]. In some forest types, larger mammals prefer significant amounts (100 to 200 m³/ha or more) and large sizes (>50 centimeter diameter) of down wood (Table 2; [365]). Needs of such species are best provided in areas where late-successional attributes are being maintained. Provision of some large trees and pieces of dead wood in forests where the dominant goal is fiber production may facilitate dispersal among areas of more favorable habitat. Given the mobility of birds and patterns of larger mammals' use of space, large snags and subsequent pieces of down wood for such species can be well distributed across large areas.

8.8. Do Not Do the Same Thing Everywhere. Retention of trees in patches reduces safety risks of snag retention and windthrow [366, 367] and facilitates retention of a range of size and decay classes. It also concentrates recruitment of down wood. Some organisms favour piles of down wood, while others favour scattered pieces (review in [307]). Dispersed retention of individual snags, or declining live trees intended to become snags, may be particularly advantageous for perching birds and for territorial secondary users, such as raptors and some small birds, but can impact shrub nesters negatively by encouraging aerial predators [368]. Any single approach will disadvantage some group of species, so a range of practices is preferable when a range of species is to be sustained in an area.

8.9. Limit Salvage Logging after Tree Mortality. Concern for biodiversity led to restrictions on green-tree harvest, particularly in the PNW. That in turn encouraged more complete harvest of dying and dead trees in areas subject to insect attack or fire kill. We learned two things. First, if all vertebrates are to be sustained in forests experiencing mortality to insects or fire, remaining standing dead trees should not all be logged. Review of more than 200 studies addressing forest harvest in beetle-killed forests [369] reported negative effects on bird, mammal, and fish species. Similar findings emerged from review of logging postfire areas [331, 370]. A second major point emerging from these reviews is that postdisturbance variation is so large that small-scale studies are unlikely to provide general principles for mitigating damage, beyond "do not take all dead trees and do not do the same thing everywhere."

Acknowledgments

This work was partially supported by the Forest Sciences Program of British Columbia. Several individuals provided published data in raw form, so that the data could be included in figures, specifically Figures 2, 3, 4, 5, 6, 7, and 9. The author is grateful to B. Booth, J. Deal, D. Huggard, W. Klenner, M. Machmer, K. Martin, K. Squires, and C. Steeger for providing data. D. J. Huggard prepared Figure 12 from data he acquired during his review [340]. I. Houde, L. Kremsater, and K. Squires helped collate the scattered data. A. Farr and three anonymous reviewers improved the paper.

References

[1] B. G. Jonsson, N. Kruys, and T. Ranius, "Ecology of species living on dead wood—lessons for dead wood management," *Silva Fennica*, vol. 39, no. 2, pp. 289–309, 2005.

[2] C. D. Oliver, C. Harrington, M. Bickford et al., "Maintaining and creating old growth structural features in previously disturbed stands typical of the Eastern Washington Cascades," *Journal of Sustainable Forestry*, vol. 2, no. 3, pp. 353–387, 1994.

[3] G. F. Wilhere, "Simulations of snag dynamics in an industrial Douglas-fir forest," *Forest Ecology and Management*, vol. 174, no. 1–3, pp. 521–539, 2003.

[4] A. Berg, B. Ehnstrom, L. Gustafsson, T. Hallingback, M. Jonsell, and J. Weslien, "Threatened plant, animal, and fungus species in Swedish forests: distribution and habitat associations," *Conservation Biology*, vol. 8, no. 3, pp. 718–731, 1994.

[5] J. Pojar, K. Klinka, and D. V. Meidinger, "Biogeoclimatic ecosystem classification in British Columbia," *Forest Ecology and Management*, vol. 22, no. 1-2, pp. 119–154, 1987.

[6] L. W. Spring, "Climbing and pecking adaptations in some North American woodpeckers," *The Condor*, vol. 67, no. 6, pp. 457–488, 1965.

[7] K. E. H. Aitken and K. Martin, "The importance of excavators in hole-nesting communities: availability and use of natural tree holes in old mixed forests of western Canada," *Journal of Ornithology*, vol. 148, supplement 2, pp. S425–S434, 2007.

[8] E. L. Bull, S. R. Peterson, and J. W. Thomas, "Resource partitioning among woodpeckers in northeastern Oregon," *USDA Forest Service—Research Note*, vol. 444, pp. 1–18, 1986.

[9] K. L. Cockle, K. Martin, and T. Wesołowski, "Woodpeckers, decay, and the future of cavity-nesting vertebrate communities worldwide," *Frontiers in Ecology and the Environment*, vol. 9, no. 7, pp. 377–382, 2011.

[10] G. C. Daily, P. R. Ehrlich, and N. M. Haddad, "Double keystone bird in a keystone species complex," *Proceedings of the National Academy of Sciences of the United States of America*, vol. 90, no. 2, pp. 592–594, 1993.

[11] D. S. Dobkin, A. C. Rich, J. A. Pretare, and W. H. Pyle, "Nest-site relationships among cavity-nesting birds of riparian and snowpocket aspen woodlands in the northwestern Great Basin," *Condor*, vol. 97, no. 3, pp. 694–704, 1995.

[12] W. L. Foster and T. Tate Jr., "The activities and coactions of animals at sapsucker trees," *Living Bird*, vol. 5, pp. 87–113, 1966.

[13] R. S. Miller and R. W. Nero, "Hummingbird- sapsucker associations in northern climates," *Canadian Journal of Zoology*, vol. 61, no. 7, pp. 1540–1546, 1983.

[14] G. D. Sutherland, C. L. Gass, P. A. Thompson, and K. P. Lertzman, "Feeding territoriality in migrant rufous hummingbirds: defense of yellow-bellied sapsucker (*Sphyrapicus varius*) feeding sites (Selasphorus rufus)," *Canadian Journal of Zoology*, vol. 60, no. 9, pp. 2046–2050, 1982.

[15] R. T. Holmes, "Ecological and evolutionary impacts of bird predation on forest insects: an overview," *Studies in Avian Biology*, no. 13, pp. 6–13, 1990.

[16] J. W. Thomas, R. G. Anderson, C. Maser et al., "Snags," in *Wildlife Habitats in Managed Forests. The Blue Mountains of Oregon and Washington*, J. W. Thomas, Ed., pp. 60–77, USDA Forest Service Agricultural Handbook, 1979.

[17] F. L. Bunnell and L. A. Dupuis, "Riparian habitats in British Columbia: their nature and role," in *Riparian Habitat Management and Research*, K. H. Morgan and M. A. Lashmar, Eds., pp. 7–21, Special Publication of the Fraser River Action Plan, Canadian Wildlife Service, Delta, Canada, 1995.

[18] R. B. Bury, "Differences in amphibian populations in logged and old growth redwood forest," *Northwest Science*, vol. 57, no. 3, pp. 167–178, 1983.

[19] J. J. Stelmock and A. S. Harestad, "Food habits and life history of the clouded salamander (Aneides ferreus) on northern Vancouver Island," *Syesis*, vol. 12, pp. 71–75, 1979.

[20] P. H. Baldwin and N. K. Zaczkowski, "Breeding biology of the Vaux Swift," *The Condor*, vol. 65, no. 5, pp. 400–406, 1963.

[21] C. M. Davis, "A nesting study of the Brown Creeper," *Living Bird*, vol. 17, pp. 237–263, 1978.

[22] P. R. Ehrlich, D. S. Dobkin, and D. Wheye, *The Birder's Handbook: A Field Guide to the Natural History of North American Birds*, Simon and Schuster, New York, NY, USA, 1988.

[23] L. de K. Lawrence, "A comparative life-history study of four species of woodpeckers," *Ornithological Monographs*, no. 5, pp. 1–156, 1967.

[24] E. L. Walters, E. H. Miller, and P. E. Lowther, "Yellow-bellied Sapsucker (*Sphyrapicus varius*)," in *The Birds of North America Online*, A. Poole, Ed., Cornell Lab of Ornithology, Ithaca, NY, USA, 2002.

[25] M. C. Kalcounis and M. R. Brigham, "Secondary use of aspen cavities by tree-roosting big brown bats," *Journal of Wildlife Management*, vol. 62, no. 2, pp. 603–611, 1998.

[26] S. E. Lewis, "Roost fidelity of bats: a review," *Journal of Mammalogy*, vol. 76, no. 2, pp. 481–496, 1995.

[27] F. L. Bunnell and A. C. Allaye-Chan, "Potential of ungulate winter-range reserves as habitat for cavity-nesting birds," in *Proceedings of the Symposium on Fish and Wildlife Relationships in Old-Growth Forests*, W. R. Meehan, T. R. Merrell, and T. A. Hanley, Eds., pp. 357–365, American Institute of Fishery Research Biologists, 1984.

[28] R. N. Conner, O. K. Miller, and C. S. Adkisson, "Woodpecker dependence on trees infected by fungal heart rots," *The Wilson Bulletin*, vol. 88, no. 4, pp. 575–581, 1976.

[29] A. S. Harestad and D. G. Keisker, "Nest tree use by primary cavity-nesting birds in south central British Columbia," *Canadian Journal of Zoology*, vol. 67, no. 4, pp. 1067–1073, 1989.

[30] L. Kilham, "Reproductive behavior of yellow-bellied sapsuckers. I. Preference for nesting in Fomes infested aspens and nest hole interrelations with flying squirrels, raccoons, and other animals," *The Wilson Bulletin*, vol. 83, no. 2, pp. 159–171, 1971.

[31] A. L. Shigo and L. Kilham, "Sapsuckers and *Fomes ignarius* var. *populinus*," *USDA Forest Service Research Note*, vol. NE-RN-84, pp. 1–2, 1968.

[32] M. O. G. Eriksson, "Clutch sizes and incubation efficiency in relation to nest box size among goldeneyes *Bucephala clangula*," *Ibis*, vol. 121, no. 1, pp. 107–109, 1979.

[33] M. R. Evans, D. B. Lank, W. S. Boyd, and F. Cooke, "A comparison of the characteristics and fate of Barrow's Goldeneye and Bufflehead nests in nest boxes and natural cavities," *Condor*, vol. 104, no. 3, pp. 610–619, 2002.

[34] L. Gustaffson and S. G. Nilsson, "Clutch size and breeding success of pied and collared flycatchers *Ficedula* spp. in nest boxes of different sizes," *Ibis*, vol. 127, no. 3, pp. 380–385, 1985.

[35] J. H. Van Balen, "The relationship between nest-box size, occupation and breeding parameters of the great tit (*Parus major*) and some other hole- nesting species." *Ardea*, vol. 72, no. 2, pp. 163–175, 1984.

[36] K. E. H. Aitken, K. L. Wiebe, and K. Martin, "Nest-site reuse patterns for a cavity-nesting bird community in interior British Columbia," *Auk*, vol. 119, no. 2, pp. 391–402, 2002.

[37] C. E. Bock, A. Cruz, M. C. Grant, C. S. Aid, and T. R. Strong, "Field experimental evidence for diffuse competition among southwestern riparian birds," *American Naturalist*, vol. 140, no. 5, pp. 815–828, 1992.

[38] L. von Haartman, "Adaptation in hole-nesting birds," *Evolution*, vol. 11, no. 3, pp. 339–347, 1957.

[39] E. L. Bull, "Longevity of snags and their use by woodpeckers," in *Proceedings of the Snag Habitat Management Symposium*, J. W. Davis, G. A. Goodwin, and R. A. Ochenfels, Eds., pp. 264–67, USDA Forest Service General Technical Report, RM-GTR-99, 1983.

[40] R. Everett, J. Lehmkuhl, R. Schellhaas et al., "Snag dynamics in a chronosequence of 26 wildfires on the east slope of the cascade range in Washington State, USA," *International Journal of Wildland Fire*, vol. 9, no. 4, pp. 223–234, 2000.

[41] S. M. Garber, J. P. Brown, D. S. Wilson, D. A. Maguire, and L. S. Heath, "Snag longevity under alternative silvicultural regimes in mixed-species forests of central Maine," *Canadian Journal of Forest Research*, vol. 35, no. 4, pp. 787–796, 2005.

[42] M. L. Morrison and M. G. Raphael, "Modeling the dynamics of snags," *Ecological Applications*, vol. 3, no. 2, pp. 322–330, 1993.

[43] R. E. Russell, V. A. Saab, J. G. Dudley, and J. J. Rotella, "Snag longevity in relation to wildfire and postfire salvage logging," *Forest Ecology and Management*, vol. 232, no. 1–3, pp. 179–187, 2006.

[44] J. K. Agee, *Fire Ecology of Pacific Northwest Forests*, Island Press, Washington, DC, USA, 2003.

[45] F. L. Bunnell, "Forest-dwelling vertebrate faunas and natural fire regimes in British Columbia: patterns and implications for conservation," *Conservation Biology*, vol. 9, no. 3, pp. 636–644, 1995.

[46] F. L. Bunnell, I. Houde, B. Johnston et al., "How dead trees sustain live organisms in western forests," in *Proceedings of the Symposium on the Ecology and Management of Dead Wood in Western Forests*, W. F. Laudenslayer Jr., P. J. Shea, B. E. Valentine, C. P. Weatherspoon, and T. E. Lisle, Eds., pp. 291–318, USDA Forest Service, General Technical Report PSW-GTR-181, 2002.

[47] W. G. Dahms, "How long do ponderosa pine snags stand?" in *Research Note*, pp. 1–4, USDA Forest Service, Pacific Northwest Forest and Range Experiment Station, Portland, Ore, USA, 1949.

[48] F. P. Keen, "How soon do yellow pine snags fall?" *Journal of Forestry*, vol. 27, no. 6, pp. 735–737, 1929.

[49] J. M. Schmid, S. A. Mata, and W. F. McCambridge, "Natural falling of beetle-killed ponderosa pine," pp. 1–3, USDA Forest Service Research Note RM-454, 1985.

[50] S. Zack, T. L. George, and W. F. Laudenslayer Jr., "Are there snags in the system? Comparing cavity use among nesting birds in "snag-rich" and "snag-poor" eastside pine forests," in *Proceedings of the Symposium on the Ecology and Management*

of Dead Wood in Western Forests USDA Forest Service, General Technical Report PSW- GTR-181, W. F. Laudenslayer Jr., P. J. Shea, B. E. Valentine, C. P. Weatherspoon, and T. E. Lisle, Eds., pp. 179–191, 2002.

[51] K. Martin, A. Norris, and M. Drever, "Effects of bark beetle outbreaks on avian biodiversity in the British Columbia interior: implications for critical habitat management," *BC Journal of Ecosystems and Management*, vol. 7, no. 3, pp. 10–24, 2006.

[52] J. Schepps, S. Lohr, and T. E. Martin, "Does tree hardness influence nest-tree selection by primary cavity nesters?" *Auk*, vol. 116, no. 3, pp. 658–665, 1999.

[53] P. Li and T. E. Martin, "Nest-site selection and nesting success of cavity-nesting birds in high elevation forest drainages," *The Auk*, vol. 108, no. 2, pp. 405–418, 1991.

[54] B. Schreiber and D. S. deCalesta, "The relationship between cavity-nesting birds and snags on clearcuts in western Oregon," *Forest Ecology and Management*, vol. 50, no. 3-4, pp. 299–316, 1992.

[55] K. Martin, K. E. H. Aitken, and K. L. Wiebe, "Nest sites and nest webs for cavity-nesting communities in interior British Columbia, Canada: nest characteristics and niche partitioning," *Condor*, vol. 106, no. 1, pp. 5–19, 2004.

[56] K. B. Aubry and C. M. Raley, "Selection of nest and roost trees by pileated woodpeckers in coastal forests of Washington," *Journal of Wildlife Management*, vol. 66, no. 2, pp. 392–406, 2002.

[57] E. L. Bull, "Ecology of the pileated woodpecker in northeastern Oregon," *Journal of Wildlife Management*, vol. 51, no. 2, pp. 472–481, 1987.

[58] P. Ohanjanian, I. A. Manley, and P. Davidson, "Williamson's sapsucker in the East Kootenay Region of British Columbia: results of 2006 inventory," Report to Tembec Industries, Forest Investment Account, BC Ministry of Environment, Victoria, Canada, 2006.

[59] R. van den Driessche, T. Chatwin, and M. Mather, "Habitat selection by bats in temperate old-growth forests, Clayoquot Sound, British Columbia," in *Proceedings of the Conference on the Biology and Management of Species and Habitats at Risk*, L. Darling, Ed., vol. 1, pp. 313–319, BC Ministry of Environment, Lands and Parks, University College of the Cariboo, Kamloops, Canada, 2000.

[60] M. J. Vonhof and J. C. Gwilliam, "Intra- and interspecific patterns of day roost selection by three species of forest-dwelling bats in Southern British Columbia," *Forest Ecology and Management*, vol. 252, no. 1–3, pp. 165–175, 2007.

[61] C. L. Mahon, J. D. Steventon, and K. Martin, "Cavity and bark nesting bird response to partial cutting in Northern conifer forests," *Forest Ecology and Management*, vol. 256, no. 12, pp. 2145–2153, 2008.

[62] D. E. Runde and D. E. Capen, "Characteristics of northern hardwood trees used by cavity-nesting birds.," *Journal of Wildlife Management*, vol. 51, no. 1, pp. 217–223, 1987.

[63] C. Savignac and C. S. Machtans, "Habitat requirements of the Yellow-bellied Sapsucker, Sphyrapicus varius, in boreal mixedwood forests of northwestern Canada," *Canadian Journal of Zoology*, vol. 84, no. 9, pp. 1230–1239, 2006.

[64] N. Nielsen-Pincus and E. O. Garton, "Responses of cavity-nesting birds to changes in available habitat reveal underlying determinants of nest selection," *Northwest Naturalist*, vol. 88, no. 3, pp. 135–146, 2007.

[65] J. A. Deal and M. Setterington, *Woodpecker Nest Habitat in the Nimpkish Valley, Northern Vancouver Island*, Canadian Forest Products Ltd., Woss, Canada, 2000.

[66] *British Columbia Nest Records Scheme*, Biodiversity Centre of Wildlife Studies, Victoria, Canada.

[67] R. T. Reynolds, B. D. Linkhart, and J. J. Jeanson, *Characteristics of Snags and Trees Containing Cavities in a Colorado Conifer Forest*, USDA Forest Service Research Note RM-RN-455, 1985.

[68] W. Klenner and D. J. Huggard, "Nesting and foraging habitat requirements of woodpeckers in relation to experimental harvesting treatments at Opax Mountain," in *Proceedings of the Dry Douglas-Fir Workshop New Information for the Management of Dry Douglas-Fir Forests*, C. Hollstedt, A. Vyse, and D. Huggard, Eds., pp. 277–291, Research Branch, BC Ministry of Forests, 1998.

[69] S. J. Madsen, *Habitat use by cavity-nesting birds in the Okanogan National Forest, Washington [M.S. thesis]*, University of Washington, Seattle, Wash, USA, 1985.

[70] M. G. Raphael and M. White, "Use of snags by cavity-nesting birds in the Sierra Nevada," *Wildlife Monographs*, no. 86, pp. 1–66, 1984.

[71] B. Winternitz and H. Cahn, "Nestholes in live and dead aspen," in *Proceedings of the Symposium in Snag Habitat Management*, J. W. Davis, G. A. Goodwin, and R. A. Ockenfels, Eds., pp. 102–106, USDA Forest Service General Technical Report RM-99, 1983.

[72] R. W. Lundquist and J. M. Mariani, "Nesting habitat and abundance of snag-dependent birds in the southern Washington Cascade Range," in *Wildlife and Vegetation of Unmanaged Douglas-Fir Forest*, L. F. Ruggiero, K. B. Aubry, A. B. Carey, and M. H. Huff, Eds., pp. 220–239, USDA Forest Service General Technical Report PNW-286, 1991.

[73] E. Miller and D. R. Miller, "Snag use by birds," in *Management of Western Forests and Grasslands for Nongame Birds*, R. M. DeGraaf, Ed., pp. 337–356, USDA Forest Service General Technical Report INT-GTR-86, 1980.

[74] E. L. Bull, *Resource partitioning among woodpeckers in northeastern Oregon [Ph.D. thesis]*, University of Idaho, Moscow, Idaho, USA, 1980.

[75] E. L. Caton, *Effects of fire and salvage logging on the cavity-nesting bird community in northwestern Montana [Ph.D. thesis]*, University of Montana, Missoula, Mont, USA, 1996.

[76] N. Hoffman, *Distribution of Picoides woodpeckers in relation to habitat disturbance within the Yellow stone area [M.S. thesis]*, Montana State University, Bozeman, Mont, USA, 1997.

[77] R. W. Campbell, N. K. Dawe, I. McTaggart-Cowan et al., *The Birds of British Columbia. Vol. 2, Diurnal Birds of Prey Through Woodpeckers*, University of British Columbia Press, Vancouver, Canada, 1990.

[78] R. N. Conner, R. G. Hooper, H. S. Crawford et al., "Woodpecker nesting habitat in cut and uncut woodlands in Virginia," *Journal of Wildlife Management*, vol. 39, no. 1, pp. 144–150, 1975.

[79] K. E. Kelleher, *A study of hole-nesting avifauna of southwest British Columbia [M.S. thesis]*, University of British Columbia, Vancouver, Canada, 1963.

[80] V. E. Scott, J. A. Whelan, and P. L. Svoboda, "Cavity-nesting birds and forest management," in *Proceedings of the Workshop on the Management of Western Forests and Grasslands for Nongame Birds*, R. M. DeGraaf, Ed., pp. 311–324, USDA Forest Service General Technical Report INT-GTR-86, 1980.

[81] K. L. Wiebe, "Microclimate of tree cavity nests: is it important for reproductive success in northern flickers?" *Auk*, vol. 118, no. 2, pp. 412–421, 2001.

[82] B. R. McClelland and P. T. McClelland, "Pileated woodpecker nest and roost trees in Montana: links with old-growth and forest "health"," *Wildlife Society Bulletin*, vol. 27, no. 3, pp. 846–857, 1999.

[83] T. K. Mellen, *Home range and habitat use of Pileated Woodpeckers, western Oregon [M.S. thesis]*, Oregon State University, Corvalis, Ore, USA, 1987.

[84] J. B. Joy, "Characteristics of nest cavities and nest trees of the Red-breasted Sapsucker in Coastal Montane forests," *Journal of Field Ornithology*, vol. 71, no. 3, pp. 525–530, 2000.

[85] B. R. McClelland and P. T. McClelland, "Red-naped Sapsucker nest trees in Northern Rocky Mountain old-growth forest," *Wilson Bulletin*, vol. 112, no. 1, pp. 44–50, 2000.

[86] C. J. Conway and T. E. Martin, "Habitat suitability for Williamsons's sapsuckers in mixed-conifer forests," *Journal of Wildlife Management*, vol. 57, no. 2, pp. 322–328, 1993.

[87] P. N. Hooge, M. T. Stanback, and W. D. Koenig, "Nest-site selection in the acorn woodpecker," *Auk*, vol. 116, no. 1, pp. 45–54, 1999.

[88] R. W. Campbell, N. K. Dawe, I. Mctaggart-Cowan et al., *The Birds of British Columbia. Vol. 3. Passerines: Flycatchers Through Vireos*, University of British Columbia Press, Vancouver, Canada, 1997.

[89] G. K. Peck and R. D. James, *Breeding Birds of Ontario, Nidiology and Distribution, Volume 2: Passerines*, Royal Ontario Museum, Toronto, Canada, 1987.

[90] A. H. Miller and C. E. Bock, "Natural history of the Nuttall Woodpecker at the Hastings Reservation," *The Condor*, vol. 74, no. 3, pp. 284–294, 1972.

[91] S. M. McEllin, "Nest sites and population demographies of White-breasted and Pygmy Nuthatches in Colorado," *The Condor*, vol. 81, no. 4, pp. 348–352, 1979.

[92] K. A. Milne and S. J. Hejl, "Nest-site characteristics of white-headed woodpeckers," *Journal of Wildlife Management*, vol. 53, no. 1, pp. 50–55, 1989.

[93] R. D. Dixon, *Ecology of White-headed Woodpeckers in the central Oregon Cascades [M.S. thesis]*, University of Idaho, Moscow, Idaho, USA, 1995.

[94] M. Haggard and W. L. Gaines, "Effects of stand-replacement fire and salvage logging on a cavity-nesting bird community in eastern Cascades, Washington," *Northwest Science*, vol. 75, no. 4, pp. 387–396, 2001.

[95] S. M. Hitchcock, *Abundanceand nesting success of cavity nesting birds in unlogged and salvage-logged burned forests in northwest Montana [M.S. thesis]*, University of Montana, Missoula, Mont, USA, 1996.

[96] C. Y. Smith, I. G. Warkentin, and M. T. Moroni, "Snag availability for cavity nesters across a chronosequence of post-harvest landscapes in western Newfoundland," *Forest Ecology and Management*, vol. 256, no. 4, pp. 641–647, 2008.

[97] M. A. Harris, *Habitat use among woodpeckers on forest burns [M.S. thesis]*, University of Montana, Missoula, Mont, USA, 1982.

[98] J. A. Sedgwick and F. L. Knopf, "Habitat relationships and nest site characteristics of cavity- nesting birds in cottonwood floodplains," *Journal of Wildlife Management*, vol. 54, no. 1, pp. 112–124, 1990.

[99] L. W. Gyug, C. Steeger, and I. Ohanjanian, "Characteristics and densities of Williamson's Sapsucker nest trees in British Columbia," *Canadian Journal of Forest Research*, vol. 39, no. 12, pp. 2319–2331, 2009.

[100] G. C. Daily, "Heartwood decay and vertical distribution of red-naped sapsucker nest cavities," *Wilson Bulletin*, vol. 105, no. 4, pp. 674–679, 1993.

[101] C. J. Conway and T. E. Martin, "Habitat suitability for Williamsons's sapsuckers in mixed-conifer forests," *Journal of Wildlife Management*, vol. 57, no. 2, pp. 322–328, 1993.

[102] L. Saari, E. Pulliainen, O. Hildén, A. Järvinen, and I. Mäkisalo, "Breeding biology of the Siberian Tit Parus cinctus in Finland," *Journal of Ornithology*, vol. 135, no. 4, pp. 549–575, 1994.

[103] D. G. Keisker, "Nest tree selection by primary cavity-nesting birds in south-central British Columbia," Wildlife Report R-13, BC Ministry of Environment, Lands and Parks, Victoria, Canada, 1987.

[104] K. M. Mazur, P. C. James, and S. D. Frith, "Barred Owl (*Strix varia*) nest site characteristics in the boreal forest of Saskatchewan," in *Proceedings of the 2nd International Symposium in Biology and Conservation of Owls of the Northern Hemisphere*, J. R. Duncan, D. H. Johnson, and T. H. Nicholls, Eds., pp. 267–271, USDA Forest Service, General Technical Report NC-190, 1997.

[105] M. R. Evans, *Breeding habitat selection by Barrow's Goldeneye and Bufflehead in the Cariboo-Chilcotin region of British Columbia: nest sites, brood-rearing habitat, and competition [Ph.D. thesis]*, Simon Fraser University, Burnaby, Canada, 2003.

[106] C. M. Davis, "A nesting study of the Brown Creeper," *Living Bird*, vol. 17, pp. 237–263, 1978.

[107] J. J. Siegel, *An evaluation of the minimum habitat quality standards for birds in old-growthponderosa pine forests, northern Arizona [M.S. thesis]*, University of Arizona, Tucson, Ariz, USA, 1989.

[108] C. Steeger and J. Dulisse, "Ecological interrelationships of three-toed woodpeckers with bark beetles and pine trees," in *Research Summary RS-035*, pp. 1–4, BC Ministry of Forests, Nelson, Canada, 1997.

[109] M. A. Stern, T. G. Wise, and K. L. Theodore, "Use of natural cavity by bufflehead nesting in Oregon," *The Murrelet*, vol. 68, no. 2, p. 50, 1987.

[110] D. P. Arsenault, "Differentiating nest sites of primary and secondary cavity-nesting birds in New Mexico," *Journal of Field Ornithology*, vol. 75, no. 3, pp. 257–265, 2004.

[111] R. W. Campbell, N. K. Dawe, I. McTaggart-Cowan et al., "The birds of British Columbia," in *Introduction, Loons Through Waterfowl*, vol. 1, University of British Columbia Press, Vancouver, Canada, 1990.

[112] D. F. Stauffer and L. B. Best, "Nest-site selection by cavity-nesting birds of riparian habitats in Iowa.," *Wilson Bulletin*, vol. 94, no. 3, pp. 329–337, 1982.

[113] W. B. Rendell and R. J. Robertson, "Nest-site characteristics, reproductive success and cavity availability for Tree Swallows breeding in natural cavities," *The Condor*, vol. 91, no. 4, pp. 875–885, 1989.

[114] E. L. Bull and C. T. Collins, "Nest site fidelity, breeding age, and adult longevity in the Vaux's swift," *North American Bird Bander*, vol. 21, no. 2, pp. 49–51, 1996.

[115] E. L. Bull and H. D. Cooper, "Vaux's swift nests in hollow trees," *Western Birds*, vol. 22, no. 2, pp. 85–91, 1991.

[116] J. E. Hunter and M. J. Mazurek, "Characteristics of trees used by nesting and roosting Vaux's Swifts in northwestern California," *Western Birds*, vol. 34, no. 4, pp. 225–229, 2003.

[117] G. D. Hayward, P. H. Hayward, and E. O. Garton, "Ecology of boreal owls in the northern Rocky Mountains, USA," *Wildlife Monographs*, no. 124, pp. 1–59, 1993.

[118] J. B. Cunningham, R. P. Balda, and W. S. Gaud, "Selection and use of snags by secondary and cavity-nesting birds of the ponderosa pine forest," USDA Forest Service Research Paper RM-RP-222, 1980.

[119] J. R. Robb and T. A. Bookhout, "Factors influencing wood duck use of natural cavities," *Journal of Wildlife Management*, vol. 59, no. 2, pp. 372–383, 1995.

[120] D. S. Gilmer, I. J. Ball, L. M. Cowardin et al., "Natural cavities used by wood ducks in north-central Minnesota," *Journal of Wildlife Management*, vol. 42, no. 2, pp. 288–298, 1978.

[121] G. J. Soulliere, "Density of suitable wood duck nest cavities in a northern hardwood forest," *Journal of Wildlife Management*, vol. 52, no. 1, pp. 86–89, 1988.

[122] H. H. Prince, "Nest sites used by wood ducks and common goldeneyes in New Brunswick," *Journal of Wildlife Management*, vol. 32, no. 3, pp. 489–500, 1968.

[123] L. L. C. Jones, M. G. Raphael, J. T. Forbes et al., "Using remotely activated cameras to monitor maternal dens of martens," in *Martes: Taxonomy, Ecology, Techniques, and Management*, G. Proulx, H. N. Bryant, and P. M. Woodard, Eds., pp. 329–349, Provincial Museum of Alberta, Edmonton, Canada, 1997.

[124] S. K. Martin and R. H. Barrett, "Resting site selection by marten at Sagehen Creek, California," *Northwestern Naturalist*, vol. 72, no. 2, pp. 37–42, 1991.

[125] M. G. Raphael and L. L. C. Jones, "Characteristics of resting and denning sites of American martens in central Oregon and western Washington," in *Martes: Taxonomy, Ecology, Techniques and Management*, G. Proulx, H. N. Bryant, and P. M. Woodward, Eds., pp. 146–166, Provincial Museum of Alberta, Edmonton, Canada, 1997.

[126] L. F. Ruggiero, D. E. Pearson, and S. E. Henry, "Characteristics of American marten den sites in Wyoming," *Journal of Wildlife Management*, vol. 62, no. 2, pp. 663–673, 1998.

[127] W. D. Spencer, "Seasonal rest-site preferences of pine martens in the northern Sierra Nevada," *Journal of Wildlife Management*, vol. 51, no. 3, pp. 616–621, 1987.

[128] B. J. Betts, "Roosting behaviour of silver-haired bats (Lasionycteris noctivagans) and big brown bats (Eptesicus fuscus) in northeast Oregon," in *Proceedings of the Bats and Forests Symposium*, R. M. R. Barclay and R. M. Brigham, Eds., pp. 61–66, British Columbia Ministry of Forests, 1996.

[129] M. J. Rabe, T. E. Morrell, and H. Green, "Characteristics of ponderosa pine snag roosts used by reproductive bats in northern Arizona," *Journal of Wildlife Management*, vol. 62, no. 2, pp. 612–621, 1998.

[130] S. A. Rasheed and S. L. Holroyd, *Roosting Habitat Assessment and Inventory of Bats in the Mica Wildlife Compensation Area*, Columbia Basin Fish and Wildlife Compensation Program, Nelson, Canada, 1995.

[131] M. J. Vonhof, *A Survey of the Abundance, Diversity, and Roost-Site Preferences of Bats in the Pend d'Oreille Valley, British Columbia*, Columbia Basin Fish and Wildlife Compensation Program, Nelson, Canada, 1996.

[132] J. Akenson, *Black Bear Den Summary NE Region—Interim Progress Report. Starkey Bear Study*, Oregon Department of Fish and Game, La Grande, Ore, USA, 1994.

[133] E. L. Bull, J. J. Akenson, B. J. Betts et al., "The interdependence of wildlife and old-growth forests," in *Proceedings of the Workshop on Wildlife Tree/Stand-Level Biodiversity*, P. Bradford, T. Manning, and B. I'Anson, Eds., pp. 71–76, BC Ministry of Environment, Lands, and Parks and Ministry of Forests, 1996.

[134] H. Davis, *Characteristics and selection of winter dens by black bears in coastal British Columbia [M.S. thesis]*, Simon Fraser University, Burnaby, Canada, 1996.

[135] D. Immell and M. C. Boulay, *Progress Report—Black Bear Ecology Research Project. Wildlife Research Project*, Oregon Department of Fish and Wildlife, Portland, Ore, USA, 1994.

[136] D. J. Lindsay, *Black Bear den Catalogue: A Listing of Coastal Black Bear Dens*, Timberwest Forest, Crofton, Canada, 1999.

[137] F. G. Lindzey and C. Meslow, "Characteristics of black bear dens on Long Island, Washington," *Northwest Science*, vol. 60, no. 4, pp. 236–242, 1976.

[138] W. O. Noble, C. E. Meslow, and M. D. Pope, *Denning Habits of Black Bears in the Central Coast Range of Oregon*, Department of Fisheries and Wildlife, Oregon State University, Corvallis, Ore, USA, 1990.

[139] R. M. Brigham, M. J. Vonhof, R. M. R. Barclay, and J. C. Gwilliam, "Roosting behavior and roost-site preferences of forest-dwelling california bats (*Myotis californicus*)," *Journal of Mammalogy*, vol. 78, no. 4, pp. 1231–1239, 1997.

[140] S. Grindal, *Upper Kootenay River Bat Survey*, Columbia Basin Fish and Wildlife Compensation Program, Nelson, Canada, 1997.

[141] T. F. Paragi, S. M. Arthur, and W. B. Krohn, "Importance of tree cavities as natal dens for fishers," *Northern Journal of Applied Forestry*, vol. 13, no. 2, pp. 79–83, 1996.

[142] J. S. Yaeger, *Habitat at fisher resting sites in the Klamath province of northern California [M.S. thesis]*, Humboldt State University, Arcata, Calif, USA, 2005.

[143] A. B. Carey, T. M. Wilson, C. C. Maguire, and B. L. Biswell, "Dens of northern flying squirrels in the pacific northwest," *Journal of Wildlife Management*, vol. 61, no. 3, pp. 684–699, 1997.

[144] R. A. Mowrey and J. C. Zasada, "Den tree use and movements of northern flying squirrels in interior Alaska and implications for forest management," in *Proceedings of the Symposium on Fish and Wildlife Relationships in Old-Growth Forests*, W. R. Meehan, T. R. Merrell, and T. A. Hanley, Eds., pp. 351–356, American Institute of Fishery Research Biologists, 1984.

[145] C. Steeger and J. Dulisse, "Characteristics and dynamics of cavity nest trees in southern British Columbia," in *Proceedings of the Symposium on the Ecology and Management of Dead Wood in Western Forests*, W. F. Laudenslayer Jr., P. J. Shea, B. E. Valentine, C. P. Weatherspoon, and T. E. Lisle, Eds., pp. 275–289, USDA Forest Service, General Technical Report PSW-GTR-181, 2002.

[146] T. J. Weller and C. J. Zabel, "Characteristics of fringed myotis day roosts in northern California," *Journal of Wildlife Management*, vol. 65, no. 3, pp. 489–497, 2001.

[147] L. H. Crampton and R. M. R. Barclay, "Relationships between bats and stand age and structure in aspen mixedwood forests in Alberta," in *Relationships Between Stand Age, Stand Structure and Biodiversity in Aspen Mixedwood Forests in Alberta*, B. Stelfox, Ed., pp. 211–225, Alberta Environmental Centre, Vegreville, Canada; Canadian Forest Service, Edmonton, Canada, 1995.

[148] M. C. Kalcounis and K. R. Hecker, "Intraspecific variation in roost-site selection by little brown bats (Myotis lucifugus)," in *Proceedings of the Bats and Forests Symposium*, R. M. R. Barclay and R. M. Brigham, Eds., pp. 81–90, British Columbia Ministry of Forests, 1996.

[149] P. C. Ormsbee and W. C. McComb, "Selection of day roosts by female long-legged myotis in the central Oregon Cascade

Range," *Journal of Wildlife Management*, vol. 62, no. 2, pp. 596–603, 1998.

[150] M. C. Caceres, *The summer ecology of Myotis species bats in the interior wet-belt of British Columbia [M.S. thesis]*, University of Calgary, Calgary, Canada, 1998.

[151] M. A. Menzel, S. F. Owen, W. M. Ford et al., "Roost tree selection by northern long-eared bat (*Myotis septentrionalis*) maternity colonies in an industrial forest of the central Appalachian mountains," *Forest Ecology and Management*, vol. 155, no. 1-3, pp. 107–114, 2002.

[152] D. L. Waldien, J. P. Hayes, and E. B. Arnett, "Day-roosts of female long-eared myotis in western Oregon," *Journal of Wildlife Management*, vol. 64, no. 3, pp. 785–796, 2000.

[153] L. A. Campbell, J. G. Hallett, and M. A. O'Connell, "Conservation of bats in managed forests: use of roosts by Lasionycteris noctivagans," *Journal of Mammalogy*, vol. 77, no. 4, pp. 976–984, 1996.

[154] M. J. Vonhof and R. M. R. Barclay, "Use of tree stumps as roosts by the western long-eared bat," *Journal of Wildlife Management*, vol. 61, no. 3, pp. 674–684, 1997.

[155] P. L. Svoboda, K. E. Young, and V. E. Scott, "Recent nesting records of Purple Martins in western Colorado," *Western Birds*, vol. 11, no. 4, pp. 195–198, 1980.

[156] A. D. M. Rayner and L. Boddy, *Fungal Decomposition of Wood*, John Wiley and Sons, Chichester, UK, 1988.

[157] F. L. Bunnell and A. C. Chan-McLeod, "Terrestrial vertebrates," in *The Rain Forests of Home. Profile of a North American Bioregion*, P. K. Schoonmaker, B. von Hagen, and E. C. Wolf, Eds., pp. 103–130, Island Press, Washington, DC, USA, 1997.

[158] J. J. Akenson and M. G. Henjum, "Black bear den site selection in the Starkey study area," *Blue Mountains Natural Resources Institute, Natural Resource News*, vol. 4, no. 2, pp. 1–2, 1994.

[159] A. T. Hamilton, *Personal Communication*, BC Ministry of Fish and Wildlife, Victoria, Canada, 2010.

[160] W. C. McComb and R. E. Noble, "Nest-box and natural-cavity use in three mid-south forest habitats," *Journal of Wildlife Management*, vol. 445, no. 1, pp. 93–101, 1981.

[161] S. J. Hejl, K. R. Newlon, M. E. Mcfadzen et al., "Brown Creeper (*Certhia americana*)," in *The Birds of North America Online*, A. Poole, Ed., Cornell Lab of Ornithology, Ithaca, NY, USA, 2002.

[162] B. Peterson and G. Gauthier, "Nest site use by cavity-nesting birds of the Cariboo Parkland, British Columbia," *The Wilson Bulletin*, vol. 97, no. 3, pp. 319–331, 1985.

[163] F. L. Bunnell, E. Wind, and R. Wells, "Dying and dead hardwoods: their implications to management," in *Proceedings of the Symposium on the Ecology and Management of Dead Wood in Western Forests*, W. F. Laudenslayer Jr, P. J. Shea, B. E. Valentine, C. P. Weatherspoon, and T. E. Lisle, Eds., pp. 695–716, USDA Forest Service, General Technical Report PSW-GTR-181, 2002.

[164] R. Goggans, R. D. Dixon, and L. C. Seminara, "Habitat use by three-toed and black- backed woodpeckers, Deschutes National Forest," USDA Forest Service Technical Report 87-3-02, 1989.

[165] W. Klenner and D. Huggard, "Three-toed woodpecker nesting and foraging at Sicamous Creek," in *Proceedings of the Workshop on the Sicamous Creek Silvicultural Systems Project*, C. Hollstedt and A. Vyse, Eds., pp. 224–233, Research Branch, BC Ministry of Forests, 1997.

[166] E. L. Bull and J. A. Jackson, "Pileated woodpecker (*Dryocopus pileatus*)," in *The Birds of North America Online*, A. Poole, Ed., Cornell Lab of Ornithology, Ithaca, NY, USA, 2011.

[167] A. B. Crockett and H. H. Hadow, "Nest site selection by Williamson's and red-naped sapsuckers," *The Condor*, vol. 77, no. 3, pp. 365–368, 1975.

[168] M. Axelrod, "Observations on a boreal chickadee nest," *The Loon*, vol. 51, pp. 135–140, 1979.

[169] C. Galen, "A preliminary assessment of the status of the Lewis' Woodpecker in Wasco County, Oregon," Tech. Rep. 88-3-01, Oregon Department of Fish and Wildlife, Portland, Ore, USA, 1989.

[170] B. P. Booth, *The effects of thinning on forest bird communities in dry, interior Douglas-fir forests [M.S. thesis]*, University of British Columbia, Vancouver, Canada.

[171] K. L. Garrett, M. G. Raphael, and R. D. Dixon, "White-headed Woodpecker (*Picoides albolarvatus*)," in *The Birds of North America Online*, A. Poole, Ed., Cornell Lab of Ornithology, Ithaca, NY, USA, 1996.

[172] W. C. Weber and S. R. Cannings, "The white-headed woodpecker (*Dendrocopus albovatus*) In British Columbia," *Syesis*, vol. 9, pp. 215–220, 1976.

[173] C. L. Hartwig, D. S. Eastman, and A. S. Harestad, "Characteristics of pileated woodpecker (*Dryocopus pileatus*) cavity trees and their patches on southeastern Vancouver Island, British Columbia, Canada," *Forest Ecology and Management*, vol. 187, no. 2-3, pp. 225–234, 2004.

[174] C. Steeger, M. Machmer, and E. Walters, "Ecology and management of woodpeckers and wildlife trees in British Columbia," in *Fraser River Action Plan*, pp. 1–23, Canadian Wildlife Service, Delta, Canada, 1996.

[175] R. L. Hutto and S. M. Gallo, "The effects of postfire salvage logging on cavity-nesting birds," *Condor*, vol. 108, no. 4, pp. 817–831, 2006.

[176] R. A. Cannings, R. J. Cannings, and S. G. Cannings, *Birds of the Okanagan Valley*, British Columbia, Royal British Columbia Museum, Victoria, Canada, 1987.

[177] J. A. Deal and D. W. Gilmore, "Effects of vertical structure and biogeoclimatic subzone on nesting locations for woodpeckers on north central Vancouver Island: nest tree attributes," *Northwest Science*, vol. 72, no. 2, pp. 119–121, 1998.

[178] S. Hågvar, G. Hågvar, and E. Mønness, "Nest site selection in Norwegian woodpeckers," *Holarctic Ecology*, vol. 13, no. 2, pp. 156–165, 1990.

[179] T. Wesołowski and L. Tomiałojć, "The breeding ecology of woodpeckers in a temperate primaeval forest—preliminary data," *Acta Ornithologica*, vol. 22, no. 1, pp. 1–21, 1986.

[180] K. Eckert, "First Minnesota nesting record of northern three-toed woodpecker," *Loon*, vol. 53, pp. 221–223, 1981.

[181] A. J. Erskine and W. D. McLaren, "Sapsucker nest holes and their use by other species," *The Canadian Field-Naturalist*, vol. 86, no. 4, pp. 357–361, 1972.

[182] C. E. Bock, "The ecology and behavior of the Lewis's Woodpecker (*Asyndesmus lewis*)," in *University of California Publication in Zoology*, vol. 92, University of California Press, Berkeley, Calif, USA, 1970.

[183] W. M. Block, "Foraging ecology of Nuttall's woodpecker," *The Auk*, vol. 108, no. 2, pp. 303–318, 1991.

[184] L. L. Short Jr., "The systematics and behavior of some North American woodpeckers, genus Picoides(Aves)," *Bulletin of the American Museum of Natural History*, vol. 145, 118 pages, 1971.

[185] J. R. Waters, *Population and habitat characteristics of cavity-nesting birds in a California oak woodland [M.S. thesis]*, Humboldt State University, Arcata, Calif, USA, 1988.

[186] V. A. Saab and J. G. Dudley, *Responses of Cavity-Nesting Birds to Stand-Replacement Fire and Salvage Logging in Ponderosa Pine/Douglas-Fir Forests of Southwestern Idaho*, USDA Forest Service Research Paper RMRS-RP-11, 1998.

[187] D. L. Leonard Jr., "Three-toed Woodpecker (*Picoides tridactylus*)," in *The Birds of North America Online*, A. Poole, Ed., Cornell Lab of Ornithology, Ithaca, NY, USA, 2001.

[188] R. W. Mannan, E. C. Meslow, and H. M. Wight, "Use of snags by birds in Douglas-fir forests, Western Oregon," *Journal of Wildlife Management*, vol. 44, no. 4, pp. 787–797, 1980.

[189] B. G. Marcot and R. Hill, "Flammulated owls in northwestern California," *Western Birds*, vol. 11, no. 3, pp. 141–149, 1980.

[190] M. L. Richmond, L. R. DeWeese, and R. E. Pillmore, "Brief observations on the breeding biology of the flammulated owl in Colorado," *Western Birds*, vol. 11, no. 1, pp. 35–46, 1980.

[191] B. Webb, "Distribution and nesting requirements of montane forest owls in Colorado—part III: flammulated owl (*Otus flammeolus*)," *Journal of the Colorado Field Ornithologists*, vol. 6, pp. 76–81, 1982.

[192] E. L. Bull, A. L. Wright, and M. G. Henjum, "Nesting habitat of flammulated owls in Oregon," *Journal of Raptor Research*, vol. 24, no. 3, pp. 52–55, 1990.

[193] R. T. Reynolds and B. D. Linkhart, "The nesting biology of flammulated owls in Colorado," in *Proceedings of the Biology and Conservation of Northern Forest Owls Symposium*, R. W. Nero, R. J. Clark, R. J. Knapton, and R. H. Hamre, Eds., pp. 239–248, USDA Forest Service Technical Report RM-GTR-42, 1997.

[194] A. M. van Woudenberg, *Integrated management of flammulated owl breeding habitat and timber harvest in British Columbia [M.S. thesis]*, University of British Columbia, Vancouver, Canada, 1992.

[195] H. E. Kingery and C. K. Ghalambor, "Pygmy nuthatch (*Sitta pygmaea*)," in *The Birds of North America Online*, A. Poole, Ed., Cornell Lab of Ornithology, Ithaca, NY, USA, 2001.

[196] T. Brush, B. W. Anderson, and R. D. Ohmart, "Habitat selection related to resource availability among cavity-nesting birds," in *Proceedings of the Snag Habitat Management Symposium*, J. W. Davis and G. A. Goodwin R .A. Ockenfels, Eds., pp. 88–98, USDA Forest Service General Technical Report RM-99, 1983.

[197] S. P. Cline, A. B. Berg, and H. M. Wight, "Snag characteristics and dynamics in Douglas-fir forests, western Oregon," *Journal of Wildlife Management*, vol. 44, no. 4, pp. 773–786, 1980.

[198] K. H. Wright and G. M. Harvey, *The Deterioration of Beetle-Killed Douglas-Fir in Western Oregon and Washington*, USDA Forest Service Research Paper PNW-RP-50, 1967.

[199] C. Steeger and C. L. Hitchcock, "Influence of forest structure and diseases on nestsite selection by red-breasted nuthatches," *Journal of Wildlife Management*, vol. 62, no. 4, pp. 1349–1358, 1998.

[200] T. S. Buchanan and G. H. Englerth, *Decay and Other Losses in Windthrown Timber on the Olympic Peninsula, Washington*, USDA Forest Service Technical Bulletin 733, Washington, DC, USA, 1940.

[201] N. T. Engelhardt, "Pathological deterioration of looper-killed western hemlock on southern Vancouver Island," *Forest Science*, vol. 3, no. 2, pp. 125–136, 1957.

[202] S. Parsons, K. J. Lewis, and J. M. Psyllakis, "Relationships between roosting habitat of bats and decay of aspen in the sub-boreal forests of British Columbia," *Forest Ecology and Management*, vol. 177, no. 1–3, pp. 559–570, 2003.

[203] P. M. Cryan, M. A. Bogan, and G. M. Yanega, "Roosting habits of four bat species in the black hills of South Dakota," *Acta Chiropterologica*, vol. 3, no. 1, pp. 43–52, 2001.

[204] R. W. Perry and R. E. Thill, "Roost selection by big brown bats in forests of Arkansas: importance of Pine snags sand open forest habitats to males," *Southeastern Naturalist*, vol. 7, no. 4, pp. 607–618, 2008.

[205] S. J. Rancourt, M. I. Rule, and M. A. O'Connell, "Maternity roost site selection of big brown bats in ponderosa pine forests of the Channeled Scablands of northeastern Washington State, USA," *Forest Ecology and Management*, vol. 248, no. 3, pp. 183–192, 2007.

[206] R. M. R. Barclay and R. M. Brigham, "Year-to-year reuse of tree-roosts by California bats (*Myotis californicus*) in southern British Columbia," *American Midland Naturalist*, vol. 146, no. 1, pp. 80–85, 2001.

[207] J. L. Boland, J. P. Hayes, W. P. Smith, and M. M. Huso, "Selection of day-roosts by Keen's myotis (*Myotis Keenii*) at multiple spatial scales," *Journal of Mammalogy*, vol. 90, no. 1, pp. 222–234, 2009.

[208] L. H. Crampton and R. M. R. Barclay, "Selection of roosting and foraging habitat by bats in different-aged aspen mixedwood stands," *Conservation Biology*, vol. 12, no. 6, pp. 1347–1358, 1998.

[209] J. M. Psyllakis and R. M. Brigham, "Characteristics of diurnal roosts used by female *Myotis bats* in sub-boreal forests," *Forest Ecology and Management*, vol. 223, no. 1–3, pp. 93–102, 2006.

[210] P. C. Ormsbee, "Characteristics, use, and distribution of day roosts selected by female *Myotis volans* (long-legged myotis) in forested habitat of the central Oregon Cascades," in *Proceedings of the Bats and Forests Symposium*, R. M. R. Barclay and R. M. Brigham, Eds., pp. 124–130, British Columbia Ministry of Forests, 1996.

[211] R. W. Foster and A. Kurta, "Roosting ecology of the northern bat (*Myotis septentrionalis*) and comparisons with the endangered Indiana bat (*Myotis sodalis*)," *Journal of Mammalogy*, vol. 80, no. 2, pp. 659–672, 1999.

[212] M. J. Lacki and J. H. Schwierjohann, "Day-roost characteristics of northern bats in mixed mesophytic forest," *Journal of Wildlife Management*, vol. 65, no. 3, pp. 482–488, 2001.

[213] M. D. Baker, M. J. Lacki, G. A. Faixa, P. L. Droppelman, R. A. Slack, and S. A. Slankard, "Habitat use of pallid bats in coniferous forests of northern California," *Northwest Science*, vol. 82, no. 4, pp. 269–275, 2008.

[214] R. M. R. Barclay, P. A. Faure, and D. R. Farr, "Roosting behavior and roost selection by migrating silver-haired bats (*Lasionycteris noctivagans*)," *Journal of Mammalogy*, vol. 69, no. 4, pp. 821–825, 1988.

[215] T. A. Mattson, S. W. Buskirk, and N. L. Stanton, "Roost sites of the silver-haired bat (*Lasionycteris noctivagans*) in the Black Hills, South Dakota," *Great Basin Naturalist*, vol. 56, no. 3, pp. 247–253, 1996.

[216] C. L. Cotton and K. L. Parker, "Winter habitat and nest trees used by northern flying squirrels in subboreal forests," *Journal of Mammalogy*, vol. 81, no. 4, pp. 1071–1086, 2000.

[217] J. S. Gerrow, *Home range, habitat use, nesting ecology and diet of the northern flying squirrel in southern New Brunswick [M.S. thesis]*, Acadia University, Wolfville, Canada, 1996.

[218] G. L. Holloway and J. R. Malcolm, "Nest-tree use by northern and southern flying squirrels in central Ontario," *Journal of Mammalogy*, vol. 88, no. 1, pp. 226–233, 2007.

[219] K. J. Martin, *Movements and habitat characteristics of northern flying squirrels in the central Oregon Cascades [M.S. thesis]*, Oregon State University, Corvallis, Ore, USA, 1994.

[220] M. D. Meyer, D. A. Kelt, and M. P. North, "Nest trees of northern flying squirrels in the Sierra Nevada," *Journal of Mammalogy*, vol. 86, no. 2, pp. 275–280, 2005.

[221] M. D. Meyer, M. P. North, and D. A. Kelt, "Nest trees of northern flying squirrels in Yosemite National Park, California," *Southwestern Naturalist*, vol. 52, no. 1, pp. 157–161, 2007.

[222] J. W. Witt, "Home range and density estimates for the northern flying squirrel, *Glaucomys sabrinus*, in western Oregon," *Journal of Mammalogy*, vol. 73, no. 4, pp. 921–929, 1992.

[223] M. J. Lacki and M. D. Baker, "Day roosts of female fringed myotis (*Myotis thysanodes*) in xeric forests of the pacific northwest," *Journal of Mammalogy*, vol. 88, no. 4, pp. 967–973, 2007.

[224] C. R. Willis and R. M. Brigham, "Physiological and ecological aspects of roost selection by reproductive female hoary bats (*Lasiurus cinereus*)," *Journal of Mammalogy*, vol. 86, no. 1, pp. 86–94, 2005.

[225] D. G. Constantine, "Ecological observations on lasiurine bats in Iowa," *Journal of Mammalogy*, vol. 47, no. 1, pp. 34–41, 1966.

[226] M. A. Vonhof and J. C. Gwilliam, *A Summary of Bat Research in the Pend d'Oreille Valley in Southern British Columbia*, Columbia Basin Fish and Wildlife Compensation Program, Nelson, Canada, 2000.

[227] D. I. Solick and R. M. R. Barclay, "Thermoregulation and roosting behaviour of reproductive and nonreproductive female western long-eared bats (*Myotis evotis*) in the Rocky Mountains of Alberta," *Canadian Journal of Zoology*, vol. 84, no. 4, pp. 589–599, 2006.

[228] S. J. Rancourt, M. I. Rule, and M. A. O'Connell, "Maternity roost site selection of long-eared myotis, Myotis evotis," *Journal of Mammalogy*, vol. 86, no. 1, pp. 77–84, 2005.

[229] M. J. Vonhof and J. C. Gwilliam, *Survey of the Roost-Site Preferences of California, Western Long-Eared, and Long-Legged Bats in the Pend d'Oreille Valley, British Columbia*, Columbia Basin Fish and Wildlife Compensation Program, Nelson, Canada, 1999.

[230] R. H. Waring and J. F. Franklin, "Evergreen coniferous forests of the Pacific Northwest," *Science*, vol. 204, no. 4400, pp. 1380–1386, 1979.

[231] S. S. Niemiec, G. R. Ahrens, S. Willits et al., *Hardwoods of the Pacific Northwest*, Research Contribution No. 8, Forest Research Laboratory, Oregon State University, Corvallis, Ore, USA, 1995.

[232] F. L. Bunnell, E. Wind, and M. Boyland, "Diameters and heights of trees with cavities: their implications to management," in *Proceedings of the Symposium on the Ecology and Management of Dead Wood in Western Forests*, W. F. Laudenslayer Jr., P. J. Shea, B. E. Valentine, C. P. Weatherspoon, and T. E. Lisle, Eds., pp. 717–738, USDA Forest Service, General Technical Report PSW-GTR-181, 2002.

[233] F. L. Bunnell, L. L. Kremsater, and E. Wind, "Managing to sustain vertebrate richness in forests of the Pacific Northwest: relationships within stands," *Environmental Reviews*, vol. 7, no. 3, pp. 97–146, 1999.

[234] D. A. Sibley, *The Sibley Field Guide to Birds of Western North America*, Alfred A. Knopf, New York, NY, USA, 2003.

[235] B. J. Putnam, *Songbird responses of precommercially thinned and unthinned stands in east- central Washington [M.S. thesis]*, Oregon State University, Corvallis, Ore, US, 1983.

[236] B. R. McClelland, *Relationships between hole-nesting birds, forest snags, and decay in Western larch-douglas-fir forests of the Northern Rocky Mountains [M.S. thesis]*, University of Montana, Missoula, Mont, USA, 1977.

[237] B. R. McClelland, S. S. Frissle, W. C. Fischer et al., "Habitat management for hole-nesting birds in forests of western larch and Douglas-fir," *Journal of Forestry*, vol. 77, no. 8, pp. 480–483, 1979.

[238] S. K. Nelson, *Habitat use and densities of cavity-nesting birds in the Oregon coast ranges [M.S. thesis]*, Oregon State University, Corvallis, Ore, USA, 1988.

[239] J. E. Zarnowitz and D. A. Manuwal, "The effects of forest management on cavity-nesting birds in northwestern Washington.," *Journal of Wildlife Management*, vol. 49, no. 1, pp. 255–263, 1985.

[240] M. A. Machmer and B. Korol, "Assessment of wildlife tree habitat in the Revelstoke Forest District," Final Technical Report, Forest Renewal BC, Victoria, Canada, 1998.

[241] E. L. Bull and C. Meslow, "Habitat requirements of the pileated woodpecker in northeastern Oregon," *Journal of Forestry*, vol. 75, no. 6, pp. 335–337, 1977.

[242] E. L. Bull, R. S. Holthausen, and M. G. Henjum, "Roost trees used by pileated woodpeckers in northeastern Oregon," *Journal of Wildlife Management*, vol. 56, no. 4, pp. 786–793, 1992.

[243] E. L. Walters, *Habitat and space use of red-naped sapsucker, Sphyrapicus nuchalis, in the Hat Creek valley, south-central British Columbia [M.S. thesis]*, University of Victoria, Victoria, Canada, 1990.

[244] C. L. Mahon, K. Martin, and J. D. Steventon, "Habitat attributes and chestnut-backed chickadee nest site selection in uncut and partial-cut forests," *Canadian Journal of Forest Research*, vol. 37, no. 7, pp. 1272–1285, 2007.

[245] K. A. Linder, *Habitat utilization and behavior of nesting Lewis's Woodpeckers (Melanerpes lewis) in the Laramie range, southeast Wyoming [M.S. thesis]*, University of Wyoming, Laramie, Wyo, USA, 1994.

[246] K. T. Vierling, "Habitat selection of Lewis' woodpeckers in southeastern Colorado," *Wilson Bulletin*, vol. 109, no. 1, pp. 121–130, 1997.

[247] V. A. Saab, R. E. Russell, and J. G. Dudley, "Nest-site selection by cavity-nesting birds in relation to postfire salvage logging," *Forest Ecology and Management*, vol. 257, no. 1, pp. 151–159, 2009.

[248] R. W. Mannan and E. C. Meslow, "Bird populations and vegetation characteristics in managed and old-growth forests, northeastern Oregon," *Journal of Wildlife Management*, vol. 48, no. 4, pp. 1219–1238, 1984.

[249] K. Viste-Sparkman, *White-breasted nuthatch density and nesting ecology in oak woodlands of the Willamette Valley, Oregon [M.S. thesis]*, Oregon State University, Corvallis, Ore, USA, 2005.

[250] R. D. Dixon, "Density, nest-site and roost-site characteristics, home-range, habitat-use, and behavior of White-headed Woodpeckers: Deshutes and Winema National Forests, Oregon," in *Nongame Project 93-3-01*, pp. 1–90, Oregon Department of Fish and Wildlife, Salem, Ore, USA, 1995.

[251] B. Fall, "Early summer warbler records and boreal chickadee nest near Itasca Park," *The Loon*, vol. 49, pp. 198–201, 1977.

[252] D. B. Hay and M. Guntert, "Seasonal selection of tree cavities by pygmy nuthatches based on cavity characteristics," in *Proceedings of the Symposium on Snag Habitat Management*, J. W. Davis, G. A. Goodwin, and R. A. Ockenfels, Eds., pp. 117–120, USDA Forest Service General Technical Report RM-99, 1983.

[253] J. F. Poulin, M. A. Villard, M. Edman, P. J. Goulet, and A. M. Eriksson, "Thresholds in nesting habitat requirements of an old forest specialist, the Brown Creeper (*Certhia americana*), as conservation targets," *Biological Conservation*, vol. 141, no. 4, pp. 1129–1137, 2008.

[254] M. A. Stern, T. G. Wise, and K. L. Theodore, "Use of natural cavity by bufflehead nesting in Oregon," *The Murrelet*, vol. 68, no. 2, p. 50, 1987.

[255] A. P. Yetter, S. P. Havera, and C. S. Hine, "Natural-cavity use by nesting wood ducks in Illinois," *Journal of Wildlife Management*, vol. 63, no. 2, pp. 630–638, 1999.

[256] G. M. Haramis, *Wood duck (Aix sponsa) ecology and management within the green-timber impoundments at Montezuma National Wildlife Refuge [M.S. thesis]*, Cornell University, Ithaca, NY, USA, 1975.

[257] L. R. Belmonte, *Home range and habitat characteristics of boreal owls in northeastern Minnesota [M.S. thesis]*, University of Minnesota, Duluth, Minn, USA, 2005.

[258] F. L. Bunnell, L. L. Kremsater, and R. W. Wells, *Likely Consequences of Forest Management on Terrestrial, Forest-Dwelling Vertebrates in Oregon*, Oregon Forest Resources Institute, Portland, Ore, USA, 1997.

[259] W. J. Sydeman and M. Guntert, "Winter communal roosting in the pygmy nuthatch," in *Proceedings of the Symposium on Snag Habitat Management*, J. W. Davis, G. A. Goodwin, and R. A. Ockenfels, Eds., pp. 121–124, USDA Forest Service General Technical Report RM-99, 1983.

[260] E. L. Bull and A. K. Blumton, "Roosting behavior of postfledging Vaux's Swifts in northeastern Oregon," *Journal of Field Ornithology*, vol. 68, no. 2, pp. 302–305, 1997.

[261] M. C. Kalcounis-Rüppell, J. M. Psyllakis, and R. M. Brigham, "Tree roost selection by bats: an empirical synthesis using meta-analysis," *Wildlife Society Bulletin*, vol. 33, no. 3, pp. 1123–1132, 2005.

[262] R. M. Brigham, "Flexibility in foraging and roosting behaviour by the big brown bat (*Eptesicus fuscus*)," *Canadian Journal of Zoology*, vol. 69, no. 1, pp. 117–121, 1991.

[263] H. G. Broders and G. J. Forbes, "Interspecific and intersexual variation in roost-site selection of northern long-eared and little brown bats in the greater fundy national park ecosystem," *Journal of Wildlife Management*, vol. 68, no. 3, pp. 602–610, 2004.

[264] C. Caceres, "Northern long-eared bat," Progress Report 014, Columbia Basin Fish and Wildlife Compensation Program, Nelson, Canada, 1997.

[265] M. D. Baker and M. J. Lacki, "Day-roosting habitat of female long-legged myotis in ponderosa pine forests," *Journal of Wildlife Management*, vol. 70, no. 1, pp. 207–215, 2006.

[266] T. C. Carter and G. A. Feldhamer, "Roost tree use by maternity colonies of Indiana bats and northern long-eared bats in southern Illinois," *Forest Ecology and Management*, vol. 219, no. 2-3, pp. 259–268, 2005.

[267] C. J. Garroway and H. G. Broders, "Day roost characteristics of northern long-eared bats (*Myotis septentrionalis*) in relation to female reproductive status," *Ecoscience*, vol. 15, no. 1, pp. 89–93, 2008.

[268] J. B. Johnson, J. W. Edwards, W. M. Ford, and J. E. Gates, "Roost tree selection by northern myotis (*Myotis septentrionalis*) maternity colonies following prescribed fire in a Central Appalachian Mountains hardwood forest," *Forest Ecology and Management*, vol. 258, no. 3, pp. 233–242, 2009.

[269] T. S. Jung, I. D. Thompson, and R. D. Titman, "Roost site selection by forest-dwelling male Myotis in central Ontario, Canada," *Forest Ecology and Management*, vol. 202, no. 1–3, pp. 325–335, 2004.

[270] M. A. Menzel, S. F. Owen, W. M. Ford et al., "Roost tree selection by northern long-eared bat (*Myotis septentrionalis*) maternity colonies in an industrial forest of the central Appalachian mountains," *Forest Ecology and Management*, vol. 155, no. 1–3, pp. 107–114, 2002.

[271] R. W. Perry and R. E. Thill, "Roost selection by male and female northern long-eared bats in a pine-dominated landscape," *Forest Ecology and Management*, vol. 247, no. 1–3, pp. 220–226, 2007.

[272] D. B. Sasse and P. J. Pekins, "Summer roosting ecology of northern long-eared bats (*Myotis septentrionalis*) in the White Mountain National Forest," in *Proceedings of the Bats and Forests Symposium*, R. M. R. Barclay and R. M. Brigham, Eds., pp. 91–101, British Columbia Ministry of Forests, 1996.

[273] J. C. Timpone, J. G. Boyles, K. L. Murray, D. P. Aubrey, and L. W. Robbins, "Overlap in roosting habits of Indiana bats (*Myotis sodalis*) and northern bats (*Myotis septentrionalis*)," *American Midland Naturalist*, vol. 163, no. 1, pp. 115–123, 2010.

[274] M. J. Evelyn, D. A. Stiles, and R. A. Young, "Conservation of bats in suburban landscapes: roost selection by *Myotis yumanensis* in a residential area in California," *Biological Conservation*, vol. 115, no. 3, pp. 463–473, 2004.

[275] V. J. Bakker and K. Hastings, "Den trees used by northern flying squirrels (*Glaucomys sabrinus*) in southeastern Alaska," *Canadian Journal of Zoology*, vol. 80, no. 9, pp. 1623–1633, 2002.

[276] H. M. Hackett and J. F. Pagels, "Nest site characteristics of the endangered northern flying squirrel (*Glaucomys sabrinus coloratus*) in Southwest Virginia," *American Midland Naturalist*, vol. 150, no. 2, pp. 321–331, 2003.

[277] L. McDonald, "Relationships between northern flying squirrels and stand age and structure in aspen mixedwood forests in Alberta," in *Relationships between Stand Age, Stand Structure, and Biodiversity in Aspen Mixedwood Forests in Alberta*, J. B. Stelfox, Ed., pp. 227–231, Alberta Environmental Centre, Vegreville, Canada; Canadian Forest Service, Edmonton, Canada, 1995.

[278] P. D. Weigl, "Study of the northern flying squirrel, *Glaucomys sabrinus*, by temperature telemetry," *American Midland Naturalist*, vol. 92, no. 2, pp. 482–486, 1974.

[279] M. J. Merrick, S. R. Bertelsen, and J. L. Koprowski, "Characteristics of mount graham red squirrel nest sites in a mixed conifer forest," *Journal of Wildlife Management*, vol. 71, no. 6, pp. 1958–1963, 2007.

[280] E. L. Bull and T. W. Heater, "Resting and denning sites of American martens in Northeastern Oregon," *Northwest Science*, vol. 74, no. 3, pp. 179–185, 2000.

[281] T. G. Chapin, D. M. Phillips, D. J. Harrison et al., "Seasonal selection of habitat by resting marten in Maine," in *Martes: Taxonomy, Ecology, Techniques, and Management*, G. Proulx, H. N. Bryant, and P. M. Woodard, Eds., pp. 166–181, Provincial Museum of Alberta, Edmonton, Canada, 1997.

[282] J. H. Gilbert, J. L. Wright, D. J. Lauten et al., "Den and rest-site characteristics of American marten and fisher in northern Wisconsin," in *Martes: Taxonomy, Ecology, Techniques, and Management*, G. Proulx, H. N. Bryant, and P. M. Woodard, Eds., pp. 135–1145, Provincial Museum of Alberta, Edmonton, Canada, 1997.

[283] T. N. Hauptman, *Spatial and temporal distribution and feeding ecology of the pine marten [M.S. thesis]*, Idaho State University, Pocatello, Idaho, USA, 1979.

[284] G. R. Ryder, "Characteristics of three natal den sites of American marten in the lower mainland region of southwestern British Columbia," *Wildlife Afield*, vol. 6, no. 1, pp. 32–35, 2009.

[285] K. M. Wynne and J. A. Sherburne, "Summer home range use by adult marten in northwestern Maine," *Canadian Journal of Zoology*, vol. 62, no. 5, pp. 941–943, 1984.

[286] S. M. Arthur, W. B. Krohn, and J. R. Gilbert, "Habitat use and diet of fishers," *Journal of Wildlife Management*, vol. 53, no. 3, pp. 680–688, 1989.

[287] A. K. Mazzoni, *Habitat use by fishers (Martes pennanti) in the southern Sierra Nevada, California [M.S. thesis]*, California State University, Fresno, Calif, USA, 2002.

[288] T. F. Paragi, S. M. Arthur, and W. B. Krohn, "Importance of tree cavities as natal dens for fishers," *Northern Journal of Applied Forestry*, vol. 13, no. 2, pp. 79–83, 1996.

[289] R. A. Powell and W. J. Zielinsky, "Fisher," in *The Scientific Basis For Conserving Forest Carnivores: American Marten, Fisher, Lynx and Wolverine in the Western United States*, K. B. Aubry, S. W. Buskirk, L. J. Lyon, and W. J. Zielinski, Eds., pp. 38–73, USDA Forest Service General Technical Report RM-254, 1994.

[290] C. M. Raley, "Ecological characteristics of fishers (*Martes pennanti*) in the Southern Oregon Cascade Range, update: July 2006. Report," in *USDA Forest Service*, pp. 1–31, Olympia Forestry Sciences Laboratory, Olympia, Wash, USA, 2006.

[291] R. D. Weir, *Diet, spatial organization, and habitat relationships of fishers in south-central British Columbia [M.S. thesis]*, Simon Fraser University, Burnaby, Canada, 1995.

[292] R. D. Weir, *Fisher Ecology in the Kiskatinaw Plateau Ecosection, Year-End Report*, Ministry of Environment of British Columbia, Victoria, Canada, 2008.

[293] R. D. Weir, F. Corbould, and A. Harestad, "Effect of ambient temperature on the selection of rest structures by fishers," in *Martens and Fishers (Martes) in HumAn-Altered Environments: An International Perspective*, D. J. Harrison, A. K. Fuller, and G. Proulx, Eds., pp. 187–197, Springer Science and Business Media, New York, NY, USA, 2004.

[294] J. J. Beecham, D. G. Reynolds, and M. G. Hornocker, "Black bear denning activities and den characteristics in west-central Idaho," *Bears: Their Biology and Management*, vol. 5, pp. 79–86, 1983.

[295] E. L. Bull, J. J. Akenson, and M. G. Henjum, "Characteristics of black bear dens in trees and logs in northeastern Oregon," *Northwest Naturalist*, vol. 81, no. 3, pp. 148–153, 2000.

[296] A. W. Erickson, B. M. Hanson, and J. J. Brueggeman, "Black bear denning study, Mitkof Island, Alaska," Project Report FRI-UW-8214, School of Fisheries, University of Washington, Seattle, Wash, USA, 1982.

[297] T. K. Fuller and L. B. Keith, "Summer ranges, cover type use, and denning of black bears near Fort. McMurray, Alberta," *The Canadian Field-Naturalist*, vol. 94, no. 1, pp. 80–83, 1980.

[298] K. G. Johnson and M. R. Pelton, "Selection and availability of dens for black bears in Tennessee," *Journal of Wildlife Management*, vol. 45, no. 1, pp. 111–119, 1981.

[299] C. J. Jonkel and I. M. Cowan, "The black bear in the spruce-fir forest," *Wildlife Monographs*, no. 27, pp. 1–55, 1971.

[300] G. B. Kolenosky and S. M. Strathearn, "Winter denning of black bears in east-central Ontario," *Bears: Their Biology and Management*, vol. 7, pp. 305–316, 1987.

[301] D. A. Martorello and M. R. Pelton, "Microhabitat characteristics of American black bear nest dens," *Ursus*, vol. 14, no. 1, pp. 21–26, 2003.

[302] Manning, Cooper and Associates, *2002 Black Bear Winter Den Inventory. TFL 37, Northern Vancouver Island, BC*, Canadian Forest Products, Woss, Canada, 2003.

[303] M. K. Oli, H. A. Jacobson, and B. D. Leopold, "Denning ecology of black bears in the White River National Wildlife Refuge, Arkansas," *Journal of Wildlife Management*, vol. 61, no. 3, pp. 700–706, 1997.

[304] C. W. Ryan and M. R. Vaughan, "Den characteristics of black bears in southwestern Virginia," *Southeastern Naturalist*, vol. 3, no. 4, pp. 659–668, 2004.

[305] W. G. Wathen, K. G. Johnson, and M. R. Pelton, "Characteristics of black bear dens in the southern Appalachian region," *Bears: Their Biology and Management*, vol. 6, pp. 119–127, 1986.

[306] T. H. White, J. L. Bowman, H. A. Jacobson, B. D. Leopold, and W. P. Smith, "Forest management and female black bear denning," *Journal of Wildlife Management*, vol. 65, no. 1, pp. 34–40, 2001.

[307] F. L. Bunnell and I. Houde, "Down wood and biodiversity—implications to forest practices," *Environmental Reviews*, vol. 8, pp. 397–421, 2010.

[308] D. Huber and H. U. Roth, "Denning of brown bears in Croatia," *Bears: Their Biology and Management*, vol. 9, pp. 79–83, 1997.

[309] K. Elgmork, "Denning behaviour of a female brown bear, *Ursus arctos* (Linne, 1758), with three young," *Säugetierkundliche Mitteilungen*, vol. 29, no. 3, pp. 59–66, 1981.

[310] S. P. Cline, *The characteristics and dynamics of snags in Douglas-fir forests of the Oregon Coast Range [Ph.D. thesis]*, Oregon State University, Corvallis, Ore, USA, 1977.

[311] S. G. Nilsson, "The evolution of nest-site selection among hole-nesting birds: the importance of nest predation and competition." *Ornis Scandinavica*, vol. 15, no. 3, pp. 167–175, 1984.

[312] A. Nappi and P. Drapeau, "Reproductive success of the black-backed woodpecker (*Picoides arcticus*) in burned boreal forests: are burns source habitats?" *Biological Conservation*, vol. 142, no. 7, pp. 1381–1391, 2009.

[313] N. Nielsen-Pincus, *Nest site selection, nest success, and density of selected cavity-nesting birds in northeastern Oregon with a method for improving accuracy of density estimates [M.S. thesis]*, University of Idaho, Moscow, Idaho, USA, 2005.

[314] W. F. Laudenslayer Jr., "Cavity-nesting bird use of snags in eastside pine forests of northeastern California," in *Proceedings of the Symposium on the Ecology and Management of Dead Wood in Western Forests*, W. F. Laudenslayer Jr., P. J. Shea, B. E. Valentine, C. P. Weatherspoon, and T. E. Lisle, Eds., pp. 223–236, USDA Forest Service, General Technical Report PSW-GTR-181, 2002.

[315] S. T. Walter and C. C. Maguire, "Snags, cavity-nesting birds, and silvicultural treatments in western Oregon," *Journal of Wildlife Management*, vol. 69, no. 4, pp. 1578–1591, 2005.

[316] R. J. Fisher and K. L. Wiebe, "Nest site attributes and temporal patterns of northern flicker nest loss: effects of predation and competition," *Oecologia*, vol. 147, no. 4, pp. 744–753, 2006.

[317] K. J. Gutzwiller and S. H. Anderson, "Multiscale associations between cavity-nesting birds and features of Wyoming streamside woodlands," *Condor*, vol. 89, no. 3, pp. 534–548, 1987.

[318] C. L. Hartwig, *Effect of forest age, structural elements, and prey density on the relative abundance of Pileated Woodpecker (Dryocopus pileatus abieticola) on southeastern Vancouver Island [M.S. thesis]*, University of Victoria, Victoria, Canada, 1999.

[319] T. K. Mellen, E. C. Meslow, and R. W. Mannan, "Summertime home range and habitat use of pileated woodpeckers in western Oregon," *Journal of Wildlife Management*, vol. 56, no. 1, pp. 96–103, 1992.

[320] R. G. Troetschler, "Acorn woodpecker breeding strategy as affected by starling nest-hole competition," *The Condor*, vol. 78, no. 2, pp. 151–165, 1976.

[321] B. G. Hill and M. R. Lein, "Ecological relations of sympatric black-capped and mountain chickadees in southwestern Alberta," *The Condor*, vol. 90, no. 4, pp. 875–884, 1988.

[322] S. M. Ramsay, K. Otter, and L. M. Ratcliffe, "Nest-site selection by female black-capped Chickadees: settlement based on conspecific attraction?" *Auk*, vol. 116, no. 3, pp. 604–617, 1999.

[323] A. E. Allin, "Nesting of the barred owl (*Strix varia*) in Ontario," *The Canadian Field-Naturalist*, vol. 58, pp. 8–9, 1944.

[324] K. R. Bevis, "Primary excavators in grand fir forests of Washington's east Cascades and forestry on the Yakima Indian Nation, Washington," in *Proceedings of the Wildlife Tree/Stand-Level Biodiversity Workshop*, P. Bradford, T. Manning, and B. I'Anson, Eds., pp. 77–86, BC Ministry of Environment, Lands, and Parks and Ministry of Forests, Victoria, Canada, 1996.

[325] D. J. Spiering and R. L. Knight, "Snag density and use by cavity-nesting birds in managed stands of the Black Hills National Forest," *Forest Ecology and Management*, vol. 214, no. 1–3, pp. 40–52, 2005.

[326] A. B. Carey, M. M. Hardt, S. P. Horton et al., "Spring bird communities in the Oregon Coast Range," in *Wildlife and Vegetation of Unmanaged Douglas-Fir Forests*, L. F. Ruggiero, K. B. Aubry, A. B. Carey, and M. F. Huff, Eds., pp. 123–144, USDA Forest Service General Technical Report, PNW-GTR-285, 1991.

[327] R. W. Mannan, *Use of snags by birds, Douglas-fir region, western Oregon [M.S. thesis]*, Oregon State University, Corvallis, Ore, USA, 1977.

[328] R. W. Mannan and E. C. Meslow, "Bird populations and vegetation characteristics in managed and old-growth forests, northeastern Oregon," *Journal of Wildlife Management*, vol. 48, no. 4, pp. 1219–1238, 1984.

[329] J. E. Zarnowitz and D. A. Manuwal, "The effects of forest management on cavity-nesting birds in northwestern Washington," *Journal of Wildlife Management*, vol. 49, no. 1, pp. 255–263, 1985.

[330] C. Steeger and H. Quesnel, "Impacts of partial cutting on old-growth forests in the Rocky Mountain trench: interim report," Tech. Rep. 9, Enhanced Forest Management Pilot Project, Invermere, Canada, 1998.

[331] R. L. Hutto, "Toward meaningful snag-management guidelines for postfire salvage logging in North American conifer forests," *Conservation Biology*, vol. 20, no. 4, pp. 984–993, 2006.

[332] E. L. Bull, C. G. Parks, and T. Torgerson, *Trees and Logs Important to Wildlife in the Interior Columbia River Basin*, USDA Forest Service General Technical Report PNW-GTR-391, 1997.

[333] W. C. McComb, S. A. Bonney, R. M. Sheffield, and N. D. Cost, "Snag resources in Florida—are they sufficient for average populations of primary cavity-nesters?" *Wildlife Society Bulletin*, vol. 14, no. 1, pp. 40–48, 1986.

[334] G. A. McPeek, W. C. McComb, J. J. Moriarty et al., "Bark-foraging bird abundance unaffected by increased snag availability in a mixed mesophytic forest," *The Wilson Bulletin*, vol. 99, no. 2, pp. 253–257, 1987.

[335] W. A. Nietro, V. W. Binkley, S. P. Cline et al., "Snags (wildlife trees)," in *Management of Wildlife and Fish Habitats in Forests of Western Oregon and Washington*, E. R. Brown, Ed., pp. 129–169, USDA Forest Service Publication R6-F&WL-192-1985, 1985.

[336] F. L. Bunnell, T. Spribille, I. Houde, T. Goward, and C. Björk, "Lichens on down wood in logged and unlogged forest stands," *Canadian Journal of Forest Research*, vol. 38, no. 5, pp. 1033–1041, 2008.

[337] F. R. Larson, *Downed Woody Material in Southeast Alaska Forest Stands*, USDA Forest Service Research Paper PNW-RP-452, 1992.

[338] P. Sollins, "Input and decay of coarse woody debris in coniferous stands in western Oregon and Washington," *Canadian Journal of Forest Research*, vol. 12, no. 1, pp. 18–28, 1982.

[339] B. G. Marcot, J. L. Ohmann, K. L. Mellen-McLean, and K. L. Waddell, "Synthesis of regional wildlife and vegetation field studies to guide management of standing and down dead trees," *Forest Science*, vol. 56, no. 4, pp. 391–404, 2010.

[340] D. J. Huggard, *Synthesis of Studies of Forest Bird Responses to Partial-Retention Forest Harvesting*, Pamphlet, Centre for Applied Conservation Research, University of British Columbia, Vancouver, Canada, 2006.

[341] F. L. Bunnell and B. G. Dunsworth, "Making adaptive management for biodiversity work—the example of Weyerhaeuser in coastal British Columbia," *Forestry Chronicle*, vol. 80, no. 1, pp. 37–43, 2004.

[342] A. J. Huggett, "The concept and utility of "ecological thresholds" in biodiversity conservation," *Biological Conservation*, vol. 124, no. 3, pp. 301–310, 2005.

[343] D. B. Lindenmayer and G. Luck, "Synthesis: thresholds in conservation and management," *Biological Conservation*, vol. 124, no. 3, pp. 351–354, 2005.

[344] J. S. Guénette and M. A. Villard, "Thresholds in forest bird response to habitat alteration as quantitative targets for conservation," *Conservation Biology*, vol. 19, no. 4, pp. 1168–1180, 2005.

[345] D. Huggard, "Forest birds and retention levels," *BC Journal of Ecosystems and Management*, vol. 8, no. 3, pp. 120–124, 2007.

[346] F. L. Bunnell, M. Boyland, and E. Wind, "How should we spatially distribute dead and dying wood?" in *Proceedings of the Symposium on the Ecology and Management of Dead Wood in Western Forests*, W. F. Laudenslayer Jr., P. J. Shea, B. E. Valentine, C. P. Weatherspoon, and T. E. Lisle, Eds., pp. 739–752, USDA Forest Service, General Technical Report PSW-GTR-181, 2002.

[347] B. G. Marcot, "Snag use by birds in Douglas-fir clearcuts," in *Proceedings of the Symposium in Snag Habitat Management*, J. W. Davis, G. A. Goodwin, and R. A. Ockenfels, Eds., pp. 134–139, USDA Forest Service General Technical Report RM-99, 1983.

[348] L. J. Bate, E. O. Garton, and M. J. Wisdom, *Estimating Snag and Large Tree Densities and Distributions on a Landscape for Wildlife Management*, USDA Forest Service General Technical Report PNW-GTR-425, Portland, Ore, USA, 1999.

[349] D. R. Petit, K. E. Petit, T. C. Grubb Jr. et al., "Habitat and snag selection by woodpeckers in a clear-cut: an analysis using artificial snags," *The Wilson Bulletin*, vol. 97, no. 4, pp. 525–533, 1985.

[350] W. Walankiewicz, "Do secondary cavity-nesting birds suffer more from competition for cavities or from predation in a primeval deciduous forest," *Natural Areas Journal*, vol. 11, no. 4, pp. 203–212, 1991.

[351] C. J. E. Welsh and D. E. Capen, "Availability of nesting sites as a limit to woodpecker populations," *Forest Ecology and Management*, vol. 48, no. 1-2, pp. 31–41, 1992.

[352] J. G. Dickson, R. N. Conner, and J. H. Williamson, "Snag retention increases bird use of clear-cut," *Journal of Wildlife Management*, vol. 47, no. 3, pp. 799–804, 1983.

[353] V. E. Scott, "Bird responses to snag removal in ponderosa pine," *Journal of Forestry*, vol. 77, no. 1, pp. 26–28, 1979.

[354] J. L. Ohmann, W. C. McComb, and A. A. Zumrawi, "Snag abundance for primary cavity-nesting birds on nonfederal forest lands in Oregon and Washington," *Wildlife Society Bulletin*, vol. 22, no. 4, pp. 607–619, 1994.

[355] F. L. Bunnell and G. B. Dunsworth, Eds., *Forestry and Biodiversity. Learning How to Sustain Biodiversity in Managed Forests*, University of British Columbia Press, Vancouver, Canada, 2009.

[356] B. Söderström, "Effects of different levels of green- and dead-tree retention on hemi-boreal forest bird communities in Sweden," *Forest Ecology and Management*, vol. 257, no. 1, pp. 215–222, 2009.

[357] K. Mellen, B. G. Marcot, J. L. Ohmann et al., "DecAID: a decaying wood advisory model for Oregon and Washington," in *Proceedings of the Symposium on the Ecology and Management of Dead Wood in Western Forests*, W. F. Laudenslayer Jr., P. J. Shea, B. E. Valentine, C. P. Weatherspoon, and T. E. Lisle, Eds., pp. 527–533, USDA Forest Service, General Technical Report PSW-GTR-181, 2002.

[358] A. J. Erskine, *Birds in Boreal Canada: Communities, Densities and Adaptations*, Canadian Wildlife Service Report Series 41, Ottawa, Canada, 1977.

[359] M. E. Harmon, J. F. Franklin, F. J. Swanson et al., "Ecology of coarse woody debris in temperate ecosystems," *Advances in Ecological Research*, vol. 15, pp. 133–302, 1986.

[360] M. C. Vanderwel, J. R. Malcolm, and S. M. Smith, "Long-term snag and downed woody debris dynamics under periodic surface fire, fire suppression, and shelterwood management," *Canadian Journal of Forest Research*, vol. 39, no. 9, pp. 1709–1721, 2009.

[361] R. L. L. Graham, *Biomass dynamics of dead Douglas-fir and western hemlock boles in mid- elevation forests of the Cascade Range [Ph.D. thesis]*, Oregon State University, Corvallis, Ore, USA, 1981.

[362] USDA Forest Service, *Ecological Characteristics of Fishers (Martes Pennanti) in the Southern Oregon Cascade Range*, USDA Forest Service, Pacific Northwest Research Station, Olympia, Wash, USA, 2006.

[363] C. S. Binkley, "Preserving nature through intensive plantation forestry: the case for forestland allocation with illustrations from British Columbia," *Forestry Chronicle*, vol. 73, no. 5, pp. 553–559, 1997.

[364] F. L. Bunnell, R. W. Wells, J. D. Nelson et al., "Effects of harvest policy on landscape pattern, timber supply and vertebrates in an East Kootenay watershed," in *Forest Fragmentation: Wildlife and Management Implications*, J. A. Rochelle, L. A. Lehmann, and J. Wisniewski, Eds., pp. 271–293, Brill, Leiden, The Netherlands, 1999.

[365] E. C. Lofroth, *Scale dependent analyses of habitat selection by marten in the sub-boreal spruce biogeoclimatic zone, British Columbia [M.S. thesis]*, Simon Fraser University, Burnaby, Canada, 1993.

[366] K. D. Coates, "Windthrow damage 2 years after partial cutting at the Date Creek silvicultural systems study in the interior Cedar-Hemlock forests of northwestern British Columbia," *Canadian Journal of Forest Research*, vol. 27, no. 10, pp. 1695–1701, 1997.

[367] J. F. Franklin, D. R. Berg, D. A. Thornburgh et al., "Alternative silvicultural approaches to timber harvesting: variable retention harvest systems," in *Creating a Forestry for the 21st Century: The Science of Ecosystem Management*, K. A. Kohm and J. F. Franklin, Eds., pp. 111–139, Island Press, Washington, DC, USA, 1997.

[368] R. M. S. Vega, *Bird communities in managed conifer stands in the Oregon Cascades: habitat associations and nest predation [M.S. thesis]*, Oregon State University, Corvallis, Ore, USA, 1993.

[369] F. L. Bunnell, L. L. Kremsater, and I. Houde, "Mountain pine beetle: a synthesis of the ecological .consequences of large-scale disturbances on sustainable forest management, with emphasis on biodiversity," Information Report BC-X-426, Natural Resources Canada, Canadian Forest Service, Pacific Forestry Centre, Victoria, Canada, 2011.

[370] J. D. McIver and L. Starr, "A literature review on the environmental effects of postfire logging," *Western Journal of Applied Forestry*, vol. 16, no. 4, pp. 159–168, 2001.

13

Using Multispectral Spaceborne Imagery to Assess Mean Tree Height in a Dryland Plantation

Michael Sprintsin,[1,2] **Pedro Berliner,**[1] **Shabtai Cohen,**[3] **and Arnon Karnieli**[1]

[1] *Jacob Blaustein Institutes for Desert Research, Ben-Gurion University of The Negev, 84990 Sede Boqer Campus, Israel*
[2] *Forest Management and GIS Department, Land Development Authority, Forest Department, Jewish National Fund (KKL), Eshtaol, M.P., 99775 Shimshon, Israel*
[3] *Institute of Soil, Water and Environmental Sciences, Agricultural Research Organization, The Volcani Center, 50250 Bet Dagan, Israel*

Correspondence should be addressed to Michael Sprintsin; michaelsp@kkl.org.il

Academic Editors: M. Kanashiro and H. Zeng

This study presents an approach for low-cost mapping of tree heights at the landscape level. The proposed method integrates parameters related to landscape (slope, orientation, and topographic height), tree size (crown diameter), and competition (crown competition factor and age), and determines the mean stand tree height as a function of tree competitive capability. The model was calibrated and validated against a standard inventory dataset collected over a dryland planted forest in the eastern Mediterranean region. The validation of the model shows a high and significant level of correlation between measured and modeled datasets ($R^2 = 0.86$; $P < 0.01$), with almost negligible (less than 1 m) levels of absolute and relative errors. The validated model was implemented for mapping mean tree height on a per-pixel basis by using high-spatial-resolution satellite imagery. The resulting map was, in turn, validated against an independent dataset of ground measurements. The presented approach could help to reduce the need for fieldwork in compiling single-tree-based inventories and to apply surface-roughness properties to hydrometeorological studies and regional energy/water-balance evaluation.

1. Introduction

Tree height is considered to be a useful structural variable in estimating wood volumes, biomass, carbon stocks, and productivity of forest stands. It also determines the light penetration into the forest canopy and is of importance for certain habitat studies. In addition, tree height plays an essential role in micrometeorological research and global climate modeling by determining forest aerodynamic roughness (i.e., zero-plane displacement and roughness length) and affecting the transport of energy and substances between the land surface and the atmosphere boundary layer [1]. Therefore, the computation and mapping of tree-height distribution in a widespread area becomes a key step in characterizing the land-surface physical processes.

Although the relationship between vegetation structure and surface reflectance obtained from satellite observation has been a focus of a great deal of research, the evaluation of mean tree height is still one of the main challenges for remote sensing applications. The most frequently used remote sensing techniques that are relevant to evaluation of tree height are (1) automated photogrammetry (e.g., [2]); (2) airborne ranging radar (e.g., [3]); and (3) laser altimetry (e.g., [4–6]). Since these methods are mainly based on airborne platforms, the data collected by them is naturally of high resolution, and usually enables observation of an individual tree within a stand. Though such data are accurate and therefore attractive for use, they are still very expensive to obtain. Thus, for operational use and for covering relatively large territories, it is more convenient to use low-cost high-frequency observations at lower spatial resolution, such as those obtained from passive optical systems (e.g., [7, 8]). However, the imagery obtained by such sensors presents significant limitations for forestry applications, because not every forest parameter has its own unique spectral response. Thus, for large-area studies, the spectral vegetation index approach, that is, a single value generated by combining data from multiple spectral bands, seems to be more appropriate.

The most basic spatial variable of the land surface that can be simply extracted from optical remote sensing data is the canopy cover (CC). From the ecological point of view, CC is an important parameter of a forest ecosystem for it is related to species richness and wildlife habitat and behavior (e.g., [9]); also, it is significant in studies of natural-hazard dynamics and understory vegetation productivity [10]. Technically, it can be assessed either by linear normalization of spectral vegetation indices [11, 12] or by supervised or unsupervised classification of multispectral imagery [13]. Both procedures are relatively simple and easily applied by users with varied levels of training. Furthermore, correlation of CC with other stand parameters could be straightforward, according to allometric relationships. It should be remembered, however, that, theoretically, the same level of CC could be found in dense stands of small trees and sparse stands of tall trees. Although in practice such similarity is unlikely to be found, the possibility highlights that CC could not be taken either as a unique or as a "stand-alone" predictor of mean tree height in the stand, and more robust combination of variables is required. Nevertheless, we assumed that stand-level canopy cover, as deduced from multispectral remote sensing imagery, can be used as a proxy for those variables, and our objective was to test this assumption in order to map mean tree height distribution on the landscape scale.

Although the majority of remote sensing applications for forestry cover a wide variety of ecoregions (e.g., [14]), the implementation of such techniques in predominantly water-limited ecosystems has rarely been reported being a subject only of some recent developments (e.g., [15–17]). However, such ecosystems occupy a significant part of the Earth's surface and are continually afforested. Because of the particular environmental problems common in such environments (e.g., low rainfall concentrated in short periods during the year, poor and shallow soils, high temperatures, and low relative humidity) and the relatively simple vegetation structure associated with these problems, landscape-level modeling is necessary to predict and optimize the benefits dryland forestry can contribute to ecosystem sustainability. Therefore, the development of applications that can be easily implemented in drylands is important from both ecological and silvicultural points of view. All in all, the major justification and motivation for our present study is specifically addressing the applicability of remote sensing in studying the structure of dryland planted forest that can be deemed as being typical of large tracts of afforested lands of the eastern Mediterranean. This assumes that little information is available and that little testing has been done until now.

The presented approach is based on a standard inventory dataset that was divided into calibration and validation data subsets. The former was used for choosing independent variables and calculating required coefficients; the latter was used for model validation. The validated model was then implemented for mapping mean tree height on a per-pixel basis, by means of multispectral high-spatial-resolution satellite imagery. The resulting map was, in turn, validated against an independent dataset of ground-based measurements.

This paper is structured as follows: first, the modeling approach is presented, then the validation data subset is analyzed, and finally, mean tree height distribution is mapped and validated within a specific studied area.

2. Theoretical Model

Although, the height to which trees can grow is still poorly understood, it is a common assumption that it is primarily limited by hydraulic factors, that is, by the tree's ability to transport water from roots to top. Consequently, Koch et al. [18] stated the following: "trees grow tall where resources are abundant, stresses are minor, and competition for light places a premium on height growth." Therefore, in water-limited environments mean tree height can be presented as a function of a tree's competitive abilities. These abilities reflect the interaction of any tree in a specific stand with other surrounding individuals (e.g., [19]) and to some extent are determined by landscape characteristics. The latter determine the water redistribution within a specific area and control the interception of incoming solar radiation.

Accordingly, we present mean tree height (H_t) as a variable related to competition (COMP) and landscape (i.e., site) factors (SITE):

$$H_t = a + (b \times \text{COMP}) + (c \times \text{SITE}), \qquad (1)$$

where COMP reflects characteristics of an individual tree and its behavior within a specific stand, SITE is related to specific plot elevation and orientation, and a, b, and c are regression coefficients.

Although the relationship between those factors is not necessarily linear, that is, of the form of (1), the lack of attention to this problem in the current literature allows us to test the applicability of this simplified intuitive approximation to exploration of the relationship between H_t and other stand variables. Moreover, many existing approaches that examine the dependency of any target stand characteristic on the combined effect of ecophysiological factors use linear relationships (e.g., [20, 21]) which support the assumption that (1) can serve as a reasonable approximation for H_t.

Since our major concern was the ability to evaluate H_t on the landscape level, in order to solve (1) we included in the final formulation only variables that were accessible—directly or indirectly—via remote sensing imagery or from existing databases. Those variables were carefully selected from the calibration data subset that initially included the entire set of inventory measures: age, diameter at breast height, tree and plot basal area, crown width, soil depth, and number of trees per plot (TPP).

Competition (COMP) has been defined as a parameter that depends on the maturity of a particular stand and involves the within-stand interaction between individuals. To characterize such relationships, we considered the crown competition factor (CCF; [22]) and stand age as independent predictors of "COMP."

The advantage of using CCF, which represents the area available to the average tree in the stand in relation to the

TABLE 1: Comparative statistics for calibration and validation data subsets.

Parameter	Calibration subset			Validation subset		
	Range	Average	STD	Range	Average	STD
H_t (m)	5–10	8	1	4–11	9	2
DBH (m)	0.12–0.2	0.16	0.02	0.10–0.19	0.16	0.03
CW (m)	3–5	4	1	3–5	4	1
CC (%)	25–91	56	20	28–90	56	19
TPP	6–10	7	1	7–10	8	2
Age (years)	32–37	35	1.5	30–38	35	2.3

H_t represents mean tree height; DBH: diameter at breast height; CW: crown width; CC: canopy cover, TPP number of trees per plot, and STD: standard deviation.

maximum area it could use if it existed in isolation [23], is that it is generally independent of site and age. A logarithmic transformation of CCF was used to reduce the effect of sampling variation in large estimates of CCF [20].

There are very few published studies of the impact of stand age and competition on simple measures of tree size. However, the age-related decline in productivity and biomass development of forests after canopy closure is well known and well documented (e.g., [24–28]). Therefore, we assume that the reason for such a decline is age-related decrease in tree competitive abilities, in which case, one would expect that any combined effect of age and CCF (that implies the influence of CC) would be representative as a measure of competition:

$$b \times COMP = b_1 + [b_2 \times \ln(CCF)] + [b_3 \times Age]. \quad (2)$$

The "SITE" factors are plot-specific variables that characterize the topography and the orientation of a plot. They were computed after Hasenauer and Monserud [20] as

$$
\begin{aligned}
d \times SITE = {}& [d_1 \times ELEV] + [d_2 \times ELEV^2] + [d_3 \times SL] \\
& + [d_4 \times SL]^2 + [d_5 \times SL \times \sin(AS)] \\
& + [d_6 \times SL \times \cos(AS)],
\end{aligned}
\quad (3)
$$

where ELEV is the elevation of the site (m), SL is the slope (°), and AS is the aspect (i.e., the direction in which a slope faces) that were calculated from the digital terrain model (DTM; [29]).

3. Materials and Methods

3.1. Study Area. The study was conducted in the Yatir forest ($31°35'$ N and $35°05'$ E, 630 m AMSL; area ~3000 ha) located in a transitional area between arid and semiarid climatic zones in southern Israel. The long-term average annual precipitation is ~285 mm, and the average total annual potential evapotranspiration is 1600 mm. The forest comprises predominantly of *Pinus halepensis* Mill. trees, mostly planted during 1964–1974. Average tree density is ~320 ± 75 trees ha^{-1} [17], mean tree height (H_t) ~9 ± 2 m, diameter at breast height (DBH) ~17 ± 4 cm, canopy cover (CC) ~53% ± 15%, and effective leaf area index (LAI) ~1.7 [30]. The trees grow on shallow Rendzina and lithosol soils, 0.2–1.5 m in depth

that overlay chalks and limestone. The understory vegetation develops during the rainy season and disappears shortly thereafter [31]. The specific study area (SSA) was set to 1 km^2 at the central most mature part of the forest, which was planted in the late 1960 s.

3.2. Sampling Design, Measurements, and Calculations. The Israeli Forest Service provided the inventory dataset. The data collection followed the traditional line-plot cruising approach, with circular plot shape [23]. Ninety-seven plots, each of 200 m^2, were established throughout the studied forest by using a forest map overlaid with the network of 250 m^2 quadrates. Training plots were located at the upper right-hand corner of every second quadrate. On each tree located within each training plot, we measured (a) DBH (measured with a caliper 1.37 m above the ground); (b) crown width (CW; measured with a measuring tape as the average of four different diameters of the canopy extension as observed from below); and (c) H_t (measured with a clinometer). Plot basal area was calculated from DBH measurements. The tree canopy was assumed to be circular, and CW that represented an equivalent canopy diameter was used to compute the crown area (CA). Canopy cover (CC) was calculated as the ratio of the sum of the crown area of all trees within a plot to the area of the plot, as adjusted for canopy overlap, according to Crookstone and Stage [32].

The entire set of 97 plots was then arbitrarily separated into calibration and validation subsets, comprising, respectively, 73 and 24 plots, that is, ~75% and ~25%, respectively, of all the plots. Both subsets were chosen with the aid of a random number generator [33, 34]. Comparative statistics for both subsets are presented in Table 1. Each subset has been classified into 5% CC classes. The mean value of each measured and calculated variable was estimated per CC class.

In addition, an independent dataset of measurements was collected over the six 1000 m^2 plots chosen within a specific research site of about 1 km^2. The exact spatial location and the perimeter of each plot were determined with a GPS (GPS) receiver with an accuracy of ±2 m. The results were then converted into a GIS polygon vector layer by using the MapInfo Professional software, Version 7.0. This additional dataset was used to validate a map of mean tree height prepared by using multispectral satellite imagery.

3.3. Multicollinearity Test. As we stated above, CCF is an age-independent parameter. However, some degree of multicollinearity between age and CCF still might be expected, because the studied forest is subject to intensive management practice that aims to reduce tree density as age increases, to provide optimal growing conditions for the remaining trees. Although there is no statistical test that can determine whether or not multicollinearity is a problem [35], its extent can be detected via the variance inflation factor (VIF; [36]) which measures the impact of multicollinearity among the predictors on the precision of estimation. The VIF is computed as $(1/(1 - R^2))$ for each independent variable. A general rule is that VIF higher than 10 indicates problems with multicollinearity [37, 38], that is, that the correlation between certain predictor variables is so large that they do not provide adequately independent information for reliable predictions.

3.4. Satellite Data Acquisition and Processing. A multispectral IKONOS image obtained on March 21, 2004 under cloud-free sky conditions was used. IKONOS has four spectral bands in the blue (0.45–0.52 μm), green (0.51–0.60 μm), red (0.63–0.70 μm), and near-infrared (0.76–0.85 μm) regions. An IKONOS image covers a nominal area of 16 km × 16 km at nadir with a spatial resolution of 4 m in all multispectral bands. The image was radiometrically and atmospherically corrected according to supplier's instructions (http://www.geoeye.com/products/imagery/ikonos/spectral.htm) and the 6S radiative transfer model [39] and registered into the UTM projection by using 20 ground control points (GCPs) obtained in the field with the dGPS receiver, resulting in average rectification errors of 0.7 and 0.85 pixels for the X and Y planes, respectively. The SSA was then extracted from the entire image with the ERDAS Imagine software.

The degree of canopy cover for an independent six-plot dataset has been described in terms of fractional vegetation cover (FVC). For that we used a simplified two-end member spectral mixture analysis model comprising a single equation that presents a surface reflectance measured by a satellite as the weighted sum of canopy and background reflectance terms [11, 30]. These two terms are further represented by minimum (min) and maximum (max) values of the normalized difference vegetation index (NDVI), obtained from in-situ reflectance measurements carried with a LICOR LI-1800 high-spectral-resolution field spectroradiometer, operating in the range 400–1100 nm with spectral resolution of 2 nm, yielding

$$\text{FVC} = \left(\frac{\text{NDVI} - \text{NDVI}_{\min}}{\text{NDVI}_{\max} - \text{NDVI}_{\min}} \right)^{0.645}. \qquad (4)$$

The FVC was then compared with the CC, as calculated from field measurements.

4. Results

4.1. Variance Inflation Factor and Multicollinearity. The statistical analysis did not support the expectation of multicollinearity between age and ln(CCF) measures; it yielded a

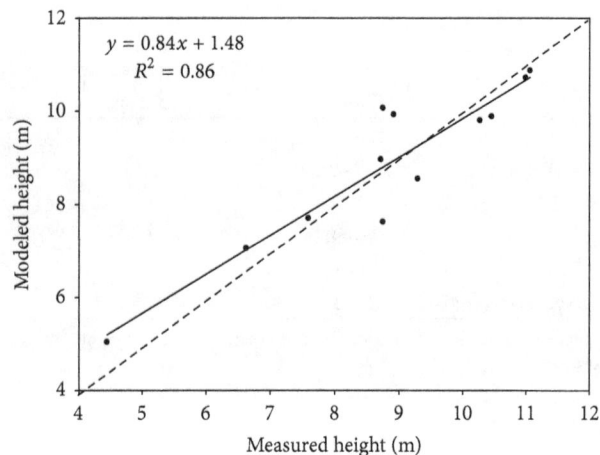

FIGURE 1: A comparison between measured and modeled mean tree height over the validation data subset.

VIF value of 1.15. In addition, the direct comparison between the two variables, which resulted in a very weak and insignificant linear correlation ($R^2 = 0.13$; $P = 0.28$), also served as a supplementary proof for the lack of multicollinearity between parameters used to represent competition.

4.2. Model Parameters and Comparisons with Measured Results. Both components of (1) had statistically significant predictive capability with almost equal positive magnitude of multipliers (slope = 0.8, $P < 0.01$ and slope = 0.6, $P < 0.05$ for COMP and SITE, resp.) showing that the variation of H_t was sufficiently explained by either measure. The overall regression's P value was 0.001, with $R^2 = 0.84$ and no recognized multicollinearity (VIF = 1.25). We suggest that the inclusion of the "age" parameter as one of the independent predictors (significant at $P < 0.004$ level as compared with $P = 0.4$ for ln(CCF)) could be speculated to be a reason for the higher significance of a COMP variable. In addition to the ecological meaning of such inclusion, in planted stands it provides a good explanation for the within-stand variation in tree sizes, because it is tightly related to stand density manipulations.

Figure 1 compares measured with predicted values of mean tree height for 12 CC classes over the validation dataset. It reveals high and significant correlation between the two ($R^2 = 0.86$; $P < 0.01$). Table 2 shows that both sets resulted in an identical average ($H_t = 9$ m) with a small standard deviation (STD) that was ~11% higher for the measured than for the predicted dataset (1.9 and 1.7 m, resp.), which highlights the higher intrinsic variability of the former.

Here it must be noted that though the effect of ln(CCF) was not statistically significant for the model represented by (2), excluding it from the calculation did not change the correlation level depicted in Figure 1, nor the resulting statistics (see below) in the specific case. Nevertheless, we decided to include it in the final model, as it is the only parameter that takes into account the "social status" of trees

Table 2: Comparison between measured and predicted values of mean tree height for twelve CC classes (each class represents an average of two plots) over the validation dataset. Each class contains two cases. Relative and absolute errors were calculated as RE = (|measured − modeled|)/measure and AE = |measured − modeled|, respectively.

CC class	COMP	SITE	Measured height (m)	Modeled height (m)	Relative Error	Absolute error (m)
<30	8.44	6.35	6.6	7.1	0.07	0.4
30–35	9.43	9.79	10.5	9.9	0.05	0.6
35–40	5.71	6.67	4.4	5.0	0.14	0.6
40–45	9.74	7.25	9.3	8.6	0.08	0.7
45–50	7.37	8.98	7.6	7.7	0.02	0.1
50–55	9.38	10.30	8.8	10.1	0.15	1.3
55–60	10.04	10.55	11.0	10.7	0.02	0.3
60–65	11.04	9.52	11.1	10.9	0.02	0.2
65–70	10.01	9.31	8.9	9.9	0.11	1.0
70–75	7.22	9.17	8.8	7.6	0.13	1.1
75–80	9.32	10.07	10.3	9.8	0.05	0.5
>80	9.96	7.89	8.7	9.0	0.03	0.3
AVG	**8.97**	**8.82**	**8.8**	**8.8**	**0.07**	**0.59**
STD	**1.53**	**1.43**	**1.9**	**1.8**	**0.05**	**0.39**
R^2	**0.2**			**0.86**		
P value	**0.15**			**<0.01**		
VIF	**1.25**					

Table 3: Comparison between measured and predicted values of mean tree height.

Plot	Measured height (m)	Predicted height (m)	Absolute error (m)
1	8.8	9.1	0.3
2	7.4	8.1	0.7
3	8.5	9.1	0.6
4	8.0	8.3	0.3
5	7.7	8.2	0.5
6	9.1	9.1	0
AVG	**8.3**	**8.7**	**0.4**
STD	**0.66**	**0.50**	**0.25**
R^2		**0.88**	
Slope		**0.71**	
Offset		**2.83**	
P value		**<0.01**	

within the plot, and including it seemed important from the silvicultural point of view.

Further analysis of the data is presented in Table 3. It highlights that for more than 75% of the classes the absolute error (AE = |measured − modeled|) was less than or equal to 1 m, whereas a higher deviation from the measured data was registered for the remaining 25% of the classes (i.e., two cases). Both groups, however, remained within the boundaries of the standard error for photogrammetric interpretation of aerial photography used by the Survey of Israel (i.e., 2 m, I. Sosnitsky, personal communication). The average AE was 0.6 m and was considered to be negligible. The average

relative error (RE = |measured − modeled|/measured) was 7% ± 5%; it was smaller than 10% for 66.67% of classes and never exceeded 15%. Hence, based on a combination of high correlation coefficient (Figure 1) and the above statistics we deduced that the overall accuracy of the model could be considered adequate, demonstrating the validity of the linear approximation (1).

In light of the results presented in Table 2, the proposed model (1) was implemented for mapping mean tree height over the SSA on a per-pixel basis, as based on a multispectral satellite image. All variables required for the SITE term were assessed at 20 m spatial resolution DTM [29]. A 4 m spatial resolution IKONOS image was then aggregated to 20 m resolution, to make it comparable with DTM.

The FVC was calculated according to (4) with values of 0.13 and 0.75 used for $NDVI_{min}$ and $NDVI_{max}$, respectively [30]. This resulted in a map of the spatial distribution of FVC that ranged from 40% to 70% over the study area within the entire image, and averaged 51.5% ± 4% for the six training plots. This value corresponds to an average CC calculated from measured crown width of 50% ± 11%, with an average absolute error less than 10% ± 6%. The FVC was then used to calculate the CCF as discussed in details by Sprintsin et al. [17] and, consequently, the COMP term.

Figure 2 is a map of spatial distribution of mean tree height over 1 km^2. This map was then overlain with the vector layer of six 1000 m^2 plots. The mean H_t value for each plot was extracted with a Zonal Statistics procedure of the ERDAS Imagine software and then compared with ground-based measurements. Those comparisons, presented in Table 3, show high and significant linear correlation between the calculated and measured values (R^2 = 0.88; slope = 0.71; P < 0.01) indicating freedom from systematic error in

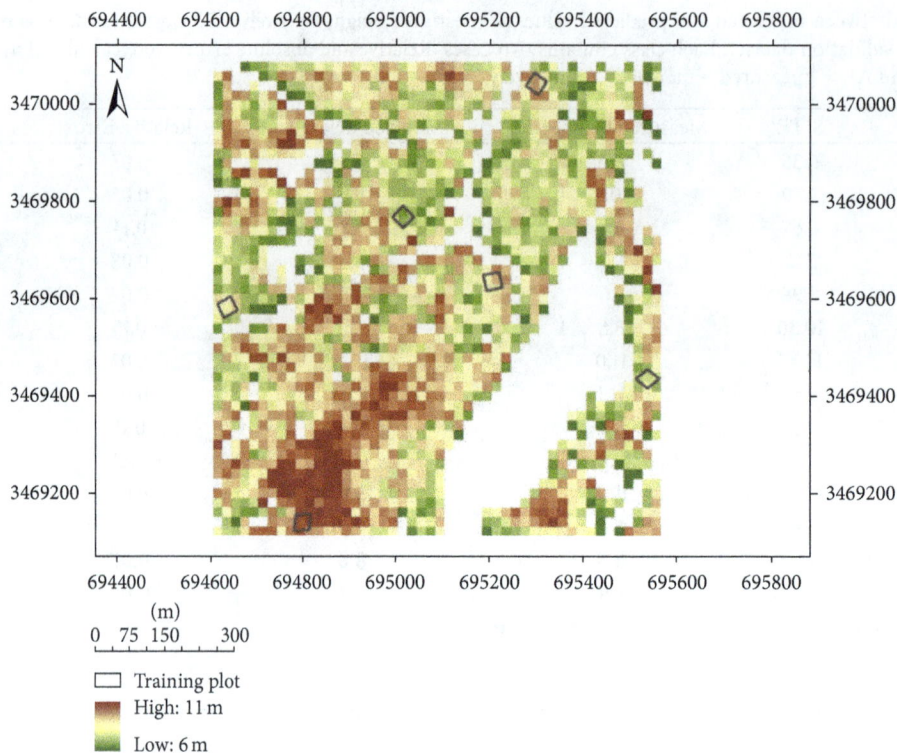

FIGURE 2: Spatial distribution of mean tree height over the specific study area. Note the location of training plots used in validation of the map accuracy (gray squares).

the calculations. It also shows that both sets were very comparable with regard to mean values (8.3 and 8.7 m for measured and modeled sets, resp.) and on a plot-to-plot basis, with an absolute error that never exceeded 0.7 m and had an average of 0.4 m, which was considered as negligible.

4.3. Practical Implementation of the Proposed Methodology.
As was mentioned earlier in this paper (Section 1), accurate approximation of the mean tree height could benefit hydrometeorological studies for which detailed description of canopy structure is required in estimation of surface aerodynamic roughness properties and consequent assessment of water and energy balances. To test the accuracy of the proposed methodology, the measured and the calculated values of H_t for each CC class of the validation data subset (see Table 2) were used as an input for surface aerodynamic resistance (r_a) calculations [40]:

$$r_a = \frac{\ln\left((z - d)/z_0\right)}{k^2 u},\qquad(5)$$

in which z is the height of meteorological measurements (15 m for Yatir); d_c is the zero-plane displacement (m); z_0 is the roughness length (m), u is the wind speed (m s^{-1}), and k is von Karman's constant ($k = 0.41$). d_c and z_0 were taken as a fraction of tree height ($d_c = 0.78 H_t$ and $z_0 = 0.075 H_t$ after [41] for conifers). Wind speed was taken to be constant ($u = 2.7$ m s^{-1}), approximated from the tower-top (18 m height) measurements that were averaged for daylight hours over six consecutive years (2000–2006).

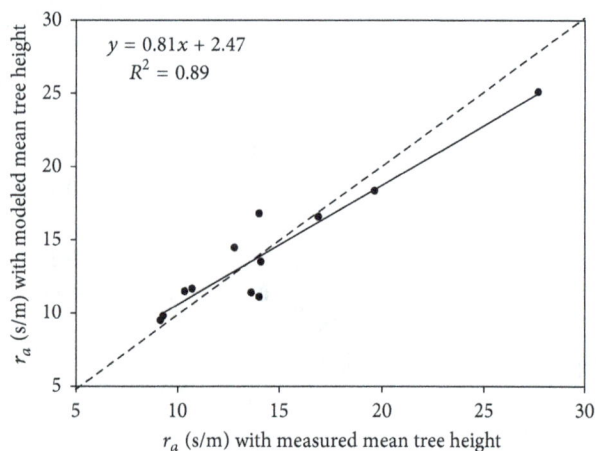

FIGURE 3: Comparison between aerodynamic resistance (r_a) calculated from field-measured values of mean tree height (H_t) and that calculated from mean tree height as estimated with the presented model (1)–(3).

Figure 3 compares resistance calculated from field-measured values of $H_t (r_{a,\text{meas}})$ with that based on estimated $H_t (r_{a,\text{mod}})$. A comparison between the two sets revealed almost equal values ($r_{a,\text{mod}} = 14.4 \pm 5.2$ s m^{-1}; $r_{a,\text{meas}} = 14.2 \pm 4.5$ s m^{-1}) and low relative error, averaging 10% ± 1%. The respective sets were highly correlated ($R^2 = 0.89$) and not significantly different from one another.

5. Discussion and Conclusions

One of the challenges currently faced by foresters is how to assess the spatial extent of standard forest characteristics. Remote-sensing methods, with their ability to cover large areas while entailing moderate costs, seem to offer a good answer to this challenge. Thus, it is important to be able to relate surface phenomena to spectral characteristics. However, as we mentioned above (Section 1), not every forest parameter has its own unique spectral response. For this reason, determining an optimal compromise between minimization of inputs and lowering spatial resolution should be a major task of remote sensing applications, and this was a major aim of the present study. This objective was achieved by adopting stand-level mean canopy cover—the only parameter that can be simply extracted from multispectral optical remote sensing imagery—as a proxy for other forest structural characteristics. Our approach determines the mean tree height of a forest stand by using a combination of landscape, stand, and canopy characteristics. Such a combination reflects the interactions among the varied factors that are responsible for tree growth and precisely describes the "social position" of an individual tree within a particular stand. Accurate estimation of the height of a vegetation layer, in turn, leads to a better determination of surface aerodynamic roughness properties and their spatial distribution, which could be easily and accurately derived from multispectral observations. These outputs can contribute to hydrometeorological studies and evaluation of the regional water and energy balance.

We should note, however, that the main challenge that one could face in applying the presented methodology in environments other than drylands is the proper estimation of CC from multispectral imagery. The main reason for this difficulty is the signal generated by understory vegetation (which is generally negligible in semiarid and arid regions) that could seriously contaminate the reflectance of a single pixel. In such cases, the number of end members of spectral mixture analysis model will be increased and an appropriate unmixing technique will be required [42]. Nevertheless, although it is technically possible to measure the crown projection area on remote sensing imagery, the actual maximum crown width cannot always be defined (even on aerial photographs) because of neighboring trees and overlapping crowns [43]. One important issue, therefore, is to examine the difficulties encountered in applying the proposed method in more complex terrain, mixed forests and milder climates that are characterized by more complex structures of vegetation layer.

The results presented here, however, indicate that the proposed combination of landscape and canopy characteristics can be exploited for mean tree height assessment on the regional scale in simply structured environments. The model presented here serves the need to estimate tree characteristics from their canopy extent; it helps to satisfy the requirement to reduce the need for fieldwork in single-tree-based forest inventories and to fill the knowledge gap caused by underrepresentation of the available applications in the current forestry literature.

Acknowledgment

The authors are grateful to the Israeli Forest Service (Keren Kayemet LeIsrael-Jewish National Fund) for the help with the field measurements.

References

[1] D. Sellier, Y. Brunet, and T. Fourcaud, "A numerical model of tree aerodynamic response to a turbulent airflow," *Forestry*, vol. 81, no. 3, pp. 279–297, 2008.

[2] G. Zagalikis, A. D. Cameron, and D. R. Miller, "The application of digital photogrammetry and image analysis techniques to derive tree and stand characteristics," *Canadian Journal of Forest Research*, vol. 35, no. 5, pp. 1224–1237, 2005.

[3] S. R. Cloude and K. P. Papathanassiou, "Polarimetric SAR interferometry," *IEEE Transactions on Geoscience and Remote Sensing*, vol. 36, no. 5, pp. 1551–1565, 1998.

[4] C. H. Hug and A. Wehr, "Detecting and identifying topographic objects in imaging laser altimetry data," *International Archives of Photogrammetry and Remote Sensing*, vol. 32, part 3-4W2, pp. 19–26, 1997.

[5] K. Kraus and W. Rieger, "Processing of laser scanning data for wooded areas," in *Photogrammetric Week 1999*, D. Fritsch and R. Spiller, Eds., pp. 221–231, Wichmann, Heidelberg, Germany, 1999.

[6] J. A. B. Rosette, P. R. J. North, and J. C. Suárez, "Vegetation height estimates for a mixed temperate forest using satellite laser altimetry," *International Journal of Remote Sensing*, vol. 29, no. 5, pp. 1475–1493, 2008.

[7] S. N. Coward and D. L. Williams, "Landsat and earth systems science: development of terrestrial monitoring," *Photogrammetric Engineering and Remote Sensing*, vol. 63, no. 7, pp. 887–900, 1997.

[8] M. A. Lefsky, W. B. Cohen, G. G. Parker, and D. J. Harding, "Lidar remote sensing for ecosystem studies," *BioScience*, vol. 52, no. 1, pp. 19–30, 2002.

[9] P. A. Zollner and K. J. Crane, "Influence of canopy closure and shrub coverage on travel along coarse woody debris by eastern chipmunks (*Tamias striatus*)," *American Midland Naturalist*, vol. 150, no. 1, pp. 151–157, 2003.

[10] F. Berger and F. Rey, "Mountain protection forests against natural hazards and risks: new french developments by integrating forests in risk zoning," *Natural Hazards*, vol. 33, no. 3, pp. 395–404, 2004.

[11] B. J. Choudhury, N. U. Ahmed, S. B. Idso, R. J. Reginato, and C. S. T. Daughtry, "Relations between evaporation coefficients and vegetation indices studied by model simulations," *Remote Sensing of Environment*, vol. 50, no. 1, pp. 1–17, 1994.

[12] J. Wang, P. M. Rich, K. P. Price, and W. D. Kettle, "Relations between NDVI and tree productivity in the central Great Plains," *International Journal of Remote Sensing*, vol. 25, no. 16, pp. 3127–3138, 2004.

[13] J. Jensen, *Introductory Digital Processing: A Remote Sensing Perspective*, Prentice-Hall, Englewood Cliffs, NJ, USA, 1996.

[14] C. J. Nichol, K. F. Huemmrich, T. A. Black et al., "Remote sensing of photosynthetic-light-use efficiency of boreal forest," *Agricultural and Forest Meteorology*, vol. 101, no. 2-3, pp. 131–142, 2000.

[15] M. F. Garbulsky, J. Peñuelas, D. Papale, and I. Filella, "Remote estimation of carbon dioxide uptake by a Mediterranean forest," *Global Change Biology*, vol. 14, no. 12, pp. 2860–2867, 2008.

[16] A. Goerner, M. Reichstein, and S. Rambal, "Tracking seasonal drought effects on ecosystem light use efficiency with satellite-based PRI in a Mediterranean forest," *Remote Sensing of Environment*, vol. 113, no. 5, pp. 1101–1111, 2009.

[17] M. Sprintsin, A. Karnieli, S. Sprintsin, S. Cohen, and P. Berliner, "Relationships between stand density and canopy structure in a dryland forest as estimated by ground-based measurements and multi-spectral spaceborne images," *Journal of Arid Environments*, vol. 73, no. 10, pp. 955–962, 2009.

[18] G. W. Koch, S. C. Stillet, G. M. Jennings, and S. D. Davis, "The limits to tree height," *Nature*, vol. 428, no. 6985, pp. 851–854, 2004.

[19] N. Liphschitz, O. Bonneh, and Z. Mendel, "Living stumps—circumstantial evidence for root grafting in *Pinus halepensis* and *P. brutia* plantations in Israel," *Israel Journal of Botany*, vol. 36, pp. 41–43, 1987.

[20] H. Hasenauer and R. A. Monserud, "A crown ratio model for Austrian Forests," *Forest Ecology and Management*, vol. 84, no. 1–3, pp. 49–60, 1996.

[21] S. Condés and H. Sterba, "Derivation of compatible crown width equations for some important tree species of Spain," *Forest Ecology and Management*, vol. 217, no. 2-3, pp. 203–218, 2005.

[22] L. E. Krajicek, K. A. Brinkman, and S. F. Gingrich, "Crown competition—a measure of density," *Forest Science*, vol. 7, pp. 35–42, 1961.

[23] T. E. Avery and H. E. Burkhart, *Forest Measurements*, McGraw-Hill, New York, NY, USA, 1994.

[24] D. M. Hyink and S. M. Zedaker, "Stand dynamics and the evaluation of forest decline," *Tree Physiology*, vol. 3, pp. 17–26, 1987.

[25] C. Deleuze, J. Herve, F. Colin, and L. Ribeyrolles, "Modelling crown shape of Picea abies: spacing effects," *Canadian Journal of Forest Research*, vol. 26, no. 11, pp. 1957–1966, 1996.

[26] A. Mäkelä and P. Vanninen, "Impacts of size and competition on tree form and distribution of aboveground biomass in Scots pine," *Canadian Journal of Forest Research*, vol. 28, no. 2, pp. 216–227, 1998.

[27] H. Ishii, J. P. Clement, and D. C. Shaw, "Branch growth and crown form in old coastal Douglas-fir," *Forest Ecology and Management*, vol. 131, no. 1–3, pp. 81–91, 2000.

[28] M. J. Ducey, "Predicting crown size and shape from simple stand variables," *Journal of Sustainable Forestry*, vol. 28, no. 1-2, pp. 5–21, 2009.

[29] J. K. Hall, "DTM project scheme of 1:50,000 topographic sheet mnemonics for Israel," *Geological Survey of Israel Current Research*, vol. 8, pp. 47–50, 1993.

[30] M. Sprintsin, A. Karnieli, P. Berliner, E. Rotenberg, D. Yakir, and S. Cohen, "The effect of spatial resolution on the accuracy of leaf area index estimation for a forest planted in the desert transition zone," *Remote Sensing of Environment*, vol. 109, no. 4, pp. 416–428, 2007.

[31] J. M. Grünzweig, T. Lin, E. Rotenberg, A. Schwartz, and D. Yakir, "Carbon sequestration in arid-land forest," *Global Change Biology*, vol. 9, no. 5, pp. 791–799, 2003.

[32] N. J. Crookstone and A. R. Stage, *Percent Canopy Cover and Stand Structure Statistics from the Forest Vegetation Simulator*, USDA Publication, 1999.

[33] Y. Tominaga, "Representative subset selection using genetic algorithms," *Chemometrics and Intelligent Laboratory Systems*, vol. 43, no. 1-2, pp. 157–163, 1998.

[34] A. Gusnanto, Y. Pawitan, J. Huang, and B. Lane, "Variable selection in random calibration of near-infrared instruments: ridge regression and partial least squares regression settings," *Journal of Chemometrics*, vol. 17, no. 3, pp. 174–185, 2003.

[35] L. D. Schroeder, D. L. Sjoquist, and P. E. Stephan, *Understanding Regression Analysis*, Sage Publications, Beverly Hills, Calif, USA, 1986.

[36] R. J. Freund, R. C. Littell, and L. Creighton, *Regression Using JMP*, SAS Institute, Cary, NC, USA, 2003.

[37] D. A. Belsley, E. Kuh, and R. E. Welsch, *Regression Diagnostics: Identifying Influential Data and Sources of Collinearity*, John Wiley & Sons, New York, NY, USA, 1980.

[38] R. D. Snee, "Some aspects of nonorthogonal data analysis—part I: developing prediction equations," *Journal of Quality Technology*, vol. 5, no. 2, pp. 67–79, 1973.

[39] E. F. Vermote, D. Tanré, J. L. Deuzé, M. Herman, and J. Morcrette, "Second simulation of the satellite signal in the solar spectrum, 6s: an overview," *IEEE Transactions on Geoscience and Remote Sensing*, vol. 35, no. 3, pp. 675–686, 1997.

[40] W. Brutsaert, *Evaporation into the Atmosphere: Theory, History, and Applications*, Kluwer Academic Publishers, Dordrecht, The Netherlands, 1982.

[41] H. G. Jones, *Plant and Microclimate*, Cambridge University Press, Cambridge, UK, 2nd edition, 1992.

[42] J. Pisek and J. M. Chen, "Mapping forest background reflectivit over North America with Multi-angle Imaging SpectroRadiometer (MISR) data," *Remote Sensing of Environment*, vol. 113, no. 11, pp. 2412–2423, 2009.

[43] J. Kalliovirta and T. Tokola, "Functions for estimating stem diameter and tree age using tree height, crown width and existing stand database information," *Silva Fennica*, vol. 39, no. 2, pp. 227–248, 2005.

14

Decomposition and Nutrient Release Dynamics of *Ficus benghalensis* L. Litter in Traditional Agroforestry Systems of Karnataka, Southern India

B. Dhanya,[1,2] Syam Viswanath,[1] and Seema Purushothaman[3]

[1] *Tree Improvement and Propagation Division, Institute of Wood Science and Technology, Malleswaram, Bangalore, Karnataka 560003, India*
[2] *Department of Environmental Science, College of Agriculture, University of Agricultural Sciences, PB No. 329, Raichur, Karnataka 584102, India*
[3] *Azim Premji University, Bangalore, Karnataka 560100, India*

Correspondence should be addressed to B. Dhanya; krupaias@gmail.com

Academic Editors: P. Robakowski and S. F. Shamoun

Decomposition and nutrient release dynamics of leaf litter of *Ficus benghalensis*, a common agroforestry species in southern dry agroclimatic zone of Karnataka, were studied using the standard litter bag technique in surface and subsurface methods of application. Results revealed a marginally higher rate of decay in subsurface placement (22.5% of initial litter mass remaining after one year of decomposition) compared to surface treatment (28.3% of initial litter mass remaining). Litter quality (lignin content and lignin/N ratio) and climatic and soil conditions of the study site (monthly rainfall and soil moisture) were found to influence the rate of decomposition. Mineralisation of litter was found to be in the order K > N > P. The paper further discusses the implications of these results for rainfed farming in Mandya and emphasises the potential of *F. benghalensis* in reducing nutrient input costs for resource-poor dryland farmers.

1. Introduction

Agroforestry systems with scattered trees in croplands have traditionally played a pivotal role in sustaining rural livelihoods in semiarid zones of the world. In Mandya district of southern dry agroclimatic zone of Karnataka, trees of the genus *Ficus* have been integral components of traditional rainfed agroecosystems with field crops like millets, pulses, maize and oil seeds. *Ficus benghalensis* L. is the major species of *Ficus* grown in these agroforestry systems, followed by *Ficus religiosa* L., *Ficus amplissima* Sm., *Ficus virens* Aiton, *Ficus racemosa* L., and *Ficus mysorensis* var. *pubescens* (Roth). [1]. In preliminary surveys held in Mandya, farmers were appreciative of the various direct benefits (fodder, firewood, small timber, and shade) and ecological services from these trees, especially soil enrichment through litterfall. Farmer interviews also revealed that litter from *Ficus* trees helped to reduce compost usage by 3 tonnes per hectare. However, scientific studies on the value of *Ficus* trees as source of litter nutrients in agroforestry systems of semiarid tropics are lacking. The relative value of litter as a source of nutrient is dependent on its decomposition rate, which in turn controls the release of the tissue-held mineral ions [2]. Further, the process of decomposition is regulated by a host of variables including physical and chemical properties (quality) of litter, climate, soil properties, and decomposer communities consisting of microorganisms and soil invertebrates [3, 4]. In the present study, leaf litter decomposition of *Ficus benghalensis*, factors influencing decomposition process, and pattern of nutrient release were assessed with the objective of eliciting information that may help resource-poor farmers to optimally exploit *Ficus* litterfall to reduce input costs for farming.

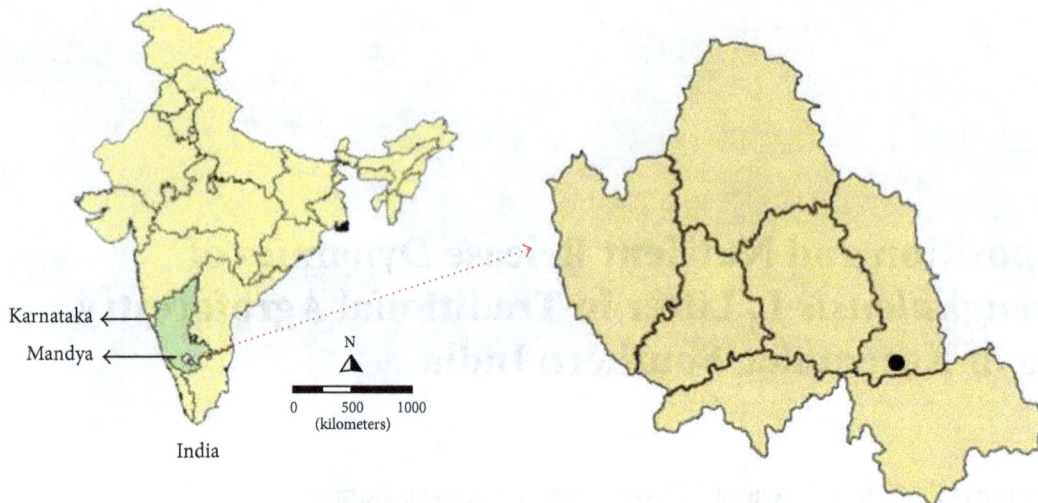

FIGURE 1: Location map of Mandya district in Karnataka state (black dot indicates the experimental site).

It is hypothesised that if utilized properly, *Ficus* litterfall can substantially contribute to nutrient needs of dryland farming in Mandya.

2. Materials and Methods

2.1. Study Area. Mandya district, spanning an area of 4961 km^3 in southern dry agroclimatic zone of Karnataka state in south India (location map in Figure 1), consists of plain lands, with an elevation of 757 m–909 m above mean sea level. A plot of area 0.8 ha, located at 12° 29' N latitude and 76° 59' E longitude, was selected for litter decomposition studies in Mandya, where *F. benghalensis* trees were grown in association with (*Eleusine coracana* (L.) Gaertn.) and horse gram (*Macrotyloma uniflorum* (Lam.) Verdc.). The mean daily minimum and maximum temperatures for 2009 were 19°C and 31.5°C, respectively, and total annual rainfall for the year was 800 mm, with most of the rain received during the premonsoon month of May and monsoon months of June–September. Soil in the study site was red and slightly acidic (pH of 6.5), and the texture ranged from gravelly sandy clay to sandy clay loam. Organic carbon content of soil was 1.33%, available N content was 204.3 kg/ha, available P was 2.5 kg/ha, and available K was 642.2 kg/ha. Exchangeable Ca and Mg were high at 2102.1 ppm and 420.4 ppm, respectively, and available S had a moderate value of 47.9 ppm.

2.2. Litter Decomposition Studies. In the present investigation, only leaf litter was studied for decomposition as this component constituted around 60% of the total litterfall of *F. benghalensis* in agroforestry systems of Mandya [5]. The standard litterbag technique [6] was employed for decomposition studies, in which nylon bags of size 30 cm × 20 cm and mesh size 2 mm were filled with 20 g of freshly fallen/senescent foliage of *F. benghalensis*, shade dried for 48 hours. Filter paper chopped into small pieces was also transferred to similar bags as control, to determine the importance of chemical constitution in influencing decomposition [7]. A total of 60 litter bags and 60 filter paper bags (total of 120 bags) were buried just below the soil surface (subsurface treatment) to simulate the litter that goes inside soil when land is ploughed. An equal number of litter bags and filter paper bags were placed on surface of soil (surface treatment) to study the effect of method of placement on decomposition. Thus, a total of 240 such bags (two substrates for 12 months and two methods of application) were laid out for decomposition experiments under *Ficus* tree canopy in substrate-wise strips to facilitate easy retrieval of bags per sampling date. Samples of air-dried litter were also kept apart for analysis of N (micro-Kjeldahl method in Kel Plus-KES 6L Automatic Distillation Unit), P (vanadomolybdophosphoric yellow colour method) K (flame photometry using Elico-CL 378 flame photometer) [8], C [9], lignin (Klason's procedure [10]), and total phenol content (Folin-Ciocalteau reagent method [11]).

Five samples each were drawn at monthly intervals from surface and subsurface treatments for one year starting from April 2009 to March 2010. The residual substrate (litter/paper) mass from bags was washed, oven-dried at 70°C, and weighed after excluding fine roots and macroarthropods penetrating the mesh. The samples were then pooled substrate-wise, powdered, and analysed for N, P and K. Decomposition rates of the substrates were estimated from the first-order exponential equation

$$e^{-kt} = \frac{L_R}{L_I},$$ (1)

where k = decomposition rate per year, t = the time interval of sampling L_R expressed in years, L_R = the litter weight remaining at a given time, and L_I = initial litter weight at time zero [15].

Half-life period ($t_{0.5}$) of the decomposing litter samples was estimated from k values using the equation

$$t_{0.5} = \frac{0.693}{-k}.$$ (2)

Decomposition and Nutrient Release Dynamics of Ficus benghalensis L. Litter in Traditional Agroforestry Systems of Karnataka, Southern India

165

Nutrient content of the decomposing leaf was derived as

$$\% \text{ nutrient remaining} = \left(\frac{C}{C_0}\right)\left(\frac{DM}{DM_0}\right) \times 10^2, \quad (3)$$

where C is the concentration of element in the leaf litter at the time of sampling, C_0 is the concentration of the initial leaf litter kept for decomposition, DM is the dry matter at the time of sampling, and DM_0 is the initial dry matter of the litter sample kept for decomposition [16].

$$\% \text{ nutrient released} = 100 - \% \text{ of original nutrient remaining;} \quad (4)$$

see [17].

2.3. Factors Influencing Litter Decomposition.
The association of litter decomposition with litter quality, climatic conditions, and soil properties was examined in the present study.

2.3.1. Litter Quality.
Litter quality determines in part rates of decomposition and release of nutrients from organic residues [18]. Initial concentrations of N, lignin (LG), and polyphenol (PP) and ratios such as C : N, LG : N, and PP : N in the biomass are some of the factors that have been shown to influence decomposition rates [19, 20]. In the present study litter quality parameters in terms of contents of nitrogen, lignin, and polyphenol and C : N and LG : N ratios were analysed following Jama and Nair [7] and Niranjana [21].

2.3.2. Collection of Weather Data.
Monthly rainfall data recorded from the rain gauge at Bharathi Nagara, located 1.5 km from the study site, was collected from the District Statistics Office, Mandya. Air temperature and humidity were measured in situ using thermohygrometer (Temp Tec brand, Mextech Company, Maharashtra, India) once monthly from April 2009 to March 2010.

2.3.3. Analysis of Soil Properties.
Soil temperature was measured in situ using a soil thermometer (Universal brand, Universal Soil Equippers, New Delhi, India). Samples from the site were carried to laboratory once in a month from April 2009 to March 2010 for gravimetrical estimation of soil moisture and measurement of soil pH.

2.4. Statistical Analyses.
Data on residual litter mass in bags and their nutrient contents after one year for surface and subsurface methods of placement were statistically analysed using two-way ANOVA technique (substrate and method of placement as two factors) in SigmaStat 3.5 statistical software, and Fisher's least square difference (LSD) values were computed. Pearson's correlation coefficients of mean monthly weight loss of litter with weather parameters and soil properties in the experimental plot were worked out following Panse and Sukhatme [22].

3. Results and Discussion

3.1. Pattern of Litter Decay.
Previous studies have shown that litter decay can follow an exponential pattern [23, 24] or a

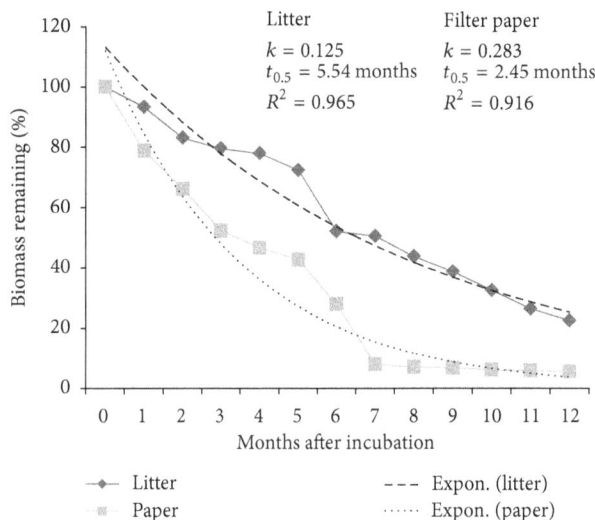

FIGURE 2: Weight loss of F. benghalensis litter and filter paper substrates in subsurface method of application in agroforestry systems of Mandya (Expon. denotes the exponential trendline fitted).

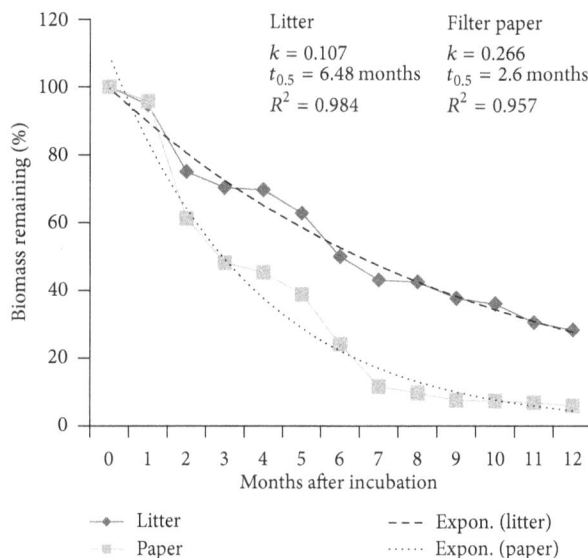

FIGURE 3: Weight loss of F. benghalensis litter and filter paper substrates in surface method of application in agroforestry systems of Mandya (Expon. denotes the exponential trendline fitted).

linear pattern [2, 14, 21] depending upon the species studied and conditions (climate, soil, method of application, etc.) under which decomposition takes place. In the case of F. benghalensis leaf litter decomposition in Mandya, a biphasic mode of decay was apparent with an initial rapid phase of mass loss in first 6-7 months, followed by a later slower phase from 8 to 12 months. Hence, a negative exponential model ($Y_t = Y_0 e^{-kx}$, where Y_0 is the initial litter mass applied and Y_t is the dry matter remaining after x months) was used to describe the weight loss pattern of litter and filter paper substrates during decomposition (Figures 2 and 3).

TABLE 1: Litter quality parameters of *F. benghalensis* in comparison with other substrates.

Species/substrate	Location	Initial N (%)	Initial lignin (%)	Initial total phenol (%)	C/N ratio	Lignin/N ratio	Half life (in months)	Reference
F. benghalensis	Mandya	1.11	30.50	6.76	23.89	27.60	5.54–6.47	Primary data
Filter paper	Mandya	0.04	0.50	0.47	70.49	12.19	2.45–2.60	Primary data
F. glomerata	Garhwal, India	1.97	13.82	7.10	19.88	7.01	8.00	[12]
F. roxburghii	Garhwal, India	0.96	12.14	11.64	39.58	12.64	12.00	[12]
Artocarpus hirsutus	Kerala, India	1.10	28.90	2.20	⋯	26.20	4.60	[13]
	Kerala, India	1.73	31.40	⋯	⋯	18.16	3.40	[14]
A. heterophyllus	Kerala, India	0.91	15.20	2.00	⋯	17.60	2.40	[13]
	Kerala, India	2.15	17.90	⋯	⋯	8.33	3.10	[14]

⋯: data not available.

Decomposition was found to be faster in subsurface method of application, with only 22.5% of initial litter mass remaining after one year of incubation, while in surface treatment, 28.3% of mass was remaining. Decay constant of litter was slightly higher for subsurface mode of application (0.125) compared to surface mode (0.107), and half-life of litter ranged from 5.54 months in subsurface treatment to 6.48 months in surface treatment indicating a modestly faster decay in subsurface treatment. But this difference in decay rate between treatments did not appear significant in two-way ANOVA ($F = 0.0187$; $P = 0.892$), probably due to the failure of placement directly below soil surface to bring litter in more contact with decomposers. Hence, for faster decomposition, deeper placement inside soil may be required which implies that in-field *Ficus* litter has to be ploughed deeply or composted separately for faster decomposition and quicker release of nutrients.

Two-way ANOVA results also revealed that between substrates, there is a significant variation in decomposition ($F = 11.640$; LSD for comparison of substrates = 2.937; $P < 0.001$). The control filter paper substrate, decomposed faster and had only 5.5%–6% of mass remaining after one year. Filter paper also had a biphasic pattern of mass decline similar to results from semiarid regions of Kenya [7]. ANOVA did not show significant differences for treatment × substrate interaction.

Decay constant of *F. benghalensis* litter as determined in the present study is lower than many tropical tree species [14], while decomposition was faster compared to *F. glomerata* and *F. roxburghii* studied in central Himalayas [12] (Table 1).

3.2. Factors Influencing Litter Decay

3.2.1. Litter Quality. Differences in decomposition rates can probably be explained by variations in litter quality [25] and in climatic and soil conditions of study sites [15]. Litters with low lignin and phenolics and higher nitrogen content are generally considered good quality material for decomposition [26]. *F. benghalensis* litter has medium N content (as per the classification of Jamaludheen and Kumar [14]) and high lignin and polyphenol contents (Table 1), making it a poor quality litter.

Comparison of litter quality of *F. benghalensis* with similar species (of the same genus or family) suggests that high lignin content and lignin/N ratio are the main attributes responsible for its slow decomposition and longer half life period compared to other species. Tian et al. [20] and Constantinides and Fownes [27] argued that initial N content of litter is the best predictor of decomposition and has a positive influence. Litter with high N content in relation to C (low C:N ratio) is known to mineralize faster. But filter paper with highest C:N ratio had rates of decomposition higher than litter, which suggests that in absence of lignin, wide C:N ratio *per se* may not be a limiting factor [7]. Lignin is highly resistant to enzymatic attack and physically interferes with decay of other chemical fractions in leaf tissue hence, slows down decomposition process. But absolute lignin concentration is not of much use as its concentration in relation to N, in predicting decay rates of litter [12]. By virtue of its high lignin content, *F. benghalensis* has the highest lignin:N ratio of all species compared in Table 1 which may explain the persistence of the litter. *Artocarpus hirsutus* litter had higher lignin content (as reported by Isaac and Nair [13]); yet, decomposition was faster due to higher N content and lower lignin/N ratio and also due to favourable *in situ* climatic conditions. Similarly, *F. roxburghii* with lower lignin and higher N contents than *F. benghalensis* had longer half-life due to the cooler environmental conditions of the study area.

3.2.2. Climatic and Soil Conditions of Study Site. Association between *F. benghalensis* litter decay and climatic and soil parameters was investigated (Table 2). As there was no significant difference in decomposition between surface and subsurface applications of litter, mean of monthly weight loss of litter in surface and subsurface treatments was correlated to various climatic parameters (temperature, humidity, and rainfall) and soil properties (moisture, temperature, and pH). Weight loss was found to be significantly and positively correlated to monthly rainfall (Pearson's correlation coefficient = 0.736; $P < 0.01$) and soil moisture (Pearson's correlation coefficient = 0.608; $P < 0.05$). This implies that litter decomposition and subsequent nutrient release are higher in high rainfall months when soil moisture is also high. Association

TABLE 2: Correlations of litter weight loss with climatic and soil parameters of study site.

Parameter	Mean monthly weight loss	Soil temperature	Soil moisture	Air temperature	Relative humidity	Rainfall	Soil pH
Mean monthly weight loss	1	−0.39	0.608*	−0.077	−0.197	0.736**	−0.234
Soil temperature		1	−0.735**	0.719**	−0.762**	−0.681*	−0.311
Soil moisture			1	−0.234	0.416	0.703*	0.268
Air temperature				1	−0.628*	−0.139	−0.446
Relative humidity					1	0.19	0.338
Rainfall						1	−0.096
Soil pH							1

*Significant at $P < 0.05$; **$P < 0.01$.

TABLE 3: Concentrations of major nutrients in *F. benghalensis* leaf litter retrieved at monthly intervals on decomposition.

Month after incubation	N (%)		P (%)		K (%)	
	Surface	Subsurface	Surface	Subsurface	Surface	Subsurface
0 (initial)	1.105		0.0845		1.004	
1	0.903	1.089	0.152	0.220	0.681	0.183
2	0.967	1.050	0.129	0.096	0.994	0.080
3	0.940	1.050	0.291	0.230	0.357	0.191
4	0.992	1.005	0.486	0.211	0.206	0.175
5	0.947	1.030	0.172	0.119	0.199	0.099
6	0.740	1.080	0.273	0.197	0.199	0.164
7	1.340	0.660	0.066	0.075	0.384	0.062
8	1.230	0.974	0.171	0.070	0.284	0.058
9	1.470	1.250	0.343	0.077	0.374	0.064
10	1.410	1.190	0.420	0.103	0.325	0.085
11	1.297	1.240	0.129	0.115	0.261	0.091
12	1.320	1.023	0.166	0.110	0.410	0.110
LSD (0.05)	T : NS	P : NS	T : 0.0774*	P : NS	T : 0.153**	P : NS

T: treatment; P: period; NS: not significant. *Significant at $P < 0.05$; **$P < 0.01$.

between rainfall and litter weight loss has been documented by Upadhyay and Singh [4] and Mugendi and Nair [15]. Jamaludheen and Kumar [14] documented a positive, yet nonsignificant, relation of soil moisture with residual litter weight. Soil pH had no significant effect on litter decay in the present study, which corroborates observations of Mugendi and Nair [15]. Soil temperature is highly negatively correlated to soil moisture (Pearson's correlation coefficient = 0.735; $P < 0.01$) and, hence, may have an indirect impact on litter decomposition due to its influence on soil moisture. Role of soil factors in litter decomposition is generally underplayed in comparison to climatic factors or litter quality [15, 28]. But results of the present study indicate that in addition to litter quality and climate, soil properties may also significantly influence litter decomposition.

3.3. Nutrient Dynamics in Litter Decay. Analysis of elemental composition of decomposing litter revealed an initial rapid loss of N for the first six-seven months, followed by an increase in the concentration in subsequent months (Table 3). The initial decline might be due to the leaching of soluble forms of nitrogen. Second phase of increase may be attributed to immobilization of N by microbial population infesting the litter. Similar pattern of N release was recorded from litters of

Ailanthus triphysa and *Swietenia macrophylla* [13]. P concentrations showed an increasing trend owing to retention of the element in microbial tissues. Similar results were reported for P release by 13, 29, and 18. Potassium concentrations showed a rapid decline. Being a nonstructural element, K is highly mobile and is easily lost by leaching [29, 30].

When the percentage of nutrients remaining after one year of decomposition was computed from residual nutrient concentrations and litter mass, it was found that proportion of residual nutrients was less in subsurface treatment compared to surface treatment, indicating comparatively faster nutrient release when the litter is incorporated to the soil. After one year of decomposition, 33.80% of initial N content was remaining in surface treatment, while in subsurface treatment only 20.83% was remaining. However, two-way ANOVA of nutrient release showed that there was no significant difference in release of N between surface and subsurface treatments ($F = 0.933; P = 0.355$). But release of P ($F = 7.724$; LSD for comparing treatments = 0.0774; $P < 0.05$) and K ($F = 15.781$; LSD = 0.153; $P < 0.01$) varied significantly between treatments. Mineralisation was fastest for K due to its highly mobile nature and susceptibility to leaching, while P was the most persistent nutrient. After one year of decomposition, 2.45% of K was remaining in subsurface treatment and 11.55% in surface treatment, while 29.29% and 55.71% of P was remaining in surface and subsurface treatments, respectively. Order of mineralization followed the pattern K > N > P in both surface and subsurface treatments. However, general trend in nutrient release showed that processes determining P and K release may not be correlated with N release [12]. Effect of months was not significant for all the elements.

4. Conclusions

The pattern of litter decomposition and nutrient release has important implications for exploiting *F. benghalensis* as an agroforestry species in Mandya. Litter quality and the timing of litterfall determine the contribution of leaf litter of agroforestry trees to soil fertility through decomposition [12]. In the case of *Ficus*, two phases of heavy litterfall occur: first during October-November and then during January-February [5]. Decomposition studies indicate that litter decomposes to half of its original mass only in six months of decay. Thus the slow release of nutrients from surge of litterfall in October-November facilitates summer cropping in April, while the second peak in February makes nutrients available for monsoon cropping in July. Higher litter weight loss during the rainy season enhances nutrient availability for monsoon crops. Results of the study also point to the need of ploughing the litter inside soil or separate composting if nutrient release is to be accelerated, while slow rate of decomposition on soil surface indicates the potential of the litter to be used as an organic mulch to conserve moisture of agricultural soils in semiarid areas. For proper utilisation of *Ficus* litter, it is also important to clearly understand its allelopathic potential, as inhibitory effects of *Ficus benghalensis* bark and leaf extracts on weeds, microbes, and crops are widely reported in the literature [31–34]. However, *Ficus* agroforestry in Mandya is a time-tested practice, and hence, negative interactions of trees with common crops in these systems may be minimal. Litter decomposition and nutrient release pattern of *Ficus* trees potentially complement the cropping pattern in rainfed farmlands of Mandya and augment nutrient availability to crops, thereby affecting considerable saving on external nutrient input costs in dryland farming and contributing to the overall sustainability of the system.

Acknowledgments

The authors wish to acknowledge the University Grants Commission, New Delhi, and Indian Council of Forestry Research and Education, Dehradun, for financial assistance and Director, Institute of Wood Science and Technology, Bangalore, for logistic support. Mr. K. M. Shivaswamy, Karadakere village, Mandya, is sincerely thanked for providing space in his farmland for litter decomposition experiment, and reviewers are thanked for critical suggestions.

References

[1] B. Dhanya, S. Viswanath, S. Purushothaman, and B. Suneeta, "*Ficus* trees as components of rainfed agrarian systems in Mandya district of Karnataka," *My Forest*, vol. 46, no. 2, pp. 161–165, 2010.

[2] S. R. Isaac and M. A. Nair, "Decomposition of wild jack (*Artocarpus hirsutus* Lamk.) leaf litter under sub canopy and open conditions," *Journal of Tropical Agriculture*, vol. 42, no. 1-2, pp. 29–32, 2004.

[3] V. Meentemeyer and B. Berg, "Regional variation in rate of mass loss of *Pinous sylvestris* needle litter in Swedish pine forest as influenced by climate and litter quality," *Canadian Journal of Forest Research*, vol. 1, pp. 167–180, 1986.

[4] V. P. Upadhyay and J. S. Singh, "Patterns of nutrient immobilization and release in decomposing forest litter in Central Himalaya, India," *Journal of Ecology*, vol. 77, no. 1, pp. 147–161, 1989.

[5] B. Dhanya, *Integrated study of a Ficus based traditional agroforestry system in Mandya district, Karnataka [Ph.D. thesis]*, Forest Research Institute Deemed University, Dehradun, India, 2011.

[6] J. M. Anderson and J. S. Ingram, *Tropical Soil Biology and Fertility: A Handbook of Methods*, CAB International, Wallingford, UK, 1993.

[7] B. A. Jama and P. K. R. Nair, "Decomposition- and nitrogen-mineralization patterns of *Leucaena leucocephala* and *Cassia siamea* mulch under tropical semiarid conditions in Kenya," *Plant and Soil*, vol. 179, no. 2, pp. 275–285, 1996.

[8] M. L. Jackson, *Soil Chemical Analysis*, Prentice Hall of India, New Delhi, India, 1973.

[9] A. Walkley and C. A. Black, "An examination of the Degtjareff method for determining soil organic matter and proposed modification of the chromic acid titration method," *Soil Science*, vol. 37, pp. 29–39, 1934.

[10] R. M. Rowell, R. Pettersen, J. S. Han, J. Rowell, and M. A. Tshabalala, "Cellwall chemistry," in *Handbook of Wood Chemistry and Wood Composites*, R. M. Rowell, Ed., p. 65, CRC & Taylor & Francis, Boca Raton, Fla, USA, 2005.

[11] S. Sadasivam and A. Manickam, *Biochemical Methods For Agricultural Sciences*, Wiley Eastern Limited and Coimbatore: Tamil Nadu Agricultural University, New Delhi, India, 1992.

[12] R. L. Semwal, R. K. Maikhuri, K. S. Rao, K. K. Sen, and K. G. Saxena, "Leaf litter decomposition and nutrient release patterns of six multipurpose tree species of central Himalaya, India," *Biomass and Bioenergy*, vol. 24, no. 1, pp. 3–11, 2003.

[13] S. R. Isaac and M. A. Nair, "Litter dynamics of six multipurpose trees in a homegarden in Southern Kerala, India," *Agroforestry Systems*, vol. 67, no. 3, pp. 203–213, 2006.

[14] V. Jamaludheen and B. M. Kumar, "Litter of multipurpose trees in Kerala, India: variations in the amount, quality, decay rates and release of nutrients," *Forest Ecology and Management*, vol. 115, no. 1, pp. 1–11, 1999.

[15] D. N. Mugendi and P. K. R. Nair, "Predicting the decomposition patterns of tree biomass in tropical highland microregions of Kenya," *Agroforestry Systems*, vol. 35, no. 2, pp. 187–201, 1997.

[16] J. G. Bockheim, E. A. Jepsen, and D. M. Heisey, "Nutrient dynamics in decomposing leaf litter of four tree species on a sandy soil in northwestern Wisconsin," *Canadian Journal of Forest Research*, vol. 21, no. 6, pp. 803–812, 1991.

[17] M. M. Giashuddin, D. P. Garrity, and M. L. Aragon, "Weight loss, nitrogen content changes, and nitrogen release during decomposition of legume tree leaves on and in the soil," *Nitrogen Fixing Tree Research Report II*, pp. 43–50, 1993.

[18] L. T. Szott, E. C. M. Fernandes, and P. A. Sanchez, "Soil-plant interactions in agroforestry systems," *Forest Ecology and Management*, vol. 45, no. 1–4, pp. 127–152, 1991.

[19] C. A. Palm and P. A. Sanchez, "Nitrogen release from the leaves of some tropical legumes as affected by their lignin and polyphenolic contents," *Soil Biology and Biochemistry*, vol. 23, no. 1, pp. 83–88, 1991.

[20] G. Tian, B. T. Kang, and L. Brussaard, "Effects of chemical composition on N, Ca, and Mg release during incubation of leaves from selected agroforestry and fallow plant species," *Biogeochemistry*, vol. 16, no. 2, pp. 103–119, 1992.

[21] K. Niranjana, *Studies on the tree-crop interactions in tea (Camellia sinensis) based shaded perennial agroforestry system in Western Ghats [Ph.D. thesis]*, Forest Research Institute Deemed University, Dehra Dun, India, 2006.

[22] V. G. Panse and P. V. Sukhatme, *Statistical Methods For Agricultural Workers*, Indian Council of Agricultural Research, New Delhi, India, 2000.

[23] M. E. Harmon, G. A. Baker, G. Spycher, and S. E. Greene, "Leaf-litter decomposition in the Picea/tsuga forests of Olympic National Park, Washington, U.S.A," *Forest Ecology and Management*, vol. 31, no. 1-2, pp. 55–66, 1990.

[24] R. L. Edmonds and T. B. Thomas, "Decomposition and nutrient release from green needles of western hemlock and Pacific silver fir in an old-growth temperate rain forest, Olympic National Park, Washington," *Canadian Journal of Forest Research*, vol. 25, no. 7, pp. 1049–1057, 1995.

[25] R. H. Waring and W. H. Schlesinger, *Forest Ecosystem: Concepts and Management*, Academic Press, New York, NY, USA, 1985.

[26] A. Young, "Agroforestry for Soil Management," CAB International and International Centre for Research in Agroforestry, 1997.

[27] M. Constantinides and J. H. Fownes, "Nitrogen mineralization from leaves and litter of tropical plants: relationship to nitrogen, lignin and soluble polyphenol concentrations," *Soil Biology and Biochemistry*, vol. 26, no. 1, pp. 49–55, 1994.

[28] V. Meentemeyer, "Macroclimate and lignin control of litter decomposition rates," *Ecology*, vol. 59, pp. 465–472, 1978.

[29] T. J. Stohlgren, "Litter dynamics in two Sierran mixed conifer forests. II. Nutrient release in decomposing leaf litter," *Canadian Journal of Forest Research*, vol. 18, no. 9, pp. 1136–1144, 1988.

[30] O. P. Toky and V. Singh, "Litter synamics in short-rotation high density tree plantations in an arid region of India," *Agriculture, Ecosystems and Environment*, vol. 45, no. 1-2, pp. 129–145, 1993.

[31] S. Shafique, R. Bajwa, A. Javaid, and S. Shafique, "Biological control of *Parthenium* IV: suppressive ability of aqueous leaf extracts of some allelopathic trees against germination and early seedling growth of *Parthenium hysterophorus* L.," *Pakistan Journal of Weed Science Research*, vol. 11, no. 1-2, pp. 75–79, 2005.

[32] M. Manikandan and M. Jayakumar, "Herbicidal effect of *Ficus bengalensis* aqueous extract on *Lpomoea pentaphylla*," *International Journal of Agriculture*, vol. 2, no. 1, pp. 35–38, 2012.

[33] S. Shafique, A. Javaid, R. Bajwa, and S. Shafique, "Effect of aqueous leaf extracts of allelopathic trees on germination and seed-borne mycoflora of wheat," *Pakistan Journal of Botany*, vol. 39, no. 7, pp. 2619–2624, 2007.

[34] S, Siddiqui, M. K. Meghvansi et al., "Efficacy of aqueous extracts of five arable trees on the seed germination of *Pisum sativum* l. Var-VRP-6 and KPM-522," *Botany Research International*, vol. 2, no. 1, pp. 30–35, 2009.

15

Simple Method of Forest Type Inventory by Joining Low Resolution Remote Sensing of Vegetation Indices with Spatial Information from the Corine Land Cover Database

Jarosław J. Zawadzki,[1] Karol Przeździecki,[1] Karol Szymankiewicz,[1] and Wojciech Marczewski[2]

[1] Department of Environmental Engineering, Warsaw University of Technology, Ulica Nowowiejska 20, 00-653 Warsaw, Poland
[2] Space Research Centre, Polish Academy of Sciences, Ulica Bartycka 18A, 00-716 Warsaw, Poland

Correspondence should be addressed to Jarosław J. Zawadzki; j.j.zawadzki@gmail.com

Academic Editors: N. Frascaria-Lacoste, G. Martinez Pastur, and J. F. Negron

The paper presents a simple, inexpensive, and effective method allowing for frequent classification of the forest type coniferous, deciduous, and mixed using medium and low resolution remote sensing images. The proposed method is based on the set of vegetation indices such as NDVI, LAI, FAPAR, and LAIxCab calculated from MODIS and MERIS satellite data. The method uses seasonal changes of the above-mentioned vegetation indices within annual cycle. The main idea was to collect and carefully analyse seasonal changes in vegetation indices in a given ecosystem type proven by a Corine Land Cover, 2006 database, and to compare them afterwards with those of a particular forest under study. Each type of a forest ecosystem has its own specific dynamics of development, thus enabling recognition of the type by comparing temporal changes of the proposed measures based on vegetation indices. Temporal measures of changes were created for selected reference stands by the ratios of particular indices determined in July and April, which are the middle and the beginning of a vegetation season in Poland, respectively. The analysed vegetation indices were additionally provided with chosen statistical measures. The statistical analyses were carried out for Poland's main national parks which represent the natural stands of temperate climate.

1. Introduction

Forests cover about 31% of the land [1], and their role in the natural environment and in human activities is essential. For example, forests have a significant influence on the composition of dust and atmospheric gases, air and soil temperature, the amount of water present, and forested areas, both in soil and in vegetation cover. Forests also play an important role in the exchange of water between the soil and the atmosphere. Forest management requires timely and accurate information on forests [2] and remote sensing methods have been used for forest inventory for decades [3–5]. A major problem in forest research is the diversity of ecosystems, which provides for the possible couplings of various physical and biological processes, especially when there is a need of mesoscale assessments. It is therefore essential how large areas can be represented by the data (satellite, terrestrial, and statistical assessment) and what time resolution measurements are performed with. For all these reasons, studies on forest ecosystem, have always required considerable effort and resources.

In remote observations of forests, spectral analysis allows for the determination of various biophysical parameters such as NDVI, LAI (leaf area index), FCover (fraction of vegetation cover), FAPAR (fraction of absorbed photosynthetically active radiation), and LAIxCab (canopy chlorophyll content of A and B types) [6–9]. These indicators and their properties are well described in the literature [10, 11], so they will not be detailed discussed here. Applying these vegetation indices, it is possible to determine quite accurately, both quantitatively and qualitatively, the state of vegetation on the Earth's surface. The vegetation indices are closely related to

Simple Method of Forest Type Inventory by Joining Low Resolution Remote Sensing of Vegetation Indices with Spatial
Information from the Corine Land Cover Database

171

the fundamental energy and biological and physical phenomena, but at the same time they are distant from the direct parametric assessments. Thus, one uses vegetation indices to evaluate more complex quantities, such as biomass and its productivity, the ability to bind water with vegetation or even with soil under the trees, diverse statistical characteristics of trees (e.g., diameter at breast height, cross-sectional area), evapotranspiration, and the ability of carbon binding.

It is worth mentioning that diverse sophisticated vegetation indices as well as their processors specialized to specific satellite imagery are still developed giving new impetus to the development of better methods for classification of forests.

High resolution observations are most commonly associated with the necessity of obtaining and processing large amounts of data [10, 11] while (contrary to popular opinion) they are usually rare, and even incidental, limited to the time of satellite flight, depending on cloud conditions, research cost, and so forth. Therefore, it is generally accepted that the satellite data with an average or low spatial resolutions are also very valuable, because such data are easily and regularly accessible, often at no cost, as open to the public.

It should be also emphasized that high spatial resolution, in observing the Earth's surface, including the forest ecosystems, is not the only or even the most important criterion of quality satellite observations. Increasing spatial resolution capabilities have caused a sharp increase in the amount of data and resources needed to process them, resulting in increased costs of research and investment of human labour. A suitable compromise in the selection of spatial and temporal resolutions is therefore essential [12, 13]. An example of a large-scale, technologically advanced satellite program designed to study the content of water in the soil, as well as the amount of the water bounded with vegetation, including forests, is the mission SMOS (soil moisture and ocean salinity), working in the range of microwave radiation (1.400–1.427 GHz), which ensures the relevance of the large-area assessments. At the same, time resolving power of this probe is relatively low, about 32 to 50 km [14]. Inventory of forest environments obviously must also involve high resolution observations, including ground-based data, but such observations are not reliable enough, if they do not enable comparisons of separate and isolated areas, such as, for instance, various national parks. Often there is not only the need for an exact assessment of detailed information as, for example, distribution of tree species, but also, especially in the climate-forest interaction studies, some essential forest features as forest type or forest state, which can sometimes be difficult to be distinguished in too small or directly neighbouring forest areas. Inexpensive and convenient observations at large scales and meso-(i.e., intermediate) scales are and will always be necessary for many regional assessments. Therefore, medium and low resolution methods of forest ecosystem observations are intensively developed together with high resolution ones.

The aim of this study was to develop an inexpensive method to distinguish between three selected classes of forest: coniferous, deciduous, and mixed by means of satellite images of low (in terms of forest applications) resolution, based on seasonal changes in selected indicators of vegetation. The

TABLE 1: Land cover class in the studied area according to the Corine Land Cover, 2006 database.

Land cover	Area [km^2]	Percentage [‰]
Anthropogenised area	10.8	4.83
Agricultural area	358.4	160.44
Deciduous forests	587.9	263.18
Coniferous forests	716.9	320.93
Mixed forests	276.8	123.91
Forests in changes	68.4	60.62
Pastures	1.6	0.72
Wetlands	212.2	95.00
Water areas	0.9	0.40
Total	2233.8	1000

results of this method could be used, for example, to validate the aforementioned SMOS data or used in the climate-forest interaction investigations. It was assumed that the classification results could be easily updated annually, which is much more frequent than, for example, the spatially detailed, and widely used in European Union, Corine Land Cover database [15], which has been updated at great expense every few years (namely, in 1990, 2000, and 2006). Therefore, the Corine Land Cover data should be considered as independent of seasonal changes and may serve for validating analysis with the use of lower resolutions satellite sensors as MERIS, MODIS, SPOT, CHRIS, and so forth.

In addition, the aim of this work was to motivate the interest in remote observations at medium and low resolution, which are inexpensive, relatively easily accessible, and very well equipped with free tools available for data analysis. Particular value of spectral analysis is an opportunity of evaluating evapotranspiration of forest areas, as it is one of the most important elements of water balance on forest area.

2. Study Area

The statistical analyses were carried out for seven national parks of Poland that represent the typical, natural stands of temperate climate, namely, the Białowieski National Park, the Biebrzański National Park, the Bolimowski National Park, the Kampinoski National Park, the Kozienicki National Park, the Roztoczański National Park, and the Świętokrzyski National Park. The location of these parks on the map of Poland is shown in Figure 1. The total area of studied parks was about 2234 km^2, including about 1582 km^2 of coniferous, deciduous, and mixed forests which is more than 70% of the examined area. The detailed information on land cover within the whole studied area is given in Table 1. As a test area, the Kampinoski National Park was selected, which is located in the vicinity of Warsaw, capital of Poland. The total area of this park is about 385 km^2.

3. Satellite and Test Data Description

The study was based on low spatial resolution data FR_1P and FR_2P of the ENVISAT/MERIS (medium resolution imaging

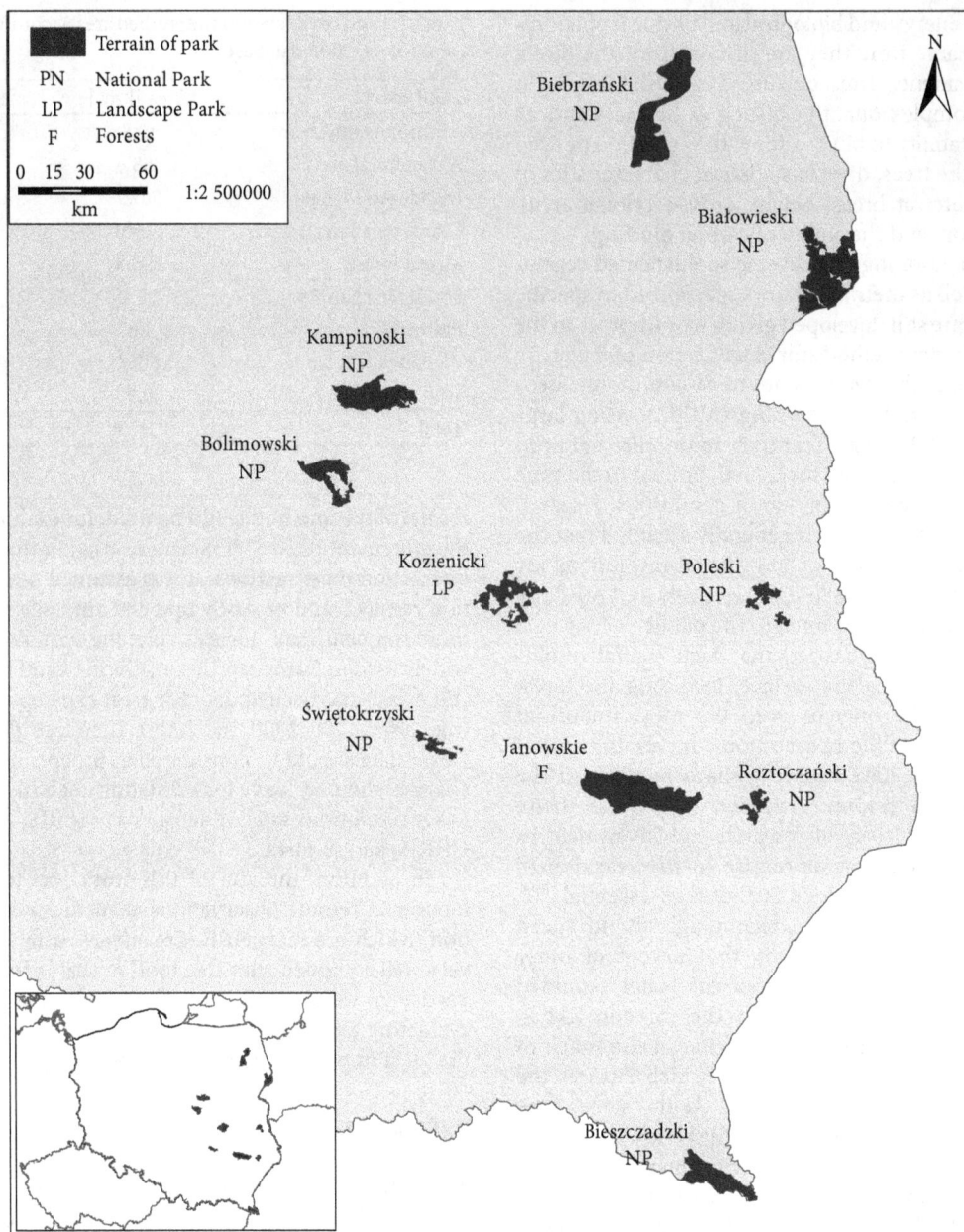

FIGURE 1: Location of study area on the map of Poland.

spectrometer, spatial resolution 300 m) satellite sensor of levels, as well as on similar MOD15A2 MOD13A2 data of the MODIS/TERRA (moderate resolution imaging spectrora-diometer, spatial resolution 1000 m). The selection of images was guided by the above-described reasons, as well as the by the evaluation of image quality in terms of the instantaneous cloud cover. MODIS images were obtained free of charge directly from the WIST site (The Warehouse Inventory Search Tool). MERIS images were also obtained free of charge in cooperation with the Space Research Centre, Polish Academy of Sciences, the Cat-1 in the frame of project AO-3275. All the preliminary operations and the analysis of vegetation indices from MERIS images were performed using the Visat (BEAM) program from Brockmann Consult and Contributors, shared

free of charge by the European Space Agency (ESA). To prepare remote images of good quality before the main analysis, it was necessary to perform carefully many time-consuming operations such as calibration, scaling, and image orthorectification. Detailed description of these necessary, but time-consuming analyses, is not possible in this paper.

Figures 2 and 3 show examples of spatial distributions of exemplary indices (FAPAR_MERIS and FAPAR_MODIS) in the area covering the whole of the satellite image.

When performing the spatial analysis, the GIS (Geographical Information Systems) tools were also used, namely ArcGIS 9.1, while the statistical calculations were made in the Statistica package. The Corine Land Cover, 2006 database, was used to take into account only natural stands from the

Simple Method of Forest Type Inventory by Joining Low Resolution Remote Sensing of Vegetation Indices with Spatial Information from the Corine Land Cover Database

173

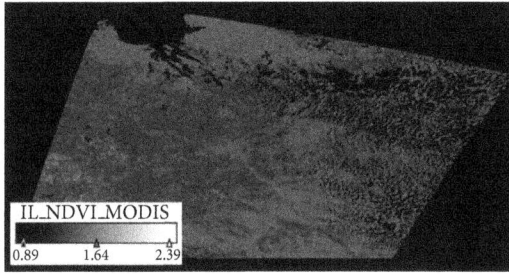

FIGURE 2: Example of spatial distribution of the IL_NDVI_MODIS vegetation index obtained from the MERIS image, on July 3, 2008, and on April 2, 2009.

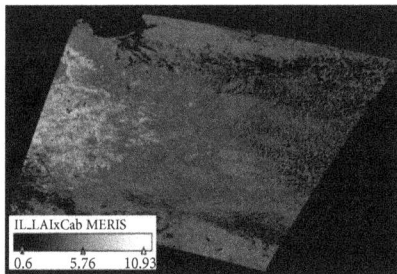

FIGURE 3: Example of spatial distribution of the IL_LAIxCab_MERIS vegetation index obtained from the MODIS image, on July 3 2008, and on April 2 2009.

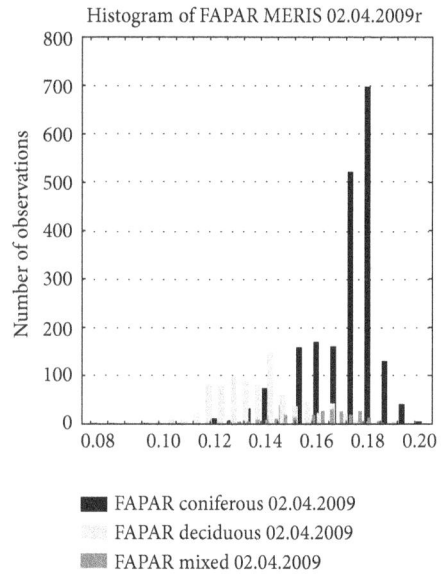

FIGURE 4: Multiple histogram of the FAPAR MERIS vegetation index for the photograph on April 2, 2009.

above-described seven national parks. To assure the best quality of results of GIS and statistical analyses, all pixels taken for calculations were carefully filtered. The area, which included the above-mentioned three types of forests and was not obscured by clouds in satellite images, had an area of approximately 875 km^2.

4. Results and Discussion

The demonstration of significant differences between the three forest types coniferous, deciduous, and mixed in remote sensing data of relatively low resolution satellite images required proper selection of vegetation indices. This was made by carefully analyzing the major seasonal changes of vegetation indices, ranging from classical vegetation indeces NDVI and LAI, to much more sophisticated, such as FAPAR, fCover, LAIxCab, and MGVI. After preliminary analysis described in the previous section, seven different vegetation indices were chosen for the study: NDVI_MERIS, FAPAR_MERIS, LAI_MERIS, LAIxCab_MERIS, NDVI_MODIS, and IL_LAI_MODIS.

In view of the fact that most conifer trees do not lose their needles, one can expect that their vegetation indices show much smaller seasonal changes than those deciduous ones [16, 17].

As an example of such behaviour, histograms presented in Figures 4 and 5 show the typical distributions of the FAPAR_MERIS index values in April 2, 2009 and July 3, 2009 over area of the Kampinoski National Park. As expected, the temporal variability of this index in the case of deciduous

forests is much higher than that in case of coniferous one. In addition, one can notice that the value of the FAPAR index in April for coniferous forests is greater than that for deciduous forests, while in July the FAPAR index value is greater for deciduous forests. As it can be seen in Figures 4 and 5, indices for mixed forests take, in both cases, intermediate values as compared with those for deciduous and coniferous ones.

This justifies the use of some indices based on seasonal changes in vegetation for the classification of forest types. In this work, the ratios of particular vegetation indices determined in July and April, which are the middle and the beginning of a vegetation season in Poland, were calculated. These ratios of vegetation indices are further known briefly as ratio indices, and indicated by the letters IL. For example,

$$\text{IL\_FAPAR\_MERIS} = \frac{\text{value of FAPAR from MERIS in July}}{\text{value of FAPAR from MERIS in April}}. \quad (1)$$

It is worth noting that in order to perform such calculations it was necessary to ensure that pixels values determined in July and April were taken exactly from the same place.

The similar histograms to those for FAPAR were obtained also for all seven above-listed vegetation indices calculated from MODIS and MERIS satellite images. For all analysed national parks, the same marked tendency has been observed: ratio indices determined for coniferous forests had the smallest mean values, those determined for deciduous forests were the largest, whereas those obtained for mixed forests were intermediate. The average values of ratio indices calculated from the whole area of national parks are presented in Figure 6, whereas Table 2 shows the mean values of ratio indices calculated both for the whole area and separately for each national park.

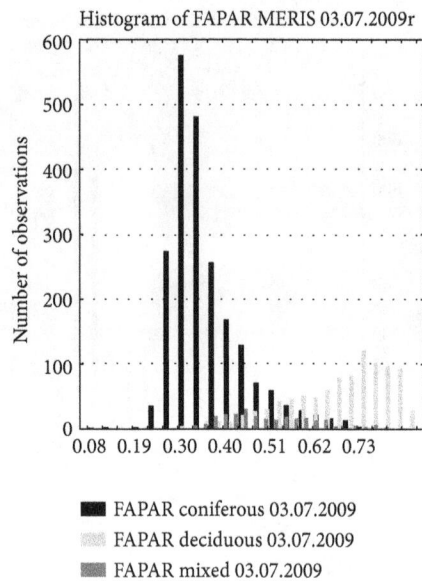

FIGURE 5: Multiple histogram of the FAPAR MERIS vegetation index for the photograph on July 3, 2009.

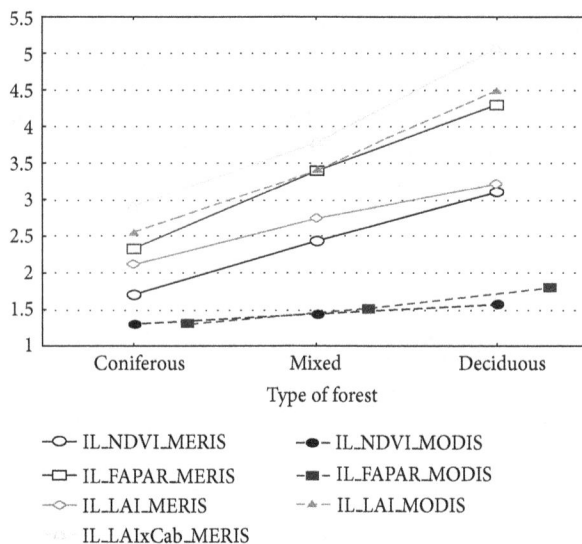

FIGURE 6: The average values of ratio indices calculated from the whole area of national parks under study.

The distinct differences in mean values of ratio indices of specific forest type made it possible to use the ratio indices for classification purposes. To do so, also the ranges of all ratio indices corresponding to different forest types were determined for the whole studied area and separately for each national park. Determination of precise limits of the ranges of all ratio indices for the selected type of forest was a difficult task because the ranges of these indices overlap each other due to natural variations in the pixel values. One could expect that the ranges of ratio indices obtained from the whole study area were more reliable; therefore they were used to obtain the spatial distributions of the forest type in the test area.

TABLE 2: The means of ratio indices for the studied types of forests in Poland.

Ratio index	Forest type		
	Coniferous	Mixed	Deciduous
IL_NDVI_MERIS	1.74	2.51	3.32
IL_FAPAR_MERIS	2.43	3.48	4.38
IL_LAI_MERIS	2.16	2.72	3.29
IL_LAIxCab_MERIS	2.88	3.85	5.21
IL_NDVI_MODIS	1.31	1.45	1.57
IL_FAPAR_MODIS	1.28	1.46	1.74
IL_LAI_MODIS	2.43	3.14	4.14
PIXEL NUMBER	7201	2385	3860

Table 3 presents the limits of ranges of ratio indices with the corresponding classification accuracy of the mixed forests.

Ranges of ratio indices when applied to the classification of forest type in the Kampinoski National Park led to surprisingly good results both for coniferous and deciduous forests. The obtained accuracy ranged from about 60% to about 90%, depending on chosen ratio index, which could be consider as good result [18, 19].

Tables 4 and 5 show the results of the classification for different types of forest for the test site, that is, the Kampinoski National Park using ratio indices calculated on the basis of MODIS and MERIS images, respectively.

The results presented in Tables 3 and 4 indicate a reasonably good classification accuracy of deciduous and coniferous forests for both types of images and apparently lower classification accuracy of mixed forests. The last effect can be attributed to too coarse spatial resolution of remote imagery. It is worth noting that the classification accuracy of the mixed forests was significantly improved when MERIS images with better spatial resolution (more than three times) were used. It should also be stressed that the highest classification accuracy of the mixed forests was obtained using IL_FAPAR_MODIS and IL_FAPAR_MERIS indices based on the FAPAR vegetation index. In order to check the proposed method spatially, also the maps at test area were drawn showing spatial distribution of the forest types obtained by means of the low resolution imagery and then compared with true maps of forests from the Corine Land Cover database. As can be seen in Figure 7, surprisingly good fit of both kinds of forest type distributions was obtained, in particular in areas with large surface and regular shape. It can be seen again that the distributions of forest types based on vegetation indices derived from satellite imagery of MERIS sensor show a much better fit than those of the MODIS sensor.

5. Summary and Conclusions

This paper describes a rapid and inexpensive method of the annual classification of forest ecosystems into three categories: coniferous forests, mixed forests, and deciduous ones, using optical satellite images of MODIS/TERRA and MERIS/ENVISAT sensors with low spatial resolution. Direct use of vegetation indices such as NDVI, FAPAR, and LAI

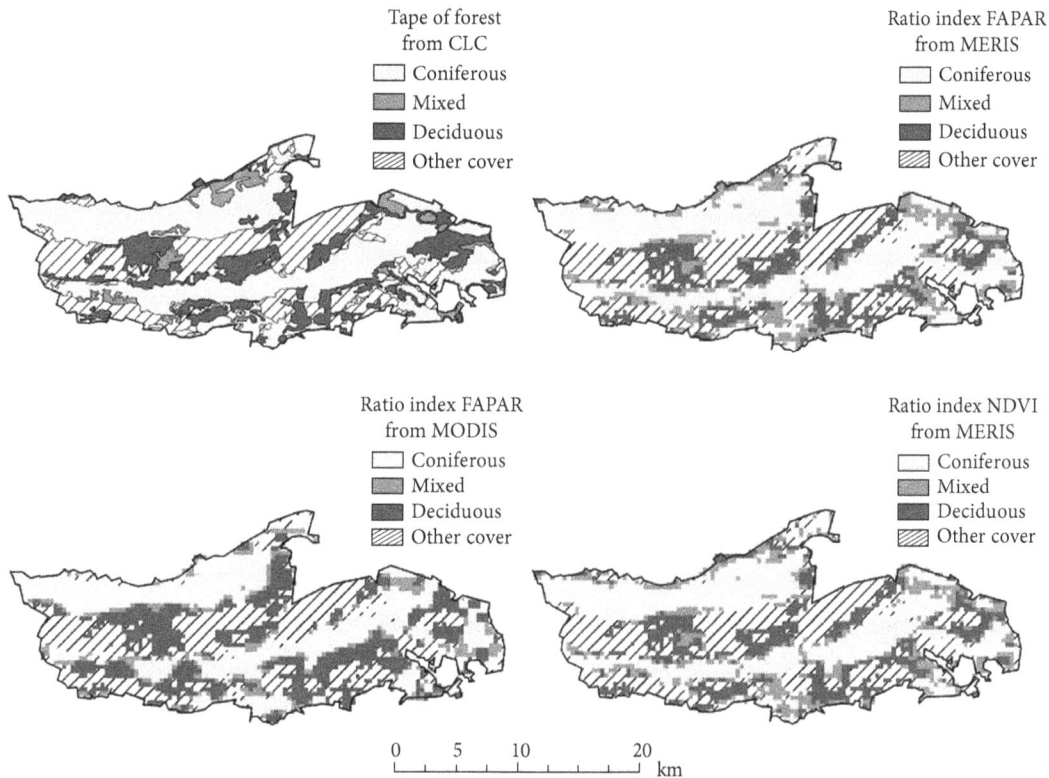

FIGURE 7: Comparison of the classification results of forest types obtained using ratio indices with the spatial distribution obtained on the basis of the accurate data from CLC database. Example based on FAPAR NDVI vegetation indices calculated from MERIS and MODIS images, with resolutions 300 m and 1000 m, respectively, in the area of the Kampinoski National Park.

TABLE 3: Limits indices for the studied types of forest and the classification accuracy for the mixed forests in Poland.

Ratio index	Forest type			Mixed forest classification accuracy
	Coniferous	Mixed	Deciduous	
IL_NDVI_MERIS	<1.75	1.75–2.60	>2.60	50.2
IL_FAPAR_MERIS	<2.35	2.35–3.75	>3.75	51.9
IL_LAI_MERIS	<2.05	2.05–2.65	>2.65	32.8
IL_LAIxCab_MERIS	<2.90	2.90–3.80	>3.80	45.6
IL_NDVI_MODIS	<1.33	1.33–1.48	>1.48	30.9
IL_FAPAR_MODIS	<1.38	1.38–1.55	>1.55	18.6
IL_LAI_MODIS	<2.55	2.55–3.25	>3.25	17.2

for this purpose did not provide sufficient accuracy, even in the case of high-resolution satellite observations. In this study, it was decided to use seasonal changes in these indices. The use of ratios of vegetation indices (the ratio indices) calculated from satellite observations performed in July and April was here proposed. The typical ranges of values of selected ratio indices for coniferous, deciduous, and mixed forests in natural stands were determined. The results of analysis performed in the work show that limits of ranges of ratio indices developed in this work may be used for efficient forest classification using low resolution satellite imagery. The method was verified at the Kampinoski National Park area using the Corine Land Cover database. Good accuracy of classification was obtained despite the relatively

low spatial resolutions used in the analysis comparing to those used commonly in the satellite observations of forest ecosystems. Lack of full agreement of the classification results with the true spatial distribution of forests may also be due to the fact that boundaries of different types of forest are not sharp but blurred. The transition between forest specified in the Corine Land Cover database as coniferous and mixed in nature takes place smoothly. Moreover, in areas where there is a large variation of the forest type, the corresponding pixel in satellite image may be located on the border between two or even three forests and thus incorrectly classified. The highest classification accuracy of the mixed forests was obtained using ratio indices based on FAPAR.

TABLE 4: The results of the classification of studied types of forest in the Kampinoski National Park using ratio indices calculated on the basis of MODIS images.

Ratio index	IL_NDVI_MODIS			IL_FAPAR_MODIS			IL_LAI_MODIS		
Forest type	Coniferous	Mixed	Deciduous	Coniferous	Mixed	Deciduous	Coniferous	Mixed	Deciduous
Classification accuracy [%]	63.6	10.4	62.5	61.8	15.6	67.8	61.8	14.4	66.4

TABLE 5: The results of the classification of studied types of forest in the Kampinoski National Park using ratio indices calculated on the basis of MERIS images.

Ratio index	IL_NDVI_MERIS			IL_FAPAR_MERIS		
Forest type	Coniferous	Mixed	Deciduous	Coniferous	Mixed	Deciduous
Classification accuracy [%]	75.1	52.4	77.3	77.7	60.0	72.8
Ratio index	IL_LAI_MERIS			IL_LAIxCab_MERIS		
Forest type	Coniferous	Mixed	Deciduous	Coniferous	Mixed	Deciduous
Classification accuracy [%]	76.7	22.0	71.3	68.7	36.4	90.0

In view of further global-scale research of the hydrological conditions and evapotranspiration at forest areas, the use of MODIS optical data and very low resolution microwave SMOS data is considered. The results obtained in this work could be applied in such studies.

Acknowledgments

The authors acknowledge the support received under a grant financed by the European Union under the European Social Fund, which is awarded by the Centre for Advanced Studies, Warsaw University of Technology in the framework of the project "Warsaw University of Technology Development Program."

References

[1] Food and Agriculture Organization of the United Nations, http://www.fao.org/news/story/en/item/40893/icode/ .

[2] M. C. Hansen, S. V. Stehman, P. V. Potapov et al., "Humid tropical forest clearing from 2000 to 2005 quantified by using multitemporal and multiresolution remotely sensed data," *Proceedings of the National Academy of Sciences of the United States of America*, vol. 105, no. 27, pp. 9439–9444, 2008.

[3] M. C. Hansen, S. V. Stehman, and P. V. Potapov, "Quantification of global gross forest cover loss," *Proceedings of the National Academy of Sciences of the United States of America*, vol. 107, no. 19, pp. 8650–8655, 2010.

[4] M. C. Hansen, S. V. Stehman, and P. V. Potapov, "Reply to Wernick et al.: Global scale quantification of forest change," *Proceedings of the National Academy of Sciences of the United States of America*, vol. 107, no. 38, p. E148, 2010.

[5] J. Zawadzki, C. J. Cieszewski, M. Zasada, and R. C. Lowe, "Applying geostatistics for investigations of forest ecosystems using remote sensing imagery," *Silva Fennica Monographs*, vol. 39, no. 4, pp. 599–617, 2005.

[6] F. Baret, K. Pavageau, Bacour et al., "Algorithm Theoretical Basis Document for MERIS Top of Atmosphere Land Prod. (TOA-VEG)," INRA-Novelties, 2009, http://www.brockmann-consult.de/beam/plugins.html.

[7] C. Bacour, F. Baret, D. Béal, M. Weiss, and K. Pavageau, "Neural network estimation of LAI, fAPAR, fCover and LAI×Cab, from top of canopy MERIS reflectance data: principles and validation," *Remote Sensing of Environment*, vol. 105, no. 4, pp. 313–325, 2006.

[8] N. Gobron, B. Pinty, F. Mélin et al., "Evaluation of the MERIS/ENVISAT FAPAR product," *Advances in Space Research*, vol. 39, no. 1, pp. 105–115, 2007.

[9] R. B. Myneni, S. Hoffman, Y. Knyazikhin et al., "Global products of vegetation leaf area and fraction absorbed PAR from year one of MODIS data," *Remote Sensing of Environment*, vol. 83, no. 1-2, pp. 214–231, 2002.

[10] J. W. Rouse, R. H. Haas, J. A. Schell, and D. W. Deering, "Monitoring vegetation systems in the great plains with ERTS," in *Proceedings of the 3rd ERTS Symposium. NASA SP-351*, pp. 309–317, NASA, Washington, DC, USA, 1973.

[11] R. Nemani, L. Pierce, S. Running, and L. Band, "Forest ecosystem processes at the watershed scale: sensitivity to remotely-sensed leaf area index estimates," *International Journal of Remote Sensing*, vol. 14, no. 13, pp. 2519–2534, 1993.

[12] A. H. Strahler, C. E. Woodcock, and J. A. Smith, "On the nature of models in remote sensing," *Remote Sensing of Environment*, vol. 20, no. 2, pp. 121–139, 1986.

[13] P. Treitz and P. Howarth, "High spatial resolution remote sensing data for forest ecosystem classification: an examination of spatial scale," *Remote Sensing of Environment*, vol. 72, no. 3, pp. 268–289, 2000.

[14] ESA—SMOS Earth Explorers, http://www.esa.int/esaLP/ESAS7C2VMOC_LPsmos_0.html.

[15] M. Bossard, J. Feranec, and J. Otahel, "Corine land cover technical guide—Addendum," Technical Report 40, EEA, Copenhagen, Denmark, 2000.

[16] A. M. Jönsson, L. Eklundh, M. Hellström, L. Bärring, and P. Jönsson, "Annual changes in MODIS vegetation indices of Swedish coniferous forests in relation to snow dynamics and tree phenology," *Remote Sensing of Environment*, vol. 114, no. 11, pp. 2719–2730, 2010.

[17] Q. Zhang, X. Xiao, B. Braswell et al., "Characterization of seasonal variation of forest canopy in a temperate deciduous broadleaf forest, using daily MODIS data," *Remote Sensing of Environment*, vol. 105, no. 3, pp. 189–203, 2006.

Simple Method of Forest Type Inventory by Joining Low Resolution Remote Sensing of Vegetation Indices with Spatial
Information from the Corine Land Cover Database

177

[18] S. Lewiński, "Object based classification of middle resolution MODIS satellite image, first results," *Archiwum Fotogrametrii, Kartografii i Teledetekcji*, vol. 21, pp. 211–219, 2010.

[19] S. Lewiński, "Applying fused multispectral and panchromatic data of Landsat ETM+ to object oriented classification," in *New Developments and Challenges in Remote Sensing*, Millpress, Rotterdam, The Netherlands, 2007.

16

Spatial Dispersal of Douglas-Fir Beetle Populations in Colorado and Wyoming

John R. Withrow,[1,2,3,4] **John E. Lundquist,**[5] **and José F. Negrón**[2]

[1] *Anadarko Industries, LLC, 500 Dallas, Suite 2750, Houston, TX 77002, USA*

[2] *USDA Forest Service, Rocky Mountain Research Station, 240 West Prospect Road, Fort Collins, CO 80525, USA*

[3] *Softec Solutions, Inc., 384 Inverness Parkwy, Ste 211, Englewood, CO 80112, USA*

[4] *USDA-FS Forest Health Technology Enterprise Team, NRRC Building A, Suite 331, 2150 Centre Avenue, Fort Collins, CO 80526, USA*

[5] *USDA Forest Service, Region 10 Forest Health Protection and Pacific Northwest Research Station, 3301 C Street, Suite 202 Anchorage, AK 99503, USA*

Correspondence should be addressed to José F. Negrón; larimer72@gmail.com

Academic Editors: D. Huber and G. Martinez Pastur

Bark beetles (Coleoptera: Curculionidae: Scolytinae) are mortality agents to multiple tree species throughout North America. Understanding spatiotemporal dynamics of these insects can assist management, prediction of outbreaks, and development of "real time" assessments of forest susceptibility incorporating insect population data. Here, dispersal of Douglas-fir beetle (*Dendroctonus pseudotsugae* Hopk.) is estimated over four regions within Colorado and Wyoming from 1994 to 2010. Infestations mapped from aerial insect surveys are utilized as a proxy variable for Douglas-fir beetle (DFB) activity and analyzed via a novel GIS technique that co-locates infestations from adjacent years quantifying distances between them. Dispersal distances of DFB infestations were modeled with a cumulative Gaussian function and expressed as a standard dispersal distance (SDD), the distance at which 68% of infestations dispersed in a given flight season. Average values of SDD ranged from under 1 kilometer for the region of northwestern Colorado to over 2.5 kilometers for infestations in Wyoming. A statistically significant relationship was detected between SDD and infestation area in the parent year, suggesting that host depletion and density-dependent factors may influence dispersal. Findings can potentially provide insight for managers—namely, likelihood of DFB infestation increase for locations within two to five kilometers of an existing infestation.

1. Introduction

The Douglas-fir beetle, *Dendroctonus pseudotsugae* Hopkins, (DFB hereafter) is a major mortality agent of Douglas-fir, *Pseudotsuga menziesii* (Mirbel) Franco, across the Western United States [1, 2]. As a native insect and a natural disturbance agent, it is always present as an endemic influence, playing an important ecological role by killing diseased or otherwise stressed trees. DFB exhibits one generation per year and attacks new hosts every year during its dispersal flight from early spring and through the summer depending on geographic location. The insect overwinters primarily in the adult stage and as larvae inside the host tree [2]. Population levels increase periodically, resulting in

widespread tree mortality [3] which can impact management resource objectives and ecosystem services. These eruptive populations usually develop after other disturbances such as fire [4], windstorms [5], or defoliation [6] events which provide an abundance of stressed trees that the insect can exploit. Once stressed trees are no longer a suitable resource, populations can disperse into surrounding stands, where the insect prefers to infest areas with an abundance of large-diameter trees growing in dense environments and exhibiting reduced growth [7, 8].

Land managers and forest health protection specialists often use models based on current stand conditions to estimate the likelihood of infestation development in a stand or the potential tree mortality levels when populations increase

[8, 9]. These models do not incorporate data about active bark beetle populations and are hampered by limited knowledge about DFB dispersal at the landscape level. Accurate ways to characterize and predict movement of Douglas-fir beetle populations across landscapes are needed to develop spatially referenced predictive models that can identify forests stands more likely to be infested due to the presence of nearby beetle populations. Such information can then be presented in the form of landscape-level assessments of the potential impact of epidemic populations. This holds the promise of directing management activities at susceptible high value areas in a timely manner. Shore and Safranyik [10] present such information for mountain pine beetle, *Dendroctonus ponderosae* Hopkins, in British Columbia.

Powers et al. [11] point out that infestation occurrence and severity depend on several factors involving multiple scales of space and time. In a multiscale study of DFB involving area wide and local spatial scales, the authors found that on larger spatial scales strong winds and prolonged drought preceded epidemics, and on more local scales, greater likelihoods of infestation were linked to drier southern facing stands, sites situated at lower elevation, and areas with greater abundance of host.

At all scales, outbreaks depend on the population dynamics of DFB. Without an abundant number of beetles, outbreaks will not occur. In many cases, what causes a beetle population to increase suddenly can be unknown and difficult to determine. However, beetle populations are a function of fecundity, mortality, immigration, and emigration, with immigration and emigration being two forms of dispersal.

Among available data sources that can be used to describe the presence of the DFB, the most prolific is the USDA Forest Service Aerial Insect and Disease Survey (IDS). Obtained from low altitude flights and conducted by regional forest health specialists, the data set provides a comprehensive and spatially explicit representation of the locations, sizes, intensities, and the geographic shapes of various forms of forest damage, including tree mortality from DFB attacks [12, 13]. Using the above data source, Dodds et al. [14] characterized patch-level characteristics and the spatial relationships between DFB infestations in northern Idaho, observing that when overall beetle populations were clearly at eruptive levels, frequencies of nearest neighbor distances between infestations in two adjacent years showed a distinct concentration at lower distances with a gradual dropoff as distances increased, becoming near zero at between 2000 and 2500 meters. The present work builds on the study by Dodds et al. [14] by examining dispersal distances of DFB. We use the locations of new infestations as a surrogate for beetle dispersal. Our objective is to examine and quantify spatial dispersal of DFB using aerially detected new infestations at the landscape scale in parts of Colorado and Wyoming.

2. Materials and Methods

2.1. Site. Our analysis focused on four areas in Colorado and Wyoming where DFB populations frequently occur—the mountainous regions of the Colorado eastern slope (ECO),

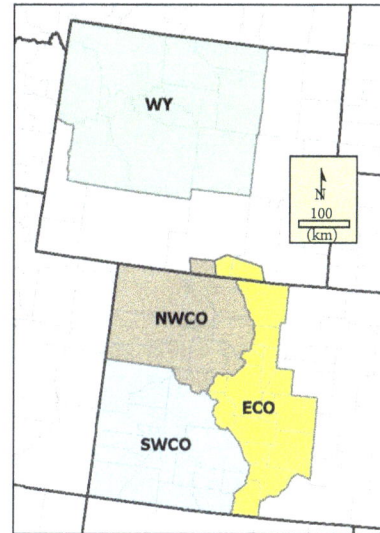

FIGURE 1: The four study areas examined in this study: the Colorado eastern slope (ECO), southwestern Colorado (SWCO), northwestern Colorado (NWCO), and forested areas of Wyoming (WY).

southwestern Colorado (SWCO), northwestern Colorado (NWCO), and Wyoming (WY) (Figure 1). The Colorado eastern slope region (approximately 37–41°N, 104–106.5°W), extending from the Continental Divide down to the urban corridor of the Colorado Front Range, includes the Roosevelt National Forest, Rocky Mountain National Park, an eastern portion of the Arapaho National Forest, and extends south to include the Pike-San Isabel National Forest. The northwestern Colorado region (approximately 39–41°N, 105.67–109°W) includes the White River and Routt National Forests as well as the western portion of the Arapaho National Forest. To the southwest, this region (approximately 37–39.4°N, 105–109°W) comprises the Gunnison, Rio Grande, Grand Mesa, San Juan, and Uncompahgre National Forests. Lastly, the Wyoming Region (approximately 42.25–45°N, 106–111°W) comprises the Bighorn and Shoshone National Forests as well as portions of the Medicine Bow and Bridger-Teton National Forests.

2.2. Aerial Survey Data Source and Background. Insect and disease survey (IDS) data is gathered by the USDA Forest Service Forest Health Protection group and state forestry agencies cooperators and is generally collected by low-flying aircrafts with aerial observers marking infestation locations in paper maps. More sophisticated methods using maps in touch screen devices linked to global positioning systems are also available but not used as frequently. Data is stored as polygons in one geographic information system shape file for each year. Observers are trained to recognize the characteristic "signature" of disturbances such as the crown fading of bark beetle-killed trees [12, 13]. Since Douglas-fir trees fade in coloration a year after attack by DFB, IDS data for one year actually represents infestations initiated the year before. Observers record delineations of affected areas by an agent, including relative size, shape, and location with the best accuracy possible. An estimate of the number of trees affected is also recorded. Each polygon shows an area of tree damage, mortality, or both, either by bark beetles, defoliators, pathogens, or abiotic factors, and each polygon has an entry in an associated attribute table that stores damage agent information.

Observations from moving aircrafts in complex terrain are inherently difficult, and even though observers are highly skilled, there is unavoidable spatial error of position, size, and shape of affected areas by the different agents. IDS data presents a picture of forest conditions at a landscape scale [15]. Johnson and Ross [16] examined the spatial accuracy of an IDS flight by using ground plots to confirm aerial observations. They reported that, when allowing tolerances in positional error of 0 m, 50 m, and 500 m, the overall accuracy of polygon location was 61%, 68%, and 79%, respectively. The authors concluded that these levels of error make the data suitable for coarse-scale landscape analysis but not for fine scales. Therefore, we assumed here that these issues of spatial accuracy produce negligible consequences with analyses designed to detect landscape-level phenomena on the scale of multiple kilometers, which is the scale at which we present our study. For future work, Wulder et al. [17] indicate that the spatial accuracy issues have improved over time with the advent of spatial mapping systems and may be improved in the future by the further use of real-time imagery during the mapping process.

2.3. Characterizing Douglas-Fir Beetle Dispersal. Quantification of dispersal requires the postulation of causational dependence between infestations in a given year and infestations nearby in a subsequent year, more specifically between all infestations in a given previous year (heretofore called "parent infestations") and all infestations in a corresponding upcoming year (heretofore called "child infestations"). This becomes a reasonable assumption when we consider that the insect is univoltine [2]. Dispersal is estimated by quantifying distances between such polygons in adjacent years. Using a 50 m grid we rasterized infestation polygons from all years and created for each polygon what in GIS terms are referred to as "allocation zones" [18], in which each 1994 infestation was then surrounded by a larger zone containing all raster grid locations having the given infestation as the nearest parent infestation (Figure 2). This was then overlaid with similarly rasterized child infestations from the following year, and specific parent-child causational dependence was assumed based upon which parent allocation zone each child infestation was located—that is, which previous-year parent infestation was closest to each child infestation (Figure 3). This procedure was performed on all pairs of consecutive infestation years beginning with 1994-1995 through 2010-2011, and independently for each of the four regional study areas.

2.4. Quantifying Dispersal Distances. In light of the complex geometric shapes inherent to many observed DFB infestations across the landscape, a systematic sampling technique was used to quantify distances of dispersal. Straight lines

FIGURE 2: An example of Douglas-fir beetle infestations in Wyoming in 2001, depicted with 50-meter grid points within allocation zones.

FIGURE 3: Same as Figure 2 but also showing infestations from 2002 in red shading and lines depicting the systematic sampling technique used in the analysis. Each line extends from a raster midpoint in a parent infestation to a similar raster midpoint in an associated child infestation in the subsequent year. The technique is systematic, sampling all possible combinations of raster midpoints in parent infestations and child infestations in the subsequent year.

from each raster midpoint in a parent infestation were connected with all possible raster midpoints in an associated child infestation in the upcoming year (Figure 3). Point-to-point connections were not allowed to cross allocation zone boundaries, since each child infestation was assumed to only be associated with the nearest parent infestation. This resulted in a histogram of sampled distances for each region and pair of consecutive years (e.g., Figure 4). These histograms were further normalized by dividing each histogram bin value by the maximum number of possible grid cells that can be located at the given distance from a single cell, a value that roughly increases by the square of the distance but can exhibit some error because of the finite raster resolution.

With the exception of the restriction of a child infestation to only a single-parent infestation, the above method approximates the more known methodology of calculating a spatial

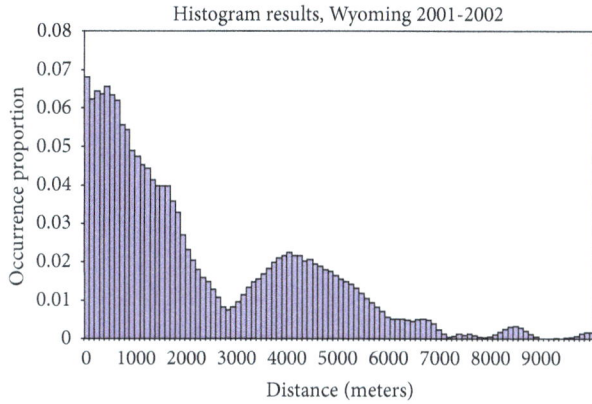

FIGURE 4: Example histogram of sampled distances in Wyoming between parent infestations in 2001 and associated child infestations in 2002.

autocorrelation [19]. The autocorrelation coefficient at a given spatial distance h is defined as follows:

$$r_h = \frac{c_h}{c_0}, \tag{1}$$

where c_h is the autocovariance function:

$$c_h = \frac{1}{N} \sum_{\langle d_{ij} \rangle = h} \left(Y_i - \overline{Y} \right) \left(Y_j - \overline{Y} \right), \tag{2}$$

and c_0 is the variance function:

$$c_0 = \frac{1}{N} \sum_{i=1}^{N} \left(Y_i - \overline{Y} \right)^2. \tag{3}$$

For our purposes, these functions are greatly simplified by the fact that we are dealing with binary data describing "detection" or "no detection":

$$Y_i \in \{0, 1\}, \qquad \overline{Y} \approx 0. \tag{4}$$

Then the equation for c_h becomes

$$c_h = \frac{1}{N} \sum_{\langle d_{ij} \rangle = h} n \left(Y_i = 1, Y_j = 1 \right) = \frac{n_h}{N}, \tag{5}$$

where n_h defines the number of instances of positive detection taking place at a given distance h. Similarly, c_0 becomes

$$c_0 = \frac{1}{N} \sum_{i=1}^{N} (1)^2 = \frac{N}{N} = 1. \tag{6}$$

Thus,

$$r_h = c_h = \frac{n_h}{N}, \tag{7}$$

which is the number of sampled distances between parent and child infestations occurring at a distance h divided by the total possible sampled distances at the same distance (note also that spatial autocovariance is here assessed only for detections, not the absence of detections, since the latter is not considered relevant and so is excluded from (5)).

2.5. Evidence for Causal Relationships. To be able to examine distances of DFB infestations between years, it was important to confirm that the distribution of new infestation across the landscape was independent from the locations of forest stands which contained the host tree and that there was indeed a relationship between the location of parent and child infestations. To confirm this, we examine if the histograms of sampled distances possessed a statistically significant departure from results expected from random spatial infestation occurrence confined exclusively within a spatially aggregated host as infestations in our study sites are restricted to locations of Douglas-fir stands or mixed stands composed of Douglas-fir and ponderosa pine (Pinus ponderosa P. & C. Lawson). Hence, histograms like the one shown in Figure 4 were compared against a null hypothesis of infestations occurring randomly throughout Douglas-fir stands. For each parent infestation year, this was done by repeating the analysis after replacing the observed child infestation layer with a rasterized layer showing locations of all nearby stands containing Douglas-fir. This procedure gave equivalent histogram results for infestations systematically distributed throughout all possible host locations, which for this analysis provides results equivalent to those of a random distribution throughout host. Rejection of the null hypothesis would support the assumption of causality between parent and child infestations.

Normalized cumulative versions of both histograms were compared using the Kolmogorov-Smirnov test, which compares the maximum differences between two cumulative histogram curves. Since the null hypothesis histogram uses allocation zones from previous year infestations and, thus is potentially compromised by observational error inherent in the IDS data, the two-sample test was used, which is more conservative than the one-sample test that assumes one distribution randomly samples from the other [20]. This analysis was performed for two locations from which vegetative data locations were obtained—Pike-San Isabel National Forest and White River National Forest.

2.6. Modeling Dispersal Distances. To describe the complex histogram distributions with a single statistic, cumulative versions of each histogram distribution were modeled using a cumulative Gaussian function [21]:

$$\Psi (z) = \frac{1}{\sigma \sqrt{2\pi}} \int_0^z e^{-x^2 / 2\sigma^2} dx, \tag{8}$$

where the σ parameter is the estimated value of a standard dispersal distance (SDD) for each flight season, from which beetle dispersal for a given region and year is collectively described with a single parameter, a parameter which will be compared with coincident values of infestation area in the parent year in an effort to investigate a relationship via linear regression [22].

2.7. Sensitivity Analyses. Optimal usage of any spatially explicit data source requires an understanding of the degree to which any spatial uncertainties in the data may or may not affect the outcome of any analytical results. A sensitivity analysis was performed by modeling the histogram results with varying degrees of positional error. This was done by modeling each raw histogram bin value as a Gaussian distribution, where negative tails were made into positive distances by taking the absolute value (Figure 5).

The procedure resulted in two adjusted histogram functions that resulted from using Gaussian standard errors of 540 and 764 meters, respectively. As the spatial errors of aerial detection survey polygons indicate locations that are reliably accurate to the landscape scale of a watershed catchment (i.e., within 2 to 3 kilometers) [16], then the above standard errors are representative of the actual spatial uncertainties associated with these data. The effect observed is one of smoothing the histogram, but not one of significantly changing the histogram's overall shape, or its estimate of SDD. Sensitivity analyses were also conducted to test our methodology with spatial resolutions other than 50 meters. From these alternate resolutions, the resulting histograms produced identical values of SDD.

All GIS procedures were performed using the ESRI ArcGIS package versions 8.3, 9.0, 9.1, 9.3, and 10.0 [23]. Model fits were estimated using PROC NLIN from the SAS statistical package v9.1 [24]. Additional statistical analyses were performed and graphs generated using the R statistical package [25].

3. Results

Standard dispersal distances are shown in Table 1 with infestation areas of their parent years (e.g., infestation area in 1997 is shown with the SDD value for 1997-1998). Regional averages for SDD ranged from about 800 meters (NWCO) to about 2,600 meters (Wyoming) with annual values ranging from 204 m to over 10 km in a Wyoming season that also displayed the highest value of overall infestation area (Table 1). Results were omitted where curve fits were not estimable.

After what appears to be a momentary episodic increase in infestation area in 1994, populations on the Colorado eastern slope reached consistently elevated levels of over 1000 ha in 1998. This was accompanied by a more long-term upward trend, with total infested area reaching 6,000 to over 7,000 hectares in 2003 and 2004, followed then by a decline and a resurgence in 2010. The largest proportional increases in total infestation area occurred at four times—in 1998, 2003, 2006, and 2009. These moments approximately coincide with the largest values of SDD occurring in 1998, 2001, 2005, and 2008. There is one infestation area peak in 1994 that for lack of data cannot be described in terms of annual growth and may be purely episodic. It is, however, approximately coincident with a similar peak SDD value in 1995.

Results for southwestern Colorado appear to show two different sets of dynamics across two separate time frames. From 1994 through about 2001 infestation areas were relatively low and values of SDD more variable across years, even momentarily reaching over 4 km in 1995 and dipping to less than 500 m just two years later. For the remaining years of 2002 through 2010, infestation areas were consistently higher than those in the previous timeframe, and values of SDD were

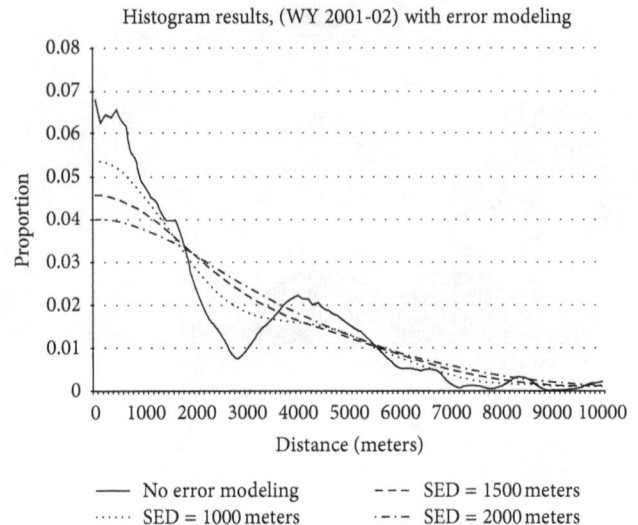

FIGURE 5: Example of the error modeling procedure applied to the histogram for the 2001-2002 DFB flight in Wyoming.

more consistent in the 1–2.5 km range, even displaying an apparent semiannual oscillation between the top and bottom of this range in the years from 2001 through 2007.

Infestation areas in northwestern Colorado were the smallest of all four study areas. Populations in the mid-1990s were small enough that no dispersal analysis was possible. In addition, when values of SDD were calculable, such values were generally smaller than those of other study areas. Infestation area appears to have increased in stages from 1999 through 2007, where two episodically high values of SDD in 2001 and 2004 preceded two separate quantum increases in infestation area in 2003 and again in 2006. The high infestation area in 1996 occurs in isolation from any other coincident phenomena.

Wyoming showed the highest values of total infestation area as well as values of SDD, with the results of 2005 showing over 60,000 ha of infestation area and having an SDD value of over 10 km. Infestation area from 1998 through 2005 showed two periods of apparently exponential growth, from 1998 through 2002 and again from 2003 through 2005. Values of SDD in the same time period show a similar pair of consistent increases from 1999 through 2011 and again from 2002 through 2005.

Rejection of the null hypothesis comparing histogram results against a null hypothesis of random infestation occurrence confined to locations of Douglas-fir host was typical, indicating that the spatial aggregation of Douglas-fir host stands cannot fully explain the location of infestations. Figure 6 illustrates an example of the results for the Kolmogorov-Smirnov test that compared cumulative histogram results with an alternate cumulative histogram computed by replacing child infestations with the entire host area, effectively showing what the results would be if the child infestations were to occur either completely throughout the host, or randomly spaced throughout the host. This latter histogram is displayed with a 95% confidence region, where any departure from this region by the dashed line would indicate a

TABLE 1: Total infestation areas for all years and regions shown with values of standard dispersal distance (SDD). Regional averages are displayed at the bottom.

Year	Colorado eastern slope		Southwestern Colorado		Northwestern Colorado		Wyoming	
	Area (ha)	SDD (m)	Area (ha)	SDD (m)	Area (ha)	SDD (m)	Area (ha)	SDD (m)
1994	7,135	903	702	1,996	0	—	0	—
1995	416	3,601	325	4,499	15	—	1,717	3,097
1996	290	1,037	2,227	1,288	5,607	—	490	656
1997	528	1,102	1,004	381	3	—	1,052	525
1998	3,843	2,718	1,218	—	255	204	214	667
1999	1,848	1,993	482	509	57	258	1,912	410
2000	2,174	1,680	3,030	792	172	342	2,977	1,542
2001	2,605	2,717	809	736	340	1,521	4,765	2,863
2002	2,350	—	3,312	1,770	175	535	10,970	894
2003	6,302	1,897	11,728	873	1,501	644	7,242	2,287
2004	7,304	1,641	8,869	2,486	1,049	2,532	27,571	3,739
2005	3,214	3,693	7,551	672	1,274	—	67,744	10,605
2006	4,644	970	5,630	2,389	3,233	1,384	6,754	1,008
2007	2,348	1,281	12,678	1,949	4,664	1,413	24,053	871
2008	555	3,530	8,122	804	2,711	315	8,535	3,817
2009	2,187	1,304	6,317	762	760	564	1,448	7,599
2010	5,108	2,124	8,983	708	757	601	2,938	2,030
Avg	3,109	2,012	4,882	1,390	1,328	859	10,022	2,663

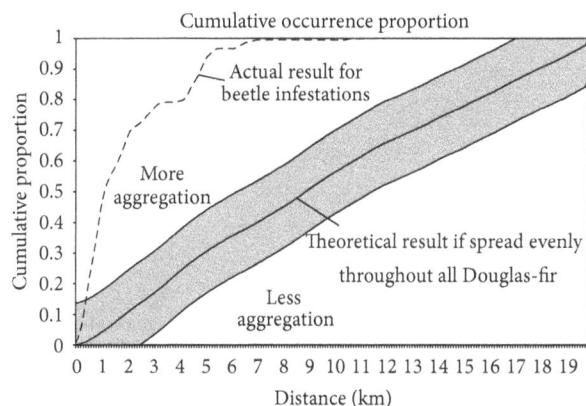

FIGURE 6: Cumulative histogram results from the 1996-1997 season of the Colorado eastern slope compared with expected results from Pike/San Isabel National Forest if 1997 infestations were spread evenly throughout instances of Douglas-fir stands, where the shaded region represents a 95% confidence interval from which the beetle result departs.

rejection of the null hypothesis. The histograms for actual child infestations almost invariably showed more aggregation than the host; therefore, host distribution solely cannot explain the spatial location of infestations. Although there are no means of directly connecting our primary histogram results to beetle flight, these comparisons against a null hypothesis imply beetle dispersal as the driver of the results. Two exceptions to this are the 2004-2005 and 2005-2006 flight seasons in the Pike and San Isabel National Forests (not shown), where the results were not statistically distinct

from results expected from infestations spread completely throughout all Douglas-fir host, which is reasonable considering the magnitude and spread of infestation in those years, which likely affected most of the susceptible stands in the landscape.

Values of SDD demonstrated a significant difference among regions ($P = .0331$), with average regional values ranging from about 1 to 2.5 kilometers. In addition, a statistically significant relationship was detected ($P = 0.0009$) between values of SDD in one year and total infestation area in the following year. In light of the process of dispersal, this relationship is intuitively reasonable, since larger dispersal distances allow beetles to have access to larger areas to infest in the following year.

Lastly, a regression analysis (Table 2, Figure 7) was performed to relate SDD and infestation area in the current year. The association

$$\log S = \beta_0 + \beta_{ECO} + \beta_{NWCO} + \beta_{SWCO} + \beta_S \log A \quad (9)$$

was detected as statistically significant ($P = 0.0006$, $R^2 = 0.296$), where S and A, respectively, refer to SDD and infestation area in a given year, and the second, third, and fourth variables are binary indicators $\{0, 1\}$ of which region is under consideration. The various trend lines show an increase in SDD with increases in infestation area. Values in the above binary variables are consistent with statistically significant differences in SDD across regions, with the Wyoming region having the largest SDD values and northwestern Colorado having the smallest.

TABLE 2: Analysis of variance table for regression parameter estimates for the model describing the relationship between standard dispersal distance and area infested by Douglas-fir beetle.

Dependent variable	= log (SDD)		$R^2 = 0.296$		
Source	DF	Sum of squares	Mean square	F value	Pr > F
Model	4	2.246	0.562	5.77	0.0006
Error	55	5.352	0.097		
Corrected total	59	7.598			
Variable	DF	Parameter estimate	Standard error	t value	Pr > |t|
Intercept (WY)	1	2.625	0.279	9.42	<.0001
Log (area)	1	0.169	0.074	2.27	0.0270
Region—ECO	1	0.070	0.112	0.62	0.5362
Region—NWCO	1	−0.292	0.133	−2.19	0.0324
Region—SWCO	1	−0.166	0.111	−1.50	0.1389

4. Discussion

Years of increasing infested area were oftentimes approximately coincidental with years of elevated values of SDD. The instances of increased dispersal after high infestations may be associated with local host depletion and the presence of distant favorable hosts, thereby fostering increased dispersal distances. In addition, density-dependent factors influencing insect population dynamics may also be indicated as triggering increased dispersal distance.

Dodds et al. [14] estimated DFB infestations to have average adjacent year nearest neighbor distances mostly residing in the 2–5 kilometer range. Our results corroborate this finding as well as that of the theoretical dispersal studies of Byers [26], who modeled dispersal of *Ips typographus* (L.) and indicated that beetles can disperse quickly over large areas but are influenced by factors such as wind, tree density, and angle of turn by the insect. Laboratory studies have shown that DFB attached to rotating mills flew continuously up to nine hours at velocities of 30 to 40 meters per minute [27], which would translate to dispersal distances potentially exceeding 20 km. The present analysis suggests that this distance, although apparently biologically possible, does not represent what commonly occurs in nature.

Dispersal distance is partially a function of resistance to movement as progeny of the parent population moves into new host environments. Although specific factors that drive beetle dispersal and establishment of new infestations in the landscape are largely outside of the scope of this study, Barclay et al. [28] demonstrated the seminal role that host connectivity can play in the landscape dispersal dynamics of the closely related mountain pine beetle. The work of Negron [8] would suggest the development of new infestations in areas of high tree density, high percentage of Douglas-fir basal area, and reduced host growth rates. At larger scales, Powers et al. [11] would stress the importance of aspect, elevation, windthrow, presence of drought, and spatial aggregation of old conifers. Resistance sources include scarcity of host, diversity of host, and distance between suitable hosts. According to Kausrud et al. [29], "mean connectivity ρ depends not only on landscape connectivity but also on factors like dispersal mortality, fat content of the insect, and attraction arrestant (windfalls and

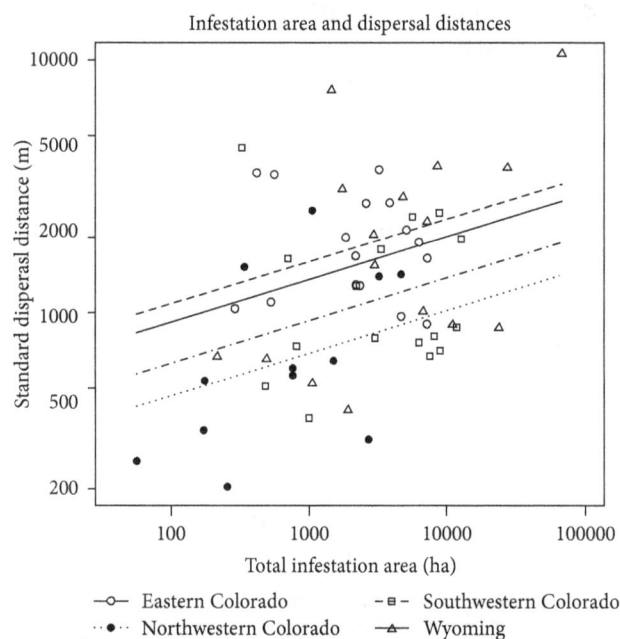

FIGURE 7: Standard dispersal distances plotted against total infestation area. A regression trend line is also shown for each region.

pheromone sources) density." The need for further study on this is clearly indicated.

The present work focuses on a representation of beetle dispersal that gives an indication of dispersal distance, but not dispersal direction. Directional anisotropies in dispersal patterns would likely be the result of wind patterns and stand structure [26], among other factors requiring subsequent analysis. Of additional consideration here would be the relationship of dispersal to differences in habitat connectivity or the ability to traverse landscape topography [30] and related aspects of habitat connectivity and fragmentation [31]. Coulson and Witter [32] divided the process of dispersal into two categories, active and passive, with active dispersal taking the form of flying or walking and passive dispersal involving windborne transport. Although the authors above classify bark beetle dispersal as active, comparing directional anisotropies in wind patterns to those of beetle dispersal may elucidate the degree that such dispersal is a mixture of processes inherent to both categories.

Ecological processes operate at a range of spatial scales, and the range of spatial scales inherent to a given analysis can have an impact on which processes are detected by that analysis [33]. In reference to the findings of Powers et al. [11] aspect is likely a particularly strong influence in Colorado as Douglas-fir occurs mostly in north facing slopes which provide more moisture. Elevation would also likely be a strong factor as Douglas-fir in Colorado has a well-defined elevation range in which it occurs. At the patch scale (tree and stand level), growth rates have also been correlated with DFB attacks [8]. In addition, the spatial distribution of DFB-caused mortality in stands with suitable diameter classes have been found to aggregate strongly around intrastand neighborhoods of high basal area [34].

Although the use of variograms in our simplified binary form (4) presents some novelty, the use of variograms to model spatially explicit ecological data is well documented [35–37], and curves used to represent the results tend to take the form of exponential, Gaussian, Whittle-Matern, or spherical. Examples of such methods being applied specifically to dispersal, however, are limited. In the context of dispersal of seeds, graphs of the form of Figure 4 would be referred to as "seed shadows" (e.g., Holbrook and Smith [38]). Further, when applying curve fits to variograms describing dispersal, the use of a Gaussian form has some biological backing, since the dispersal process can be thought of (and modeled as) a sequence of a large number of random smaller-scale movements [39], and the Central Limit Theorem in statistics indicates that distributions resulting from sums of many independent random processes theoretically approach a Gaussian distribution [40].

Schröder and Seppelt [41] suggest four foundational core questions to any model-based analysis of pattern-process interactions. These questions involve (1) the detection and quantification of spatiotemporal patterns, (2) the analysis of such patterns for underlying ecological processes, (3) the understanding of the scale dependencies of such analyses and processes, and (4) the development of methodologies for simulation of such processes. To these four, we suggest adding a fifth core question, namely, (5) the influence on management decision making. The present analysis addresses the first question by quantifying spatial patterns at the landscape level and demonstrates that spatial distribution of infestations is statistically distinct from those that would arise from spatially random infestation occurrence. Powers et al. [11] and Dodds et al. [14] begin the explanation of underlying processes of insect dispersal (question 2), yet the problem requires further investigation. The third question is addressed by clearly defining the spatial scales inherent to the analysis—namely, those of landscape dispersal. With underlying ecological processes not addressed in the present work, the question of simulating such processes (question 4) is similarly left to future analysis. Lastly, question five, management decision making, can be addressed by providing this information to forest health specialists who can monitor potential outbreak development by examining the locations

of current infestations in light of these dispersal distances. The information can also be of use in planning documents by forest managers such as Forest Plans. Ultimately, combining DFB dispersal knowledge with stand susceptibility models can provide a "real-time" assessment of stands at risk of infestation as presented for mountain pine beetle in Canada [10].

5. Conclusions

Our four study areas showed four different populations of DFB which together explained a wide array of population dynamics. The results among these four study areas, though quantitatively different, retain some apparent qualitative similarities. A model is presented which, although lacking in predictive capability ($R^2 = 0.296$), indicates that increases in total infestation area are associated with increases in SDD. In the infestation risk model of Shore and Safranyik [10], threats of mountain pine beetle infestation show a dependence on the proximity of any neighboring infestations, with threats reaching lowest values when the nearest infestations are more than 4 km away. With average values of SDD found to be in the range of 1–2.5 km, the results of this study can provide more quantified estimation of preventive management costs, by indicating that such management, to be successful, would require management and monitoring efforts to span at least 3–5 kilometers from existing beetle outbreaks (assuming an appropriate cutoff to be two standard dispersal distances which would theoretically encompass 95% of all beetle dispersal). Eventually such estimates of dispersal can be incorporated into spatially explicit landscape models to more effectively predict the risk of future infestations [31]. Lastly, the above technique can be very feasibly applied to other beetle species whose presence is described by aerial detection surveys or similar data sources such as aerial photography or other remote sensing applications.

Acknowledgments

Sincere thanks are extended to Rudy King at the USDA-FS RMRS facility in Fort Collins, CO, USA, for providing ongoing statistical consultation for this project. This project was supported by funding from the USDA Forest Service National Fire Plan Project 01.RMS.C4, Rocky Mountain Research Station, Fort Collins, CO, USA.

References

[1] R. L. Furniss and V. M. Carolin, "Western Forest Insects," Tech. Rep. 1339, U.S. Department of Agriculture, Forest Service, Miscellaneous Publication, 1977.

[2] R. F. Schmitz and K. E. Gibson, *Douglas-Fir Beetle*, Forest Insect and Disease Leaflet 5, USDA Forest Service, Washington, DC, USA, 1996.

[3] M. M. Furniss, M. D. McGregor, M. W. Foiles, and A. D. Partridge, "Chronology and characteristics of a Douglas-fir beetle outbreak in northern Idaho," General Technical Report LNT-GTR 59, USDA Forest Service, Intermountain Forest and Range Experiment Station, Ogden, Utah, USA, 1979.

[4] M. M. Furniss, "Susceptibility of fire-injured Douglas-fir to bark beetle attack in southern Idaho," *Journal of Forestry*, vol. 63, no. 1, pp. 8–11, 1965.

[5] J. A. Rudinsky, "Host selection and invasion by the Douglas-fir beetle, Dendroctonus pseudotsugae Hopkins, in coastal Douglas-fir forests," *Canadian Entomologist*, vol. 98, no. 1, pp. 98–111, 1966.

[6] L. C. Wright, A. A. Berryman, and B. E. Wickman, "Abundance of the fir engraver, Scolytus Íentralis, and the Douglas-fir beetle, Dendroctonus pseudotsugae, following tree defoliation by the Douglas-fir tussock moth, Orgyia pseudotsugata," *Canadian Entomologist*, vol. 116, no. 3, pp. 293–305, 1984.

[7] M. M. Furniss, R. L. Livingston, and M. D. McGregor, "Development of a stand susceptibility classification for Douglas-fir beetle," in *Hazard-Rating Systems in Forest Insect Pest Management: Symposium Proceedings*, R. L. Hedden, S. J. Barras, and J. E. Coster, Eds., USDA Forest Service, Washington, DC, USA, General Technical Report, no. WO-27, pp. 115–128, Society of American Foresters, Entomology Working Group, and USDA Forest Service, and The University of Georgia, Department of Entomology, 1981.

[8] J. F. Negron, "Probability of infestation and extent of mortality associated with the Douglas-fir beetle in the Colorado Front Range," *Forest Ecology and Management*, vol. 107, no. 1–3, pp. 71–85, 1998.

[9] J. F. Negron, W. C. Schaupp Jr., K. E. Gibson et al., "Estimating extent of mortality associated with the douglas-fir beetle in the central and northern rockies," *Western Journal of Applied Forestry*, vol. 14, no. 3, pp. 121–127, 1999.

[10] T. L. Shore and L. Safranyik, "Susceptibility and risk rating systems for the mountain pine beetle in lodgepole pine stands," Forestry Canada, Pacific and Yukon Region, Victoria, BC, Canada, Information Report BC-X-336, 1992.

[11] J. S. Powers, P. Sollins, M. E. Harmon, and J. A. Jones, "Plant-pest interactions in time and space: a Douglas-fir bark beetle outbreak as a case study," *Landscape Ecology*, vol. 14, no. 2, pp. 105–120, 1999.

[12] T. McConnell, E. Johnson, and B. Burns, "A guide to conducting aerial sketchmapping surveys," USDA Forest Service, Forest Health Technology Enterprise Team, Fort Collins, CO, Report FHTET 00-01, 2000.

[13] W. V. Ciesla, "Aerial signatures of forest insect and disease damage in the Western United States," USDA Forest Service, Forest Health Technology Enterprise Team, Fort Collins, CO, Report 01-06, 2006.

[14] K. J. Dodds, S. L. Garman, and D. W. Ross, "Landscape analyses of Douglas-fir beetle populations in northern Idaho," *Forest Ecology and Management*, vol. 231, no. 1–3, pp. 119–130, 2006.

[15] E. W. Johnson and D. Wittwer, "Aerial detection surveys in the United States," *Australian Forestry*, vol. 71, no. 3, pp. 212–215, 2008.

[16] E. W. Johnson and J. Ross, "Quantifying error in aerial survey data," *Australian Forestry*, vol. 71, no. 3, pp. 216–222, 2008.

[17] M. A. Wulder, C. C. Dymond, J. C. White, D. G. Leckie, and A. L. Carroll, "Surveying mountain pine beetle damage of forests: a review of remote sensing opportunities," *Forest Ecology and Management*, vol. 221, no. 1–3, pp. 27–41, 2006.

[18] D. M. Theobald, *GIS Concepts and ArcGIS Methods*, Conservation Planning Technologies, Fort Collins, Colo, USA, 1st edition, 2003.

[19] N. D. Le and J. V. Zidek, *Statistical Analysis of Environmental Space-Time Processes (Springer Series in Statistics)*, Springer, New York, NY, USA, 1st edition, 2006.

[20] W. W. Daniel, *Applied Nonparametric Statistics*, PWS-KENT Publishing Company, Boston, Mass, USA, 2nd edition, 1990.

[21] M. Abramowitz and I. A. Stegun, Eds., *Handbook of Mathematical Functions With Formulas, Graphs, and Mathematical Tables*, Dover Publications, New York, NY, USA, 1972.

[22] R. L. Ott, *An Introduction To Statistical Methods and Data Analysis*, Duxbury Press, Belmont, Calif, USA, 4th edition, 1993.

[23] T. Ormsby, E. Napoleon, R. Burke, C. Groessl, and L. Feaster, *Getting to Know ArcGIS Desktop: Basics of ArcView, ArcEditor, and ArcInfo*, ESRI Press, Redlands, Calif, USA, 2001.

[24] S. A. S. Institute, *SAS/STAT 9. 1 USer'S Guide*, SAS Institute, Cary, NC, USA, 2004.

[25] R Development Core Team. R: A language and environment for statistical computing. R Foundation for Statistical Computing, Vienna, Austria, 2010, http://www.r-project.org/.

[26] J. A. Byers, "Wind-aided dispersal of simulated bark beetles flying through forests," *Ecological Modelling*, vol. 125, no. 2-3, pp. 231–243, 2000.

[27] J. A. Chapman, "Flight of Dendroctonus pseudotsugae in the laboratory," Bi-Monthly Progress Report 10, Canada Department of Agriculture, Division of Forest Biology, 1954.

[28] H. J. Barclay, T. Schivatcheva, C. Li, and L. Benson, "Effects of fire return rates on traversability of lodgepole pine forests for mountain pine beetle: implications for sustainable forest management," *British Columbia Journal of Ecosystems and Management*, vol. 10, no. 2, pp. 115–122, 2009.

[29] K. Kausrud, B. Økland, O. Skarpaas, J. C Grégoire, N. Erbilgin, and N. C. Stenseth, "Population dynamics in changing environments: the case of an eruptive forest pest species," *Biological Reviews*, vol. 87, no. 1, pp. 34–51, 2012.

[30] D. M. Johnson, O. N. Bjørnstad, and A. M. Liebhold, "Landscape mosaic induces traveling waves of insect outbreaks," *Oecologia*, vol. 148, no. 1, pp. 51–60, 2006.

[31] J. Hof and M. Bevers, *Spatial Optimization For Managed Ecosystems, Complexity in Ecological Systems Series*, Columbia University Press, New York, NY, USA, 1998.

[32] R. N. Coulson and J. A. Witter, *Forest Entomology: Ecology and Management*, John Wiley and Sons, New York, NY, USA, 1984.

[33] P. Aplin, "On scales and dynamics in observing the environment," *International Journal of Remote Sensing*, vol. 27, no. 11, pp. 2123–2140, 2006.

[34] J. F. Negrón, J. A. Anhold, and A. S. Munson, "Within-stand spatial distribution of tree mortality caused by the douglas-fir beetle (coleoptera: Scolytidae)," *Environmental Entomology*, vol. 30, no. 2, pp. 215–224, 2001.

[35] A. M. Liebhold, X. Zhang, M. E. Hohn et al., "Geostatistical analysis of gypsy moth (Lepidoperta: Lymantriidae) egg mass populations," *Environmental Entomology*, vol. 20, no. 5, pp. 1407–1417, 1991.

[36] N. D. Le and J. V. Zidek, *Statistical Analysis of Environmental Space-Time Processes*, Series in Statistics, Springer, New York, NY, USA, 2006.

[37] P. Cízek, W. Härdle, and J. Symanzik, "Spatial statistics," in *Statistical Methods For Biostatistics and Related Fields*, W. Härdle, Y. Mori, and P. Vieu, Eds., Springer, New York, NY, USA, 2007.

[38] K. M. Holbrook and T. B. Smith, "Seed dispersal and movement patterns in two species of *Ceratogymna* hornbills in a West

African tropical lowland forest," *Oecologia*, vol. 125, no. 2, pp. 249–257, 2000.

[39] R. Kitching, "A simple simulation model of dispersal of animals among units of discrete habitats," *Oecologia*, vol. 7, no. 2, pp. 95–116, 1971.

[40] A. Papoulis, *Probability, Random Variables, and Stochastic Processes*, McGraw-Hill, New York, NY, USA, 2nd edition, 1984.

[41] B. Schröder and R. Seppelt, "Analysis of pattern-process interactions based on landscape models-Overview, general concepts, and methodological issues," *Ecological Modelling*, vol. 199, no. 4, pp. 505–516, 2006.

Permissions

The contributors of this book come from diverse backgrounds, making this book a truly international effort. This book will bring forth new frontiers with its revolutionizing research information and detailed analysis of the nascent developments around the world.

We would like to thank all the contributing authors for lending their expertise to make the book truly unique. They have played a crucial role in the development of this book. Without their invaluable contributions this book wouldn't have been possible. They have made vital efforts to compile up to date information on the varied aspects of this subject to make this book a valuable addition to the collection of many professionals and students.

This book was conceptualized with the vision of imparting up-to-date information and advanced data in this field. To ensure the same, a matchless editorial board was set up. Every individual on the board went through rigorous rounds of assessment to prove their worth. After which they invested a large part of their time researching and compiling the most relevant data for our readers. Conferences and sessions were held from time to time between the editorial board and the contributing authors to present the data in the most comprehensible form. The editorial team has worked tirelessly to provide valuable and valid information to help people across the globe.

Every chapter published in this book has been scrutinized by our experts. Their significance has been extensively debated. The topics covered herein carry significant findings which will fuel the growth of the discipline. They may even be implemented as practical applications or may be referred to as a beginning point for another development. Chapters in this book were first published by Hindawi Publishing Corporation; hereby published with permission under the Creative Commons Attribution License or equivalent.

The editorial board has been involved in producing this book since its inception. They have spent rigorous hours researching and exploring the diverse topics which have resulted in the successful publishing of this book. They have passed on their knowledge of decades through this book. To expedite this challenging task, the publisher supported the team at every step. A small team of assistant editors was also appointed to further simplify the editing procedure and attain best results for the readers.

Our editorial team has been hand-picked from every corner of the world. Their multi-ethnicity adds dynamic inputs to the discussions which result in innovative outcomes. These outcomes are then further discussed with the researchers and contributors who give their valuable feedback and opinion regarding the same. The feedback is then collaborated with the researches and they are edited in a comprehensive manner to aid the understanding of the subject.

Apart from the editorial board, the designing team has also invested a significant amount of their time in understanding the subject and creating the most relevant covers. They scrutinized every image to scout for the most suitable representation of the subject and create an appropriate cover for the book.

The publishing team has been involved in this book since its early stages. They were actively engaged in every process, be it collecting the data, connecting with the contributors or procuring relevant information. The team has been an ardent support to the editorial, designing and production team. Their endless efforts to recruit the best for this project, has resulted in the accomplishment of this book. They are a veteran in the field of academics and their pool of knowledge is as vast as their experience in printing. Their expertise and guidance has proved useful at every step. Their uncompromising quality standards have made this book an exceptional effort. Their encouragement from time to time has been an inspiration for everyone.

The publisher and the editorial board hope that this book will prove to be a valuable piece of knowledge for researchers, students, practitioners and scholars across the globe.

List of Contributors

Björn Berg
Department of Forest Sciences, University of Helsinki, FIN-00014 Helsinki, Finland

C. T. Mumbi
Environment Department, York Institute for Ecosystem Dynamics (KITE), University of York, Heslington, York, YO10 5DD, UK
Tanzania Wildlife Research Institute (TAWIRI), P.O. Box 661, Arusha, Tanzania

P. Lane
Department of Archaeology, University of York, King's Manor, York YO 17 EP, UK

R. Marchant
Environment Department, York Institute for Ecosystem Dynamics (KITE), University of York, Heslington, York, YO10 5DD, UK

Niamjit Das
Department of Forestry and Environmental Science, Shahjalal University of Science and Technology, Sylhet-3114, Bangladesh

Juliano André Bogoni
Instituto de Ciencias Biologicas, Universidade de Passo Fundo (UPF), Campus de Passo Fundo, Rodovia BR 285, Bairro Sao Jose, Caxia Postal 611, 99052-900 Passo Fundo, RS, Brazil
Departamento de Ecologia e Zoologia, Universidade Federal de Santa Catarina (UFSC), Campus de Florianopolis, Campus Universitario Reitor Joao David Ferreira Lima-Trindade, 88040-970 Florianopolis, SC, Brazil

Maurício Eduardo Graipel
Departamento de Ecologia e Zoologia, Universidade Federal de Santa Catarina (UFSC), Campus de Florianopolis, Campus Universitario Reitor Joao David Ferreira Lima-Trindade, 88040-970 Florianopolis, SC, Brazil

Talita Carina Bogoni
Instituto Federal Catarinense (IFET), Campus de Concordia-SC, Curso de Medicina Veterinaria, Rodovia SC 283, Distrito de Santo Antonio, 89900-000 Concordia, SC, Brazil

Jorge Reppold Marinho
Programa de Pos Graduacao em Ecologia, Universidade Regional Integrada do Alto Uruguai e das Missoes (URI), Campus de Erechim-RS, Avenida Sete de Setembro, 1621, Caxia Postal 743, 99700-000 Erechim, RS, Brazil

Vandana Sharma and Smita Chaudhry
Institute of Environmental Studies, Kurukshetra University, Kurukshetra 136119, Haryana, India

Patrick Addo-Fordjour
School of Biological Sciences, Universiti Sains Malaysia, 11800 Pulau Penang, Penang, Malaysia
Department of Theoretical and Applied Biology, College of Science, Kwame Nkrumah University of Science and Technology (KNUST), Kumasi, Ghana

Zakaria B. Rahmad
School of Biological Sciences, Universiti Sains Malaysia, 11800 Pulau Penang, Penang, Malaysia

Stefanie Fischer
Department of Geography, University of Bonn, Meckenheimer Allee 166, 53115 Bonn, Germany

Burkhard Neuwirth
De La Wi Tree-Ring Analysis, Preschlinallee 2, 51570 Windeck, Germany

Renato Vinícius Oliveira Castro
Department of Forestry, Faculty of Technology, University of Brasília, Campus Darcy Ribeiro, 70904-970 Brasília, DF, Brazil

Carlos Pedro Boechat Soares, Helio Garcia Leite and Agostinho Lopes de Souza
Department of Forestry, Federal University of Vicosa, Campus UFV, 36570-000 Vicosa, MG, Brazil

Gilciano Saraiva Nogueira
Department of Forestry, Federal University of the Valleys of Jequitinhonha and Mucuri, Campus Diamantina, 39100-000 Diamantina, MG, Brazil

Fabrina Bolzan Martins
Natural Resources Institute, Federal University of Itajuba, Campus Itajuba, 37500-903 Itajuba, MG, Brazil

Edward Missanjo, Gift Kamanga-Thole and Vidah Manda
Malawi College of Forestry and Wildlife, Private Bag 6, Dedza, Malawi

Pere Riera
Department of Applied Economics, Autonomous University of Barcelona, 08193 Bellaterra, Spain

Joan Mogas
Department of Economics, Rovira i Virgili University, Avinguda Universitat 1, 43204 Reus, Spain

Raul Brey
Department of Economics, Pablo de Olavide University, Ctrretera de Utrera km 1, 41013 Sevilla, Spain

A. D. Agbelade
Department of Forestry, Wildlife and Fisheries Management, Ekiti State University, P.M.B. 5363, Ado Ekiti, Ekiti State 300001, Nigeria

J. C. Onyekwelu
Department of Forestry and Wood Technology, Federal University of Technology, P.M.B. 704, Akure, Ondo State 340001, Nigeria

Fred L. Bunnell
Forest Sciences Department, University of British Columbia, 3041-2424 Main Mall, Vancouver, BC, Canada

Michael Sprintsin
Jacob Blaustein Institutes for Desert Research, Ben-Gurion University of The Negev, 84990 Sede Boqer Campus, Israel
Forest Management and GIS Department, Land Development Authority, Forest Department, Jewish National Fund (KKL), Eshtaol, M.P., 99775 Shimshon, Israel

Pedro Berliner and Arnon Karnieli
Jacob Blaustein Institutes for Desert Research, Ben-Gurion University of The Negev, 84990 Sede Boqer Campus, Israel

Shabtai Cohen
Institute of Soil, Water and Environmental Sciences, Agricultural Research Organization, The Volcani Center, 50250 Bet Dagan, Israel

Syam Viswanath
Tree Improvement and Propagation Division, Institute of Wood Science and Technology, Malleswaram, Bangalore, Karnataka 560003, India

B. Dhanya
Tree Improvement and Propagation Division, Institute of Wood Science and Technology, Malleswaram, Bangalore, Karnataka 560003, India
Department of Environmental Science, College of Agriculture, University of Agricultural Sciences, PB No. 329, Raichur, Karnataka 584102, India

Seema Purushothaman
Azim Premji University, Bangalore, Karnataka 560100, India

Jaros Baw J. Zawadzki, Karol Prze Fdziecki and Karol Szymankiewicz
Department of Environmental Engineering, Warsaw University of Technology, Ulica Nowowiejska 20, 00-653Warsaw, Poland

Wojciech Marczewski
Space Research Centre, Polish Academy of Sciences, Ulica Bartycka 18A, 00-716Warsaw, Poland

John R. Withrow
Anadarko Industries, LLC, 500Dallas, Suite 2750, Houston, TX 77002, USA
USDA Forest Service, Rocky Mountain Research Station, 240West Prospect Road, Fort Collins, CO 80525, USA
Softec Solutions Inc., 384 Inverness Parkwy, Ste 211, Englewood, CO 80112, USA
USDA-FS Forest Health Technology Enterprise Team, NRRC Building A, Suite 331, 2150 Centre Avenue, Fort Collins, CO 80526, USA

José F. Negrón
USDA Forest Service, Rocky Mountain Research Station, 240West Prospect Road, Fort Collins, CO 80525, USA

John E. Lundquist
USDA Forest Service, Region 10 Forest Health Protection and Pacific Northwest Research Station, 3301 C Street, Suite 202 Anchorage, AK 99503, USA